INTERNATIONAL UNION OF CRYSTALLOGRAPHY
BOOK SERIES

IUCr Monographs on Crystallography

1 *Accurate molecular structures*
A. Domenicano, I. Hargittai, editors
2 *P. P. Ewald and his dynamical theory of X-ray diffraction*
D. W. J. Cruickshank, H. J. Juretschke, N. Kato, editors
3 *Electron diffraction techniques, Vol. 1*
J. M. Cowley, editor
4 *Electron diffraction techniques, Vol. 2*
J. M. Cowley, editor
5 *The Rietveld method*
R. A. Young, editor
6 *Introduction to crystallographic statistics*
U. Shmueli, G. H. Weiss
7 *Crystallographic instrumentation*
L. A. Aslanov, G. V. Fetisov, J. A. K. Howard
8 *Direct phasing in crystallography*
C. Giacovazzo
9 *The weak hydrogen bond*
G. R. Desiraju, T. Steiner
10 *Defect and microstructure analysis by diffraction*
R. L. Snyder, J. Fiala and H. J. Bunge
11 *Dynamical theory of X-ray diffraction*
A. Authier
12 *The chemical bond in inorganic chemistry*
I. D. Brown
13 *Structure determination from powder diffraction data*
W. I. F. David, K. Shankland, L. B. McCusker, Ch. Baerlocher, editors

14 *Polymorphism in molecular crystals*
J. Bernstein
15 *Crystallography of modular materials*
G. Ferraris, E. Makovicky, S. Merlino
16 *Diffuse X-ray scattering and models of disorder*
T. R. Welberry
17 *Crystallography of the polymethylene chain: an inquiry into the structure of waxes*
D. L. Dorset
18 *Crystalline molecular complexes and compounds: structure and principles*
F. H. Herbstein
19 *Molecular aggregation: structure analysis and molecular simulation of crystals and liquids*
A. Gavezzotti
20 *Aperiodic crystals: from modulated phases to quasicrystals*
T. Janssen, G. Chapuis, M. de Boissieu
21 *Incommensurate crystallography*
S. van Smaalen
22 *Structural crystallography of inorganic oxysalts*
S. V. Krivovichev
23 *The nature of the hydrogen bond: outline of a comprehensive hydrogen bond theory*
Gastone Gilli, Paola Gilli

IUCr Texts on Crystallography

1 *The solid state*
A. Guinier, R. Julien
4 *X-ray charge densities and chemical bonding*
P. Coppens
6 *Crystal structure analysis: principles and practice*
W. Clegg, editor
7 *Fundamentals of crystallography, second edition*
C. Giacovazzo, editor
8 *Crystal structure refinement: a crystallographer's guide to SHELXL*
P. Müller, editor
9 *Theory and techniques of crystal structure determination*
U. Shmueli
10 *Advanced structural inorganic chemistry*
Wai-Kee Li, Gong-Du Zhou, Thomas Mak
11 *Diffuse scattering and defect structure simultations: a cook book using the program DISCUS*
R. B. Neder, T. Proffen
12 *The basics of crystallography and diffraction, third edition*
C. Hammond
13 *Crystal structure analysis: principles and practice, second edition*
W. Clegg, editor

The Nature of the Hydrogen Bond

Outline of a Comprehensive Hydrogen Bond Theory

GASTONE GILLI AND PAOLA GILLI

University of Ferrara

OXFORD UNIVERSITY PRESS

OXFORD
UNIVERSITY PRESS

Great Clarendon Street, Oxford OX2 6DP

Oxford University Press is a department of the University of Oxford.
It furthers the University's objective of excellence in research, scholarship,
and education by publishing worldwide in

Oxford New York

Auckland Cape Town Dar es Salaam Hong Kong Karachi
Kuala Lumpur Madrid Melbourne Mexico City Nairobi
New Delhi Shanghai Taipei Toronto

With offices in

Argentina Austria Brazil Chile Czech Republic France Greece
Guatemala Hungary Italy Japan Poland Portugal Singapore
South Korea Switzerland Thailand Turkey Ukraine Vietnam

Published in the United States
by Oxford University Press Inc., New York

First published 2009

British Library Cataloguing in Publication Data

Data available

Library of Congress Cataloging in Publication Data

Gilli, Gastone.
The nature of the hydrogen bond : outline of a comprehensive hydrogen
bond theory / Gastone Gilli and Paola Gilli.
p. cm.— (IUCr monographs on crystallography; no. 23)
ISBN 978–0–19–955896–4 (hardback)
1. Hydrogen bonding. I. Gilli, Paola. II. Title.

QD461.G546 2009
541′.226—dc22 2009008669

Typeset by Newgen Imaging Systems (P) Ltd., Chennai, India
Printed and bound in the UK
on acid-free paper by
the MPG Books Group, Bodmin and Kings Lynn.

ISBN 978–0–19–955896–4 (Hbk)

1 3 5 7 9 10 8 6 4 2

To Anna Maria

CONTENTS

ACKNOWLEDGEMENTS

The authors wish to thank the following for permission to reproduce published copyrighted material:

American Association for the Advancement of Science (AAAS)
Science. Fig. 8.4

American Chemical Society (ACS)
Accounts of Chemical Research. Figs. 3.37, 4.2, 4.3, 5.1;
Color Plates 1, 2, 3, 4
Biochemistry. Fig. 8.1
Journal of Medicinal Chemistry. Figs. 7.1, 7.2, 7.3
Journal of the American Chemical Society. Figs. 3.3, 3.6, 3.8, 3.11, 3.13, 3.14, 3.15, 3.16, 3.27, 3.28, 3.29, 3.30, 7.7
The Journal of Physical Chemistry A. Fig. 3.17

Biophysical Society
Biophysics Journal. Fig. 8.2

Elsevier Science Publishers
Biochemical Pharmacology. Fig. 7.8
Journal of Molecular Structure. Figs. 3.35, 3.36, 4.1, 6.1, 6.2, 6.3, 6.4, 6.5, 6.6; Color Plate 5

International Union of Crystallography (IUCr)
Acta Crystallographica, Section B. Figs. 2.2, 3.19, 3.22, 3.31, 3.32

JAI Press
Advances in Molecular Structure Research, vol. 2. Figs. 3.10, 3.12

John Wiley & Sons Limited
Drug-Receptor Thermodynamics: Introduction and Applications. Figs. 7.4, 7.5

Macmillan Publishers Ltd
Nature. Fig. 8.3

Plenum Press
Fundamental Principles of Molecular Modeling. Figs. 3.9, 3.20, 3.21

Wiley-VCH Verlag GmbH & Co. KGaA
Angewandte Chemie, International Edition in English. Fig. 3.2

INTRODUCTION

The concepts of chemical bonding and electronic structure of molecules cross all the twentieth century and the understanding of the forces deputed to link the nearly hundred atoms of the Periodic Table into molecules has greatly contributed to the rationalization of chemistry and its transformation from a mostly empirical discipline into a real science. Molecules thus become the new basic constituents of matter (updating the old idea of atoms, the previously 'indivisible' unit) whose various aggregation forms (gases, liquids, glasses, crystals) are now explained through *van der Waals interactions*, a type of *non-bonded interaction* due to electric attractions of permanent or induced molecular multipoles. This conception of weakly interacting molecules cannot explain, however, chemical reactivity and we have to introduce new *intermediate interactions* that are deputed to exchange, among molecules, the normal chips of the reactivity game, namely protons and electrons, and that are, respectively, called *hydrogen bonds (H-bonds)* and *charge-transfer or donor–acceptor interactions.* While ordinary chemical bonds occur in strict ranges of distances and energies and non-bonded interactions only beyond the sum of van der Waals radii, these intermediate interactions must cover all the forbidden range between bonding and non-bonding and it would not be surprising if the laws governing them turn out to be particularly complicated.

This book is specifically dedicated to the H-bond, a molecular interaction discovered in the early 1920s and promptly recognized to play a very important role in many chemical and biological phenomena but for which it has been impossible, till very recently, to formulate theoretical models general enough to fully account for all its unique features. The difficulties come from the fact that, while energies and geometries of normal bonds are characteristic of the interacting atoms, H-bonds undergo quite large variations of their energetic and geometrical parameters even when formed by the same couple of donor and acceptor atoms, as well represented by the O–H···O bonds which can display, without any self-evident reason, a continuous of bonding energies from less than 1 to more than 30 kcal mol^{-1} while the corresponding O···O contact distances change from some 3.00 to 2.38 Å. We have found it useful to give a name to this singular phenomenon by calling it the *H-bond puzzle* (Gilli and Gilli, 2000).

Possible solutions of the H-bond puzzle have been actively discussed for more than fifty years without finding an answer that was certain and uneqivocal or, at least, unanimously shared by the scientific community. The reason that has delayed for so long the solution of the problem is very simple, though apparently paradoxical: experimental data collected on the H-bond have been insufficient, until the mid-1980s, to solve the problem and, as a consequence, any theory developed, even if correct, has been practically undemonstrable, being at the mercy of all the most trivial paradigms of the age. And, in fact, accurate H-bond geometries and energies have become accessible only with the introduction, in the mid-1970s, of two new sophisticated techniques, namely neutron diffraction, which made eventually possible a reliable localization of H-bonded protons, and high-pressure mass spectrometry, which made it possible to measure solvent-independent gas-phase H-bond enthalpies. These techniques have opened up a fast and practicable way for the study of really strong H-bonds that, for various technical reasons, were hardly accessible by solution thermodynamics and IR and NMR spectroscopic methods, making it unquestionable that a significant fraction of H-bonds can reach dissociation enthalpies as large as 20–30 kcal mol^{-1} and definitely confuting the legend that 'the hydrogen bond is a weak interaction of electrostatic origin with energies of no more than 4–5 kcal mol^{-1}', a scientific paradigm that had monopolized H-bond studies for many decades. Not dissimilar has been the evolution of theoretical methods. The first calculations at the Hartree–Fock limit on strong H-bonds date back to the early 1970s, providing the first evidence for their three-center–four-electron covalent nature, but only the great theoretical and computational advances of the following decades has made it possible to perform *ab-initio* or DFT calculations that fully account for electron-correlation effects and can correctly emulate H-bonds of any strength formed by molecules of reasonable complexity.

This book takes advantage of the recent experimental and computational advances to perform a comprehensive analysis of the H-bond phenomenon in both its phenomenological and interpretative aspects. At variance with many other similar works, it is only partially grounded on the critical comparison of the many hypotheses and ideas expressed by different authors of different ages, but rather adopts an essentially empirical *case-study* strategy that consists in: (*i*) suspending, or better making temporary *tabula rasa* of, all previous theoretical models on the nature of the H-bond; (*ii*) selecting specific H-bond problems (the cases to be studied); (*iii*) collecting sets of related experimental (and sometimes computer-emulated) data as complete, accurate and reliable as possible; and (*iv*) trying to infer internally coherent and consistent conclusions on the very nature of the H-bond from the ensembles of data collected. The aim is to collect a number of case studies sufficient to outline the full set of empirical rules controlling the process of H-bond formation that, in turn, will become the practical foundation of any subsequent interpretative or theoretical work. For these reasons, the present analysis makes extensive use of computer- or internet-based crystallographic and thermodynamic databases and, in a sense, can also be considered as an attempt to exploit the potentialities of modern chemical information technologies to the unbiased assessment, or reassessment, of chemical laws.

Chapter 1 goes through the history of the H-bond from the beginning of the twentieth century by reviewing its first discovery, the developing of new experimental methods for its study and the theoretical models invoked for its interpretation.

Chapter 2 introduces the H-bond concept as defined in the work of previous authors, providing basic nomenclature, H-bond definitions and two different classification systems based on chemical criteria or bond physical properties. Here, the H-bond is defined as a subset of the wider family of proton-shared interactions that is characterized by the formation of three-center–four-electron interactions and follow the general scheme R–D–H···:A–R', where D is the proton donor (an electronegative atom) and :A the proton acceptor or electron pair carrier (a second electronegative atom or a π-bond). The definition is essentially borrowed from Latimer and Rodebush (1920) and Vinogradov and Linnell (1971) and has the advantage of being neutral with respect to any hypothesis on the electrostatic or covalent nature of the bond, a problem that will be introduced only later when a sufficient amount of accurate experimental information will be available.

Some fifteen main-group elements and a relevant (but presently not well-known) number of transition metals can combine together as donors and acceptors to give three-center–four-electron bonds that can be classified as true H-bonds according to the definition given above. All these bonds can be organized according to the *H-bond chemical classification*, based on atomic electronegativity criteria and essentially derived from the work of Desiraju and Steiner (1999) on weak H-bonds with some additions from the recent Brammer (2003) review on H-bonds involving metal centers. The second classification is based on the thermodynamic, spectroscopic and structural properties of the H-bond and follows the scheme proposed by Jeffrey (1997), who was the first to point out that a small number of intercorrelated parameters (H-bond energies and geometries, IR D–H stretching frequencies and ^{1}H NMR proton chemical shifts) are sufficient to classify all H-bonds in the three classes of *strong, moderate and weak H-bonds*. The observed intercorrelation among the physical quantities that characterize the H-bond leads to the introduction of the problem that is at the very core of the H-bond puzzle, that is: 'Which physical quantity, out of the many intercorrelated ones, is the *independent variable* that drives the H-bond strength and, in turn, all its other associated properties?'.

Chapters 3 and 4 survey all the case studies that have been actually performed to try to assess the empirical laws governing H-bond, laws eventually summarized in Chapter 5.

Chapter 3 summarizes the results deriving from the extraordinary wealth of structural information collected in the Cambridge Structural Database (CSD) (containing, at present, some 370 000 crystal structures of organic and organometallic compounds) and, to a minor extent, in the Inorganic Crystal Structure Database (ICSD) (including some 90 000 structures of inorganic salts and minerals). Data analysis finally leads to three important conclusions: (*i*) in spite of their extraordinary variety, all H-bonds can be grouped into only six chemical classes, called the *six H-bond chemical leitmotifs (CLs)*, which include four classes of strong, one of moderate, and only one of weak bonds; (*ii*) the process of H-bond strengthening can be conceived as a continuous transition from weak, long, dissymmetric and proton-out-centered bonds to strong, short, symmetric and proton-centered ones; (*iii*) this transition of physical properties can be interpreted, in classical VB terms, as a parallel transition from bonds of purely electrostatic nature to bonds that are rather to be classified as true three-center–four-electron covalent bonds; (*iv*) these considerations can be summarized in a simple model, called the *electrostatic-covalent H-bond model (ECHBM)*.

Chapter 4 performs a similar systematic analysis on thermodynamic, instead of structural, data retrieved from the NIST (National Institute of Standards and Technology) databases and including: (*i*) proton affinities (PAs) and acid–base dissociation constants (pK_as) of the molecules participating to H-bond formation; and (*ii*) thermodynamic data ($\Delta G°$, $\Delta H°$, and $\Delta S°$) of the H-bond association equilibrium. Systematic data analysis leads to the conclusion that the PA or pK_a differences between the H-bond donor and acceptor moieties (ΔPA or ΔpK_a) are the *independent variables* driving the H-bond strength for which we were searching, a result summarized in what has been called the *PA/pK_a equalization principle*.

In Chapter 5 the various concepts previously worked out (chemical leitmotifs, electrostatic–covalent H-bond model, and PA/pK_a equalization principle) are connected and integrated to constitute the *empirical (or natural) laws governing H-bond formation*, that is a set of definitions intended to summarize the H-bond rules that can be directly derived from experiments.

The possibility of deriving, starting from the empirical laws, higher-level treatments that could legitimately be considered *H-bond theories* is briefly discussed at the beginning of Chapter 6, maintaining the general idea that scientific theories can be defined as attempts to encode the empirical laws of a given phenomenon in terms of the principles of a more fundamental discipline, a point of view that naturally leads to the conclusion that there may be *several* H-bond theories relying on different more basic theories along the reductionistic chain. In this sense, the association of the empirical laws with the VB theory or the acid–base theory leads to the formulation of two equally legitimate theories on the same subject (the H-bond), which could be called *the H-bond VB theory* or *the H-bond acid–base theory*, a mechanism we may take advantage of again to formulate other H-bond theories.

Following these lines, Chapter 6 tries to show that the way for laying the foundations of a new comprehensive H-bond theory is to recognize that the H-bond is nothing more than a stationary point along the proton-transfer pathway of the S_N2 reaction D–H···:A ⇆ D···H···A ⇆ D:···H–A transmitting the proton from the donor to the acceptor atom. The more fundamental discipline is therefore the transition-state or activated-complex theory with all its traditional paraphernalia (potential-energy surfaces, Leffler–Hammond postulate, Marcus rate-equilibrium theory, state-correlation or avoided-crossing diagrams...) which become available for a better understanding of the H-bond phenomena and can be taken advantage of to develop a novel H-bond theory that can be called the *transition-state H-bond theory* (*TSHBT*).

Chapters 3–6 conclude the attempts to elucidate the nature of the H-bond, which remains the main goal of this book. Many other problems remain, however, to be considered, some of which are of great practical relevance. One of these arises from the fact that bonding theories consider H-bond strengths in terms of enthalpies ($\Delta H°$) or, which makes little difference at room pressure, energies ($\Delta E°$ or $\Delta U°$), while the real effects produced should be measured in terms of association constants, that is of free enthalpies or Gibbs free energies ($\Delta G°$). The relationships between $\Delta H°$ and $\Delta G°$ in gas-phase, non-polar solvents, molecular crystals and, in particular, aqueous solutions are discussed in detail in Chapter 7, where they are shown to be directly connected with another topic of great thermodynamic relevance, i.e. *enthalpy–entropy compensation*. Given the present outstanding importance of pharmacological and biological studies, the general problem of the role played by the H-bond in aqueous solutions is treated in detail for the binding of drugs to both membrane and cytoplasmic receptors.

Finally, Chapter 8 surveys a number of interesting applications of the H-bond to chemistry, biology and material science. This exposition is a traditional topic of all H-bond books that is maintained here but paying much greater attention to strong H-bonds and the role they may play in nature by exerting specific functions of spatial organization or thermodynamic or kinetic control in complex chemical and biochemical systems. These strong bonds, called here *functional H-bonds* because they apparently play functional roles in nature, have been the object of a widespread bibliographic search that has singled out a relevant number of cases where single and particularly strong H-bonds have important, and sometimes crucial, implications in many phenomena, including organic reactivity, enolization, tautomerism, H-bond-driven crystal packing, enzymatic catalysis, as well as proton transmission in bulk water and across biological membranes. These first results suggest that functional H-bonds may be a novel promising field in the systematic investigation of increasingly complex molecular systems.

1

A CENTURY OF THE HYDROGEN BOND (H-BOND)

1.1 The discovery of the H-bond

The concept of attractive and repulsive molecular interactions was introduced in the second half of the nineteenth century to account for the deviations from the laws of ideal gases, finally leading to the formulation of the more general van der Waals's law of corresponding states (1881) that can account for the behavior of real gases in a very wide range of temperatures and pressures. The parallel discovery of the laws of ideal solutions in terms of both vapour tension (F.M. Raoult, 1887–1888) and osmotic pressure (J.H. van't Hoff, 1886) provided the means to study the deviations from ideality in the liquid state, where they are often much stronger than in the gas phase and can be involved in establishing real association equilibria among the different chemical species present. It was soon recognized that association effects are more likely to occur among molecules having certain functional groups rather than others and, in particular, that are considerably enhanced with respect to simple hydrocarbons by hydrogenated functional groups, such as alcohols and phenols. It was also observed that the effects produced by these hydrogen-driven interactions on the physical properties of gases, liquids, and solids can be quite significant, as exemplified by water itself, whose actual melting and boiling points (0 and 100 °C) are in complete disagreement with those extrapolated from the properties of the other Group 6 hydrides (some −90 and −80 °C, respectively).

The concept of H-bond emerges from this scientific climate and can hardly be attributed to any one single person, though a number of authors are recognized by Pimentel and McClellan (1960) as the most probable precursors of the H-bond idea. The first is probably Alfred Werner (1902) who suggested that the properties of ammonium salts can be better explained by assuming that the proton lies in between the ammonia molecule and the anion, as symbolized by his $(H_3N{\cdots}H)X$ notation. He called this new binding situation *Nebenvalenz* (secondary valence), a term used also by Hantzsch (1910) in a further study on the ammonium salts of acetic acid. Two years later Moore and Winmill (1912), in a paper entitled 'The state of amines in aqueous solution', concurred with Werner in assigning to trimethylammonium hydroxide the *weak-union* structure $Me_3N{-}H{-}OH$, so accounting for the weakness of this substance as a base when compared with tetramethylammonium hydroxide. The first intramolecular H-bond was suggested (with the name *mesohydric form*) by Oddo and Puxeddu (1906) to explain the properties of some *o*-hydroxyazo derivatives of eugenol and, some years later, Pfeiffer *et al.* (1913) made a similar proposal for the intramolecular link, called by him *inner Komplexsalzbindung,* between the hydroxyl and carbonyl functions of 1-hydroxy-anthraquinone.

The final assessment of the H-bond concept is ascribed to Huggins and, independently, to Latimer and Rodebush, three young men working in the laboratory of G.N. Lewis. Latimer and Rodebush (1920) made use of it for rationalizing the properties of highly associated liquids, such as water and hydrogen fluoride, the small basicity of ammonium hydroxide and the dimerization of acetic acid, and their definitions, based on the Lewis' dot formalism, appear to be quite lucid and accurate even in modern terms: '...a free pair of electrons on one water molecule might be able to exert sufficient force on a hydrogen held by a pair of electrons on another water molecule to bind the two molecules together...Such an explanation amounts to saying that the hydrogen nucleus held between 2 octets constitutes a *weak bond.*' Less documented is the role played by Huggins who claimed later (1971) that he was the first: 'The hydrogen bond was proposed by me in 1919 and shortly after by Latimer and Rodebush...' quoting a previous thesis at the University of California. Though this thesis has disappeared, his claim is substantiated by Lewis (1923, p. 109) in his book on the chemical valence: 'The idea was first suggested by Dr. M.L. Huggins, and was also advanced by Latimer and Rodebush, who showed the great value of the idea in their paper...'.

The descriptor *hydrogen bond* appears a few years later when it is first used by Pauling (1931) to interpret the structure of the $[F{:}H{:}F]^-$ ion in its first general paper on the nature of the chemical bond and by Huggins (1931) to account for the ionic conduction produced by hydrogen and hydroxyl ions in water. Some perplexities on the use of the term *bond* should have persisted, however, because Huggins, in two important papers on the structure of ice and water (Huggins, 1936a) and on the H-bonded interactions in organic compounds (Huggins, 1936b), introduced the new expression *hydrogen bridge*, which is still widely used today, while Bernal and Megaw (1935) used *hydroxyl bond* to describe $^-O{-}H{\cdots}^-O{-}H$ interactions in minerals. However, it was the chapter on the H-bond in *The Nature of the Chemical Bond* (Pauling, 1939, 1940, 1960) that eventually made known the H-bond to the wider chemical community.

1.2 The theoretical understanding of the H-bond

In the third edition of his book, Pauling (1960) gave a unique definition for weak and dissymmetric and strong and symmetric H-bonds: '...under certain conditions an atom of hydrogen is attracted by rather strong forces to two atoms, instead of only one, so that it may be considered to be acting as a bond between them. This is called the *hydrogen bond.*' (p. 449), but opposite reasons for their formation. For weak H-bonds an electrostatic origin is called for: 'It is recognized that the hydrogen atom, with only one stable orbital (the 1*s* orbital), can form only one covalent bond, that the *hydrogen bond* is largely ionic in character, and that it is formed only between the most electronegative atoms.' Conversely, among the very few examples of strong and symmetric H-bonds known at the time, Pauling quotes $[F\cdots H\cdots F]^-$ in KHF_2 ($d(F\cdots F) = 2.26$ (1) Å), $[O\cdots H\cdots O]^-$ in nickel dimethylglyoxime ($d(O\cdots O)$ = 2.44(2) Å), and $[O\cdots H\cdots O]^+$ in acetamide hemihydrochloride ($d(O\cdots O)$ = 2.40(2) Å) and, for all of them, invokes resonance between two VB covalent forms X–H···X and X···H–X to explain both proton centering and shortness of the X···X distance, as in the statement: '...the hydrogen atom in the $[HF_2]^-$ ion lies midway between the two fluorine atoms and may be considered to form a half-bond with each.' (p. 484, 3rd edn).

It is then quite clear that Pauling never stated that 'all H-bonds must be electrostatic' but rather that the covalent character becomes predominant only in very short and symmetric bonds that, being extremely rare, 'are exceptions' (p. 485) and then 'can be reasonably neglected in the treatment of the much more copious H-bonds of normal strength', a position also embraced by Coulson, as reported by McWeeny in his *Coulson's Valence* (1979): 'If we discount such exceptions, hydrogen bonding may be quite well understood at a qualitative level using simple electrostatic models.' This point of view was no longer justified some twenty years later because of the very large number of strong H-bonds discovered but, at that time, it has become difficult to convince many workers in the field that covalency (or delocalization, or charge-transfer) was an important factor in strengthening the H-bond, so causing a sensible delay in the developing of a more complete and comprehensive interpretation of the H-bond phenomenon and giving rise to a *querelle des anciens et des modernes* that has incredibly lasted more than fifty years.

Not surprisingly, the first theoretical models of the H-bond were purely electrostatic and assumed that the H-bonded molecules could be suitably represented by a small number of positive and negative point charges (Bernal and Fowler, 1933; Lennard-Jones and Pople, 1951; Pople, 1951) or multipoles (Rowlinson, 1951) located on the three interacting atoms. Coulson (1959) was the first to re-examine this electrostatic model by VB methods and discounted it as insufficient to account for many of the physical phenomena associated with H-bond formation. He suggested that a proper understanding of the H-bond behavior requires four distinct concepts: (*a*) *electrostatic interaction* derived from the purely electrostatic forces acting among undeformed charge clouds; (*b*) *delocalization effects* associated with permanent large-scale deformations of these clouds, that is with charge-transfer effects able to induce a partial covalent character in the H-bond formed; (*c*) *repulsive forces* associated with the Pauli exclusion principle and produced by the physical overlapping of closed-shell charge clouds; and (*d*) *dispersion forces*, that is London forces associated with small-scale coordinate motion of the electrons. Coulson and Danielsson (1954a,b) tried to evaluate the amount of the covalent contribution to the O–H···O bond by considering it a mixture of three main VB structures

Ψ_{COV1}	–O–	––H	:O<	covalent I, no charge transfer
Ψ_{IONIC}	$–O:^-$	H^+	:O<	pure ionic, no charge transfer
Ψ_{COV2}	$–O:^-$	H–––	$^+$O<	covalent II, with charge transfer

and giving *per cent* weights ω_1, ω_2, and ω_3 to the three wavefunctions, where the covalent character is clearly measured by ω_3. The authors carried out their calculations for the two O···O distances of 2.8 and 2.5 Å finding weights of 64.9 : 30.6 : 4.5 and 60.3 : 28.4 : 11.3, respectively, and concluding that long bonds are essentially electrostatic but that 'with the short hydrogen bond this covalency is beginning to be appreciable', so that 'It seems probable that a purely electrostatic model for this situation would be hardly fair.' This line of thought was also embraced by Pimentel and McClellan (1960) who wrote 'At the 1957 Ljubljana Conference one of the important points of fairly general accord was that the electrostatic model does not account for all of the phenomena associated with H-bond formation'.

Pimentel and McClellan (1960) were also the first to propose a MO model of the H-bond based on the combination of three AOs (two $2p_z$s from the donor and the acceptor atoms and one $1s$ from the hydrogen) to give three MOs, out of which the bonding and non-bonding orbitals accommodate the four electrons and the last antibonding one remains empty. This was the beginning of the quantum-mechanical approach to H-bond studies by MO methods at the Hartree–Fock and, later on, post Hartree–Fock level of theory, an approach that has developed over the years according to three different lines: (*i*) H-bond energies and population analysis; (*ii*) H-bond energy decomposition methods; and (*iii*) topological analysis of the H-bond electron densities.

The first high-level Hartree–Fock (HF) calculations are due to Noble and Kortzeborn (1970) who calculated a H-bond energy of some 40 kcal mol^{-1} for HF_2^- and Kollman and Allen (1972) who obtained 52 and 35 kcal mol^{-1} for HF_2^- and $H_5O_2^+$, respectively. These last authors were among the first to make extensive use of density-difference maps to enlighten the electronic rearrangements associated with H-bond formation, to single out the characteristic 'decrease of charge (node) around the hydrogen during H-bond formation' occurring in both strongly and weakly H-bonded systems and to notice how these 'charge density shifts... [were] compatible with the simple four-electron, three-center view of the hydrogen bond.' From there on, an ever-growing number of H-bond cases were treated by the most advanced methods available at the time in a crescendo proceeding from simple Hartree–Fock methods with fixed molecular geometry to the present *ab initio* or DFT methods accounting for electron correlation and basis set superposition errors and including full geometry optimization, frequency calculations and evaluation of full proton-transfer pathways. For detailed surveys the reader is addressed to the books by Scheiner (1997) and Grabowski (2006) and, for a short but pregnant account, to Vanquickenborne (1991).

Other important studies were performed by energy-decomposition methods. In the 1970s Morokuma (1971, 1977) and Kitaura and Morokuma (1976) developed, within the HF approximation, a new original decomposition scheme where the H-bond interaction energy was divided into the five *electrostatic* (*ES*), *polarization* (*PL*), *exchange-repulsion* (*EX*), *charge-transfer* (*CT*), and *coupling* (*MIX*) components, obtained by permitting or forbidding that the corresponding terms appear in the HF evaluation of the product $\psi_{\mathrm{A}}^{\circ}\psi_{\mathrm{B}}^{\circ}$ of the two wavefunctions representing the isolated subunits of the complex. This KM method was applied to a number of neutral and relatively weak H-bonds confirming their prevalently electrostatic nature with small but significant contributions of charge transfer, and rapidly became a reference method for MO applications to H-bond. Drawbacks were that dispersion contributions cannot be appreciated at the HF level (because the method does not take into account excited states of the molecule) and that attempts to extend the calculations to stronger H-bonds have been generally disappointing.

A different decomposition scheme (Reed, Curtiss and Weinhold, 1988) is based on natural bond orbitals (NBOs), which can be defined as the optimal orthonormal set of localized orbitals corresponding to the traditional concepts of lone and bonding pairs. At variance with traditional methods, the filled NBOs, σ, of the 'natural Lewis structure' are associated with the virtual set of non-Lewis antibonds, σ^*, the latter being essentially empty and representing the unused valence-shell capacity of the molecule. In this framework, the energy of the molecule is decomposed into the terms $E = E_{\sigma\sigma} + E_{\sigma\sigma^*}$ (respectively, the Lewis and non-Lewis contributions) and the H-bond complexation energy into a charge-transfer, ΔE_{CT} (essentially a $\mathrm{n} \rightarrow \sigma^*$ lone pair–antibond interaction) and a no-charge-transfer part, ΔE_{NCT}, the latter being related to the Heitler–London energy including exchange repulsion and electrostatic (induction and polarization) interactions. Detailed calculations on the water dimer have shown that, at variance with the KM method, the Heitler–London energy (electrostatic plus exchange) is now repulsive and that the main stabilization energy comes from the charge-transfer term as a consequence of a very small amount of charge transferred (less than 0.01 electrons). In an updated version of the NBO scheme called NEDA (natural energy decomposition analysis) Glendening and Streitwieser (1994) have further divided the no-charge-transfer part into the electrostatic, charge transfer and deformation terms that, in spite of the name, bear little relationship to the corresponding KM ones. Their results for the water dimer assign strong attractive contributions to electrostatic and charge-transfer terms and a large repulsive one to electron density deformation (charge-density redistribution). Though other energy-decomposition schemes are known, the few discussed above are enough to underline the limits of the method. These are due to the fact that, while the total interaction energy is a true quantum-mechanical observable, the components of the decomposition are not but depend on their somewhat arbitrary definition. This does not mean that these methods are not useful and efficient tools for dissecting the H-bond but that the different decomposition schemes create separate worlds inside (but not outside) which the H-bonds can be usefully compared.

All other theoretical methods so far used for interpreting the H-bond have made direct reference to the electron density, ED or $\rho(\boldsymbol{r})$, of the total system or, sometimes, to a derived quantity, the electrostatic potential, EP or $\Phi(\boldsymbol{r})$. The ED can be either derived from theoretical calculations or experimentally measured by diffraction methods (see below) and, being directly computable from the wavefunction of the system, it may be said that it is a *function* of the wavefunction or Hamiltonian of the system itself. Its importance has greatly increased since it has been proved (Hohenberg and Kohn, 1964) that the converse is also true, that is that the Hamiltonian and the ground-state wavefunction of the system are *a functional of the ED*, though the exact form of such a functional is unknown, opening the fascinating perspective that *all the properties of the system studied* (in our case, the H-bonded system) *can be directly derived from the only knowledge of its ED*. Moreover, the concept of functional has been exploited (Kohn and Sham, 1965) into the KS-DFT theory, i.e. the *density functional theory* within the KS framework (Parr and Yang, 1989), which has become a widespread method for theoretical calculations in contrast to more traditional *ab initio* methods. DFT is inclusive of electron-correlation effects directly by means of functionals of the electron density and has proved to be highly competitive with respect to many post-HF methods, providing similar benefits at essentially HF computing costs and permitting treatment of rather larger molecular systems. For this reason DFT methods have found ever-increasing applications in the study of H-bond complexes based on the supramolecular approach. The results are largely encouraging (Guo *et al.*, 1997) suggesting that DFT will become the pre-eminent method of electronic structure calculation including correlation to be used for H-bond studies.

In this context, methods of ED analysis assume particular importance and the method that has perhaps led to the greatest advances in the study of chemical bonding in general, and H-bonds in particular, is the *topological analysis of the electron density function* $\rho(\boldsymbol{r})$. Topological methods arise from the work of Bader (1990) who was the first to show that the ED of a many-electron system can be uniquely partitioned into basins (Ω) that are referred to as *atoms in molecules* (*AIM*). The concept of AIM allows a more rigorous definition of atomic charges by redefining the atoms as finite regions of space enclosed by a zero-flux surface (the surface for which there is no electron density gradient perpendicularly to the surface itself). The most interesting aspect of Bader's AIM theory is that it redefines the concept of the chemical bond in terms of the topological properties of $\rho(\boldsymbol{r})$, namely its gradient field, $\nabla\rho(\boldsymbol{r})$, and its curvature or Laplacian, $\nabla^2\rho(\boldsymbol{r})$. So, a *bond critical point* (*BCP*) is a point along the trajectory of the gradient path (*bond path, BP*) connecting two local electron density maxima with $\nabla\rho(\boldsymbol{r}) = 0$ (nuclei) and lying at the borderline of the two atomic basins involved. The theory shows that the properties of $\nabla^2\rho(\boldsymbol{r})$ (the Laplacian) at this bond critical point can discriminate among the different types of chemical bonds. *Shared interactions* (covalent bonds) are associated with relatively large values of $\rho(\boldsymbol{r}_{\mathbf{BCP}})$ and negative values of the Laplacian ($\nabla^2\rho(\boldsymbol{r}_{\mathrm{BCP}}) < 0$), i.e. with a charge concentration in the internuclear region. Conversely, *closed-shell interactions* (ionic bonds and van der Waals interactions) are characterized by small values of $\rho(\boldsymbol{r}_{\mathbf{BCP}})$ and a positive curvature of the density along the bond path ($\nabla^2\rho(\boldsymbol{r}_{\mathrm{BCP}}) > 0$), both reflecting the charge depletion at the critical point.

Applications of the AIM theory to H-bond have been various and fruitful, involving the analysis of the EDs both theoretically computed and experimentaly measured by diffraction methods (for recent accounts see Koritsanszky and Coppens, 2001; Koritsanszky, 2006). The opening of the 'H-bond topological era' is marked by new AIM-based empirical criteria for identifying D–H···A interactions as true H-bonds (Koch and Popelier, 1995), the topological classification of the experimental EDs of the most common H-bonds (Espinosa *et al.*, 1998, 1999, 2002) and the topological redefinition of the H-bond energy (Grabowski, 2003). Moreover, AIM can, in principle, distinguish between electrostatic or covalent D–H···A bonds because only the latter can have a negative Laplacian at the *H-bond critical point (HBCP)* between H and A. However, though negative $\nabla^2\rho(\boldsymbol{r})$ have been calculated for very strong H-bonds (e.g. –8.392 e Å^{-5} for the $[F\cdots H\cdots F]^-$ bond; Espinosa *et al.*, 2002), calculations have mostly shown positive Laplacians for both weak and strong H-bonds, though with smaller values in the second case, adding new fuel to the legend of purely electrostatic H-bonds. Only recently (Gilli *et al.*, 2002) was it made clear that the HBCP Laplacian is certainly negative only at the D···H···A transition state of the proton-transfer (PT) reaction pathway and that the so-called covalent H-bonds with negative Laplacian are, in fact, strong and symmetric bonds having, in the ground state, the properties of their PT transition state (see Section 3.2.2.5).

Other indicators of *electron localization* have been proposed and widely applied to the interpretation of H-bond phenomena. The first (Becke and Edgecombe, 1990) starts from the consideration that electron localization is related to the *same-spin pair probability* and the associated *Fermi hole function* reflecting the effects of the Pauli exchange repulsion. The smaller the probability of finding a second like-spin electron near a reference electron, the more highly localized this last electron turns out to be. These considerations have led to the definition of a new quantity called the *electron localization function* (ELF; $0 \leq \text{ELF} \leq 1$) where ELF = 1 or 0 represent complete or no electron localization, respectively, and ELF = ½ the reference electron-gas-like pair probability. A new ELF-based topological approach of the H-bond has been developed by Fuster and Silvi (2000) showing that it can provide unambiguous criteria to distinguish between weak and strong H-bonds. In weak and moderate bonds the proton donor D–H and the proton acceptor :A keep their individuality, so that the topology of the D–H···A complex is just the addition of the two separate moieties. Conversely, the setting up of strong and symmetric H-bonds corresponds to the formation of a new molecular entity characterized by incomplete proton transfer from the donor to the acceptor.

A new approach has been proposed by Bader and Gatti (1998) by showing that it is possible to see the electron density $\rho(\boldsymbol{r})$ at any point $\boldsymbol{r}$ within the molecule as the sum of the contributions of a *source function* $S(\boldsymbol{r};\Omega)$ from all the atomic basins Ω of the total system, according to the equation $\rho(\boldsymbol{r}) = \Sigma^{\Omega} S(\boldsymbol{r};\Omega)$. This source function represents a quantitative appreciation of the relative importance of atomic or group contributions to the density at any specific point of the system and, in this sense, has been used to appreciate the contributions to the HBCP in a number of H-bonds of very different strength (Gatti *et al.*, 2003). The results seem conclusive in assessing the substantial difference between weak and strong H-bond. In the first, the proton acts as a sink, rather than a source, of electron density and nearly 50% of the density at the HBCP actually comes from the atoms *outside* the D–H···A group, showing that the bond, consistent with its electrostatic nature, is far from being a three-center interaction. In the second, there is a strong density increment at the HBCP, whose Laplacian becomes negative, while more than 90% of this density comes from the three D, H, and A atoms directly involved in the H-bond interaction, clearly showing the electron localization typical of the three-center–four-electron covalent bond.

In summary, it can be said that quantum-chemical methods have been so far extremely successful and convincing in giving a very detailed characterization of the structural, energetic and electron-density features of both weak and strong H-bonds and in confirming the Pauling and Coulson's seminal intuition of a continuous change of properties when going from weak electrostatic to strong covalent interactions. Perhaps theoretical methods have not been equally successful in solving the H-bond puzzle, that is in explaining why bonds formed by a same donor–acceptor couple can display such different interaction energies. We have to consider, however, that the great expansion of computational H-bond studies is quite a recent phenomenon, suggesting that future developments will certainly be of great interest, even because DFT emulation can provide, today, molecular properties (geometries, vibrational spectra, PT pathways, etc.) that are challenging in accuracy the results of the most sophisticated experimental techniques.

1.3 The experimental approach to the H-bond

In spite of its relevant theoretical aspects, the H-bond remains a mostly experimental field because of the many experimental techniques used for its study and because of the many practical implications and applications of this bond in chemistry, biochemistry and material science.

The H-bond has been studied by *thermodynamic methods* from its very beginning, having been actually discovered from the anomalies it may induce in gases or solutions, such as anomalous boiling points, deviations from Raoult's law or Trouton's rule, or abnormal pK_as of acids and bases. Later, more advanced methods for the study of H-bond energetics were developed, in particular direct *calorimetry* or van't Hoff methods applied to association *equilibrium constants* measured at different temperatures. In their book on the H-bond, Pimentel and McClellan (1960) list 32 different properties that can be taken advantage of to evaluate equilibrium constants, together with more than 150 values of these constants measured for one-, two-, and three-component systems. The amount of work done in the field can be appreciated from the book by Joesten and Schaad (1974) who report an impressive list of 2551 association constants and

1555 ΔH° enthalpy values, as determined by IR, NIR, NMR, and UV spectroscopy, calorimetry, gas-liquid chromatography, dielectric constant, molar polarization, and vapor-pressure studies. The use of the different methods was quite uneven, however, IR methods covering 63% of cases (IR 47 and NIR 16%) and calorimetry, NMR and UV, respectively, 15, 14, and 7% of cases, which leaves only 1% left for all other techniques. As for the reaction environment, only 15 measurements were carried out in vapor and all others in organic solvents with a large dominance of CCl_4 and C_6H_{12}. Over the 1555 ΔH° measurements only 132 were greater than 8 kcal mol^{-1}, 67 being in the range 10–16 kcal mol^{-1} and just one reaching 37 kcal mol^{-1} (the usual $[F\cdots H\cdots F]^-$ ion in the vapor phase). These measurements in organic solvents mostly concern rather weak bonds of neutral D–H···A or zwitterionic $D^-\cdots H{-}A^+$ type because stronger positive $[D\cdots H\cdots A]^+$ or negative $[D\cdots H\cdots A]^-$ charge-assisted bonds were not accessible, as a rule, by the techniques used.

This difficulty was overcome by the introduction of new techniques designed to study *gas-phase* ion–molecule association at different temperatures, namely ion cyclotron resonance (ICR) spectroscopy and pulsed high-pressure mass spectomery (PHPMS), whose applications to strong ionic H-bonds date back to the early 1970s (Beauchamp, 1971; Kebarle, 1977) and were systematically pursued all through the 1980s (Larson and McMahon, 1982; Meot-Ner (Mautner), 1987). Presently, the computer-based NIST database (Meot-Ner (Mautner) and Lias, 2005) collects all ΔG°, ΔH°, and ΔS° values that have been so far determined for ionic H-bonds. The dichotomy between neutral and ionic bonds still persists, however, because these gas-phase techniques can deal with charged but not with neutral or zwitterionic species.

Equilibium thermodynamics is not the only important application of *vibrational spectroscopy*, which has been for long the main tool for H-bond characterization in all states of matter. At the beginning of the 1930s (Liddel and Wulf, 1933; Hilbert *et al.*, 1936; Hendricks *et al.*, 1936) it was realized that D–H···A bond formation could be monitored from the IR red shift, $\Delta\nu = \nu^\circ - \nu$, induced by the H-bond in the D–H stretching frequency, ν, with respect to the standard value in the non-bonded state, ν°. This was the starting point for IR (and, at a lesser extent, Raman) to become the primary method for H-bond detection in liquid, solid, and gas phases and an invaluable tool for measuring the concentrations of the H-bonded species from the integrated intensity of the ν(D–H) band and the H-bond strengths from the many $\Delta\nu/\nu$ versus ΔH° correlations proposed (for reviews see: Pimentel and McClellan, 1960; Murthy and Rao, 1968; Joesten and Schaad, 1974). Rapidly, IR methods developed into an invaluable tool for the study of most of the H-bond properties (strength, geometry and shape of the PT pathway) and the quality and quantity of the results was significantly improved by the introduction of NIR (or IR-overtone; Luck and Ditter, 1968; Durocher and Sandorfy, 1967; Luck, 1991) and FIR vibrational spectroscopies (Jakobsen and Brasch, 1965; Hurley, Kuntz and Leroi, 1966) that, having complementary features with respect to fundamental IR, can integrate it to give a more complete oversight of the bonds studied. Another quantity introduced (Singh and Wood, 1968; Sokolov and Savel'ev, 1977) was the hydrogen/deuterium isotopic frequency ratio, $\nu(\mathrm{D}-{}^1\mathrm{H})/\nu(\mathrm{D}-{}^2\mathrm{H})$, which was shown to be in some way able to discriminate the shape of the PT pathways of weak and strong H-bonds. Detailed accounts of the conspicuous results obtained from vibrational spectroscopy in the early times are available (Novak, 1974; Schuster, Zundel, and Sandorfy, 1976; Vinogradov and Linnell, 1971).

New significant results were obtained by combining vibrational spectroscopy with the *matrix-isolation* technique, which consists in trapping isolated molecules and complexes in an inert solid, called the matrix, usually consisting of argon or nitrogen at cryogenic temperatures of 10–15 K. A large excess (100 to 10 000 times) of the inert material is used to assure complete isolation. The technique was simultaneously proposed by Norman and Porter (1954) and Pimentel (Whittle *et al.*, 1954) to study unstable species such as free radicals, and was largely pioneered by Pimentel and coworkers for studying H-bonded species (van Thiel *et al.*, 1957a,b). The first results obtained for intermolecular H-bonds between simple molecules were reviewed by Hallam (1976) and an important compilation of IR stretching frequencies for hydrogen halide complexes is due to Barnes (1983). For a more recent review on H-bonded and donor–acceptor interactions, see Maes (1991).

Matrix isolation vibrational spectroscopy is also important because, just to interpret its findings, Ault, Steinback and Pimentel (1975) introduced the use of the *proton affinities (PAs)* in H-bond studies, showing that D–H···A interactions become much stronger when PA(A) matches PA(D^-), so that the bond can be written as $^{½-}\mathrm{D}\cdots\mathrm{H}^+\cdots\mathrm{A}^{½-}$. From there on, the concept that the H-bond strength is enhanced by the smallness of the $\Delta\mathrm{PA} = \mathrm{PA}(\mathrm{D}^-) - \mathrm{PA}(\mathrm{A})$ term has become a recurrent leitmotif in the studies of positive $[\mathrm{D}\cdots\mathrm{H}\cdots\mathrm{A}]^+$ or negative $[\mathrm{D}\cdots\mathrm{H}\cdots\mathrm{A}]^-$ charge-assisted H-bonds whose enthalpies in the gas phase were being determined by ICR and PHPMS techniques. Accordingly, a number of extensive PA scales were constructed over the years, from the early one compiled by Yamdagni and Kebarle (1971, 1976) to the recent NIST database (Hunter and Lias, 2005). Many $\Delta H°$ versus ΔPA correlations for charge-assisted $[\mathrm{D}\cdots\mathrm{H}\cdots\mathrm{A}]^+$ or $[\mathrm{D}\cdots\mathrm{H}\cdots\mathrm{A}]^-$ bonds have been reported to be essentially linear, at least within restricted ΔPA intervals (Davidson *et al.*, 1979; Meot-Ner (Mautner), 1984; Meot-Ner (Mautner) and Sieck, 1985). Generalization to broader ΔPA ranges, however, seems rather to indicate an exponential dependence (Zeegers-Huyskens, 1986, 1988).

ΔPA correlations are parallelled by analogous studies on the role played by the difference of *acid–base dissociation constants*, $\Delta\mathrm{p}K_\mathrm{a} = \mathrm{p}K_\mathrm{AH}(\mathrm{D-H}) - \mathrm{p}K_\mathrm{BH}(\mathrm{A-H}^+)$, in strengthening the H-bonds associated with the proton-transfer acid–base equilibrium

$$\mathrm{D-H}\cdots{:}\mathrm{A} \rightleftharpoons {}^{½-}\mathrm{D}{:}\cdots\mathrm{H}^+\cdots{:}\mathrm{A}^{½-} \rightleftharpoons {}^{-}\mathrm{D}{:}\cdots\mathrm{H-A}^+,$$

for which it was supposed that the condition $\Delta\mathrm{p}K_\mathrm{a} \cong 0$ (called pK_a matching or pK_a equalization) could stabilize the doubly charged form at the center giving rise to strong and proton-centered H-bonds of mostly covalent nature. Following a first thermodynamic treatment by Huyskens and Zeegers-Huyskens (1964), this hypothesis was successfully investigated by dielectric measurements (Ratajczak and Sobczyk, 1969) and IR spectroscopy in non-polar organic solvents (Malarski, Rospenk and Sobczyk, 1982; Brycki and Szafran, 1982). At the beginning of the 1990s, the concept of pK_a matching was resumed with the proposal that pK_a-matched low-barrier H-bonds (LBHBs), that is strong double-well bonds with a low PT barrier, may exert a leading role in the reaction mechanism of serine proteolytic enzymes (Cleland, 1992; Cleland and Kreevoy, 1994; Frey, Whitt and Tobin, 1994).

In addition to IR and Raman, the H-bond has been studied by many other spectroscopic techniques (including electronic absorption and fluorescence spectroscopy, neutron inelastic scattering, deuteron nuclear quadrupolar coupling and gas-phase microwave rotational spectroscopy) but none of them has had an impact comparable with *proton NMR (^{1}H NMR) spectroscopy* that has become one of the most sensitive methods for detecting the presence of H-bonds in solution, as firstly recognized by Liddel and Ramsey (1951) and then exploited by Huggins (Huggins *et al.*, 1955, 1956), who were also the first to discuss the considerable differences between NMR and IR spectroscopies.

One of these differences is that ^{1}H chemical shifts of H-bonded protons, δ(D–H), undergo much larger variations than the corresponding $\Delta\nu/\nu$ IR stretching frequency shifts. In fact, the NMR shifts of N–H and O–H donors range from 8–10 ppm for weak up to 20–22 ppm for very strong H-bonds, making ^{1}H NMR the ideal probe for detecting strong H-bonds in solution, even because the signals do not overlap with those of the other molecular protons, which rarely exceed 11–13 ppm. Moreover, the great resolution of the NMR spectral lines allows the detection of very small chemical shifts, which makes the method ideal also for studying weak donors, such as C–H and S–H, or multiple-bond or aromatic-ring acceptors (Foster and Fyfe, 1969). For C–H···A bonds, in particular, Pimentel (Pimentel and McClellan, 1960), commenting on his previous paper with Huggins (Huggins *et al.*, 1955) on the NMR spectra of nitrogen bases dissolved in chloroform, remarked that ‘this study furnishes corroborative evidence that the chloroform–base interaction can be classified as a H-bond’, a statement that, considered with the benefit of hindsight, indicates that the successive long controversy on the existence of C–H···A bonds had been perhaps solved before it started. A second difference is that, since the radiation interaction time is much longer in NMR ($10^{-1}-10^{-8}$ s) than in IR or Raman (10^{-13} or 10^{-14} s, respectively), ^{1}H NMR may have difficulties in detecting the single partners of the D–H + A $\rightleftharpoons$ D–H···A tautomeric equilibrium at room temperature, particularly in water, a problem that can be partially circumvented at cryogenic temperatures.

The advances in NMR instrumentation since Pimentel's times have been tremendous, including high magnetic field, multiprobe, pulsed Fourier transform, multidimensional, and nuclear Overhauser effect techniques applied in a wide interval of experimental temperatures. A large number of quite interesting studies on H-bond strength, geometry and dynamics performed by ^{1}H, ^{2}H, ^{3}H, ^{13}C, ^{15}N, ^{17}O, ^{31}P, and ^{19}F NMR spectroscopy have been reported in the last 20 years. In addition to normal chemical shifts, these works make extensive use of other methods that are particularly suited for H-bond characterization, in particular primary and secondary isotope effects produced by hydrogen/deuterium exchange on chemical shifts (Hansen, 1996; Bolvig and Hansen, 2000; Dziembowska *et al.*, 2004) and deuterium fractional factors, that is equilibrium constants for the exchange of hydrogen with deuterium in D_2O solutions as determined by NMR methods (Kreevoy and Liang, 1980; Harris and Mildvan, 1999). A review of these methods applied to strong H-bonds in chemistry and biology is available (Perrin and Nielson, 1997) and some results obtained for strong charge-assisted H-bonds important in enzymatic catalysis are discussed in Section 8.5.2.

Solid-state NMR spectroscopy has considerably extended the NMR application range, even because its results are directly comparable with X-ray crystallographic ones. The problem that solids produce broad NMR signals because of the hindered molecular motion is overcome by a kind of mechanical tumbling, the *magic-angle spinning*, which was firstly applied in the ^{13}C CP-MAS (cross-polarization magic-angle spinning) method. A first review is due to Fyfe (1983) and another, specifically dedicated to H-bond applications, to Etter *et al.* (1988). Solid-state ^{1}H NMR would be more directly connected to the H-bond problem but presents considerable difficulties due to the great abundance of the ^{1}H isotope. Several solutions have been proposed, such as CRAMPS (combined rotational and multiple pulse spectroscopy) (Gerstein *et al.*, 1977; Ryan *et al.*, 1980), deuterium spin dilution of the protons, and high-speed magic-angle spinning. Note that, at variance with solution measurements, solid-state NMR provides the full tensor, σ_{ij}, of the chemical shift, whose components bear a strict relationship with H-bond geometry and orientation (Berglund and Vaughan, 1980; Rohlfing *et al.*, 1983). A $d(\mathrm{H}\cdots\mathrm{O})$ versus σ correlation over a wide range of proton chemical shifts and neutron-determined H···O distances ($21.0 \leq$ average $\sigma_{ij} \leq 5.0$ ppm; $1.24 \leq \mathrm{H}\cdots\mathrm{O} \leq 1.92$ Å) has been reported by Jeffrey and Yeon (1986).

The first reviews of *X-ray diffraction methods* applied to the study of H-bonds in crystals date back to the early 1950s (Donohue, 1952; Hunter, 1944; Nakamoto *et al.*, 1955; Pimentel and Sederholm, 1956) and provided the first reliable distributions of D···A distances for the most common H-bonds (Nyburg, 1961). The information obtained, however, was rather poor because standard errors were normally unknown (and never better than 0.02 Å) and the position of the H-bonded proton was completely undetermined, so that the occurrence of bent or straight bonds could only be matter of guessing. It can be asserted that, as a general fact, X-ray diffraction has always fallen short of H-bond needs and, for this reason, relevant advances in the diffraction field have always been followed by steep improvements in H-bond understanding. For example, the concomitant introduction of computer-controlled X-ray diffractometers and new direct methods able to solve practically all organic structures produced, at the beginning of the 1970s, a true explosion of new and better H-bond studies without, however, the problem of the *exact* proton location being completely solved.

The proton problem found its final solution with the appearance of *neutron diffraction methods*, due to the fact that neutrons are scattered by nuclei and not, as X-rays, by electrons and then can locate the exact proton position instead of the centroid of the electron density (Section 2.4.2.1). In 1976 Olovvson and Jönsson published a summary of all neutron diffraction studies prior to August 1973 where standard deviations of the bond distances involving proton were smaller than ~0.02 Å. These were only 8 in the years 1965–66 and 13, 21, and 36 in the following two-year periods, for a total of 76 structures for the full period considered. Several structures dealt with bonds of charge-assisted type (mostly $[O-H\cdots O]^-$ but also two $[O-H\cdots O]^+$ bonds) with the clear aim of assessing the existence of these strong and proton-centered H-bonds, while the presence of as many as 14 structures of amino acids (surveyed by Koetzle and Lehmann, 1976) well witnesses the pressure that biochemistry was starting to exert on structural crystallography. The neutron structures of biomolecules were eventually reviewed by Jeffrey and Saenger (1991) in their book *Hydrogen Bond in Biological Molecules*. Though X-ray analyses still outnumber the neutron ones by 300:1, the few thousands of neutron structures collected have certainly given a central contribution to establishing an unquestionable standard of H-bond geometry, so setting the bases for the comprehensive H-bond classifications reported in the next chapters. For a recent review on neutron diffraction techniques that pays particular attention to the H-bond problem see Wilson (2000).

Another important achievement of X-ray diffraction is the experimental determination of molecular electron densities and electrostatic potentials (Coppens, 1997; Koritsanszky and Coppens, 2001). The function of interest is the *deformation density*, $\Delta\rho(\boldsymbol{r})$, which represents the difference between the *total electron density* of the molecule, $\rho(\boldsymbol{r})$, and that of its *promolecule*, consisting of the superposition of the spherical electron densities of non-bonded atoms centered at their nuclear positions. $\Delta\rho(\boldsymbol{r})$ reveals the effects of chemical bonding, essentially the distribution of shared valence electrons associated with single, double, and triple bonds together with that of non-bonded electron pairs. A number of reviews on the chemical applications, including H-bonding, of electron distributions are available (Dunitz, 1979; Coppens and Hall, 1981; Klein and Stevens, 1988; Spackman and Brown, 1994). Other density-related properties of great chemical interest are the *electrostatic potential*, $\Phi(\boldsymbol{r})$, and the *electric field*, $E(\boldsymbol{r})$, which can be calculated in different ways from the total electron density $\rho(\boldsymbol{r})$. The electrostatic potential of a molecule of M atoms can be defined as the sum of its nuclear and electronic contributions, $\Phi(\boldsymbol{r}) = \Phi_N(\boldsymbol{r}) + \Phi_e(\boldsymbol{r})$, where $\Phi_N(\boldsymbol{r}) = \Sigma_{k=1,M} Z_k/|\mathbf{R}_k - \boldsymbol{r}|$ and $\Phi_e(\boldsymbol{r}) = -\int d\boldsymbol{r}'\rho(\boldsymbol{r}')/|\boldsymbol{r}' - \boldsymbol{r}|$, and the electric field as the gradient of such a potential, $E(\boldsymbol{r}) = -\nabla\Phi(\boldsymbol{r})$. Also, their applications to chemistry have been reviewed (Politzer and Truhlar, 1981; Klein and Stevens, 1988).

Approximate maps of valence-electron distributions have been observed from the beginning of diffraction studies from the final difference Fourier maps carried out after refinement of accurate crystal structures, particularly when determined at low temperature. The real problem was to transform these qualitative into quantitative maps having sufficient accuracy to be compared with the theoretical ones and to be analyzable by the same methods (AIM, ELF, source-function, etc.) described above. The first deformation densities were simply computed as the difference between the experimental electron density of the crystal (i.e. derived from the observed X-ray structure factors) and the density of a kind of *procrystal* made of spherical atoms with the same positional and vibrational parameters of the real crystal, sometimes separately determined by neutron (X–N maps) instead of X-ray diffraction (X–X maps) to maximize the precision of the nuclear positions. The main drawback of these *dynamic* deformation densities was that they were *thermally averaged* by the atomic vibrations and then not directly comparable with theoretical ones. To overcome the problem, the deformation density was expressed in parametric form and refined by least-squares analysis, along with the other positional and thermal parameters of the structure (Hirshfeld, 1971; Stewart, 1976; Hansen and Coppens, 1978). This was achieved by expressing the local density around each atom as a sum of Slater-like atomic orbitals (i.e. products of a radial function with a spherical harmonic) whose parameters were to be optimized, together with the positional and vibrational ones, in a so-called least-squares *multipolar* refinement. Since this procedure adds a lot of new parameters to be refined, it is mandatory that the diffraction experiment can collect a maximum number of reflections that is obtained, apart from any diffraction consideration, by lowering the temperature at the minimum values achievable.

The great advantage of this multipolar refinement is that it provides deformation densities that do not include thermal-smearing effects. Hence, what is obtained is a *static density* that can be directly compared with the theoretical one and analyzed by similar methods. The results obtained by applying AIM topological methods to the experimental electron densities of a number of H-bonded systems have been recently surveyed by Koritsanszky and Coppens (2001), Lyssenko and Antipin (2006), and, for weak intermolecular bonds, by Munshi and Guru Row (2005).

1.4 Significant books and reviews

Of course, the H-bond history is too intricate and complex to be accounted for by the short outline above. Theory and experiments have developed for more than 80 years in strict mutual interdependence creating a kind of Gordian knot that, to be inspected, should be cut by the legendary Alexander's sword. The correct way of doing that is to cut perpendicularly to the time coordinate, that is respecting a strict chronological order, in such a way to be sure that, at any given date, future and past developments can never mix. To help the interested reader to build up his/her own temporal scale, this section reports, in chronological order, a putatively complete list of books dedicated, in full or in significant part, to the H-bond problem together with a selection of the most important reviews published in the field.

1. Lewis (1923). *Valence and the structure of atoms and molecules.*
2. Pauling (1939, 1940, 1960). *The nature of the chemical bond and the structure of molecules and crystals: An introduction to modern structural chemistry.*
3. Hadži and Thompson eds. (1959). *Hydrogen bonding.*
4. Pimentel and McClellan (1960). *The hydrogen bond.*
5. Coulson (1961). *Valence.*
6. Kavanau (1964). *Water and solute-water interactions.*
7. Hamilton and Ibers (1968). *Hydrogen bonding in solids: Methods of molecular structure determination.*
8. Vinogradov and Linnell (1971). *Hydrogen bonding.*
9. Franks (1972–82). *Water: A comprehensive treatise.*
10. Joesten and Schaad (1974). *Hydrogen bonding.*
11. Schuster, Zundel and Sandorfy eds. (1976). *The hydrogen bond: Recent developments in theory and experiments.*
12. McWeeny (1979). *Coulson's valence.*
13. Ratajczak and Orville-Thomas eds. (1980, 1981, 1982). *Molecular interactions.*
14. Kaplan (1986). *Theory of molecular interactions.*
15. Huyskens, Luck and Zeegers-Huyskens eds. (1991). *Intermolecular forces: An introduction to modern methods and results.*
16. Jeffrey and Saenger (1991). *Hydrogen bonding in biological structures.*
17. Bountis ed. (1992). *Proton transfer in hydrogen-bonded systems.*
18. Smith ed. (1994). *Modeling the hydrogen bond.*
19. Jeffrey (1997). *An introduction to hydrogen bonding.*
20. Scheiner (1997). *Hydrogen bonding: A theoretical perspective.*
21. Hadži ed. (1997). *Theoretical treatments of hydrogen bonding.*
22. Nishio, Hirota and Umezawa (1998). *The CH/π interaction. Evidence, nature, and consequences.*
23. Desiraju and Steiner (1999). *The weak hydrogen bond in structural chemistry and biology.*
24. Stone (2000). *The theory of intermolecular forces.*
25. Kubas (2001). *Metal dihydrogen and σ-bond complexes – Structure, theory, and reactivity.*
26. Bernstein (2002). *Polymorphism in molecular crystals.*
27. Elsaesser and Bakker eds. (2002). *Ultrafast hydrogen bonding dynamics and proton transfer processes in the condensed phase.*
28. Grabowski ed. (2006). *Hydrogen bonding – New insights.*
29. Maréchal (2007). *The hydrogen bond and the water molecule. The physics and chemistry of water, aqueous and bio media.*

30. Bakhmutov (2008). *Dihydrogen bond: Principles, experiments, and applications*

A. Huggins (1971). 50 Years of hydrogen bond theory.
B. Pimentel and McClellan (1971). Hydrogen bonding.
C. Kollman and Allen (1972). The theory of the hydrogen bond.
D. Speakman (1972). Acid salts of carboxylic acids, crystals with some 'very short' hydrogen bonds.
E. Novak (1974). Hydrogen bonding in solids. Correlation of spectroscopic and crystallographic data.
F. Morokuma (1977). Why do molecules interact? The origin of electron donor-acceptor complexes, hydrogen bonding, and proton affinity.
G. Emsley (1980). Very strong hydrogen bonding.
H. Barnes (1983). Molecular complexes of the hydrogen halides studied by matrix isolation infrared spectroscopy.
I. Emsley (1984). The composition, structure and hydrogen bonding of the β-diketones.
J. Etter, Hoye, and Vojta (1988). Solid-state NMR and X-ray crystallography: Complementary tools for structure determination.
K. Kuhs and Lehmann (1986). The structure of ice I_h.
L. Legon and Millen (1987). Directional character, strength, and nature of the hydrogen bond in gas-phase dimers.
M. Staab and Saupe (1988). 'Proton sponges' and the geometry of hydrogen bonds: Aromatic nitrogen bases with exceptional basicities.
N. Hibbert and Emsley (1990). Hydrogen bonding and chemical reactivity.
O. Desiraju (1991). The C–H···O hydrogen bond in crystals: What is it?
P. Llamas-Saiz, Foces-Foces and Elguero (1994). Proton sponges.
Q. MacDonald and Whitesides (1994). Solid-state structures of hydrogen-bonded tapes based on cyclic secondary diamines.
R. Bernstein *et al.* (1995). Patterns in hydrogen bonding: Functionality and graph set analysis in crystals.
S. Jeffrey (1995). Hydrogen-bonding: An update.
T. Brammer *et al.* (1995). Hydrogen bonds involving transition metal centers – A brief review.
U. Gilli *et al.* (1996). A novel approach to hydrogen bonding theory.
V. Crabtree *et al.* (1996). A new intermolecular interaction: Unconventional hydrogen bonds with element-hydride bonds as proton acceptors.
W. Perrin and Nielson (1997). 'Strong' hydrogen bonds in chemistry and biology.

X. Fuster and Silvi (2000). Does the topological approach characterize the hydrogen bond?
Y. Custelcean and Jackson (2001). Dihydrogen bonding: Structures, energetics, and dynamics.
Z. Koritsanszky and Coppens (2001). Chemical applications of X-ray charge-density analysis.
AA. Northrop (2001). Follow the protons: A low-barrier hydrogen bond unifies the mechanisms of the aspartic proteases.
BB. Desiraju (2002). Hydrogen bridges in crystal engineering: Interactions without borders.
CC. Epstein and Shubina (2002). New types of hydrogen bonding in organometallic chemistry.
DD. Fillaux *et al.* (2002). Quantum proton transfer and interconversion in the benzoic acid crystals: vibrational spectra, mechanism and theory.
EE. Roux (2002). Computational studies of the gramicidin channel.
FF. Steiner (2002). The hydrogen bond in the solid state.
GG. Brammer (2003). Metals and hydrogen bonds.
HH. Katrusiak (2003). Macroscopic and structural effects of hydrogen-bond transformations.
II. Belkova, Shubina and Epstein (2005). Diverse world of unconventional hydrogen bonds.
JJ. Meot-Ner (Mautner) (2005). The ionic hydrogen bond.
KK. Munshi and Guru Row (2005). Evaluation of weak intermolecular interactions in molecular crystals *via* experimental and theoretical charge densities.
LL. Wilson (2007). The evolution of hydrogen atom parameters under changing external conditions by time-of-flight single crystal neutron diffraction.
MM. Ball (2008). Water as an active constituent in cell biology.
NN. Gilli *et al.* (2009). Predicting hydrogen-bond strengths from acid-base molecular properties. The pK_a slide rule: Toward the solution of a long-lasting problem.

2

GENERALITIES, DEFINITIONS AND PRELIMINARY CLASSIFICATION

2.1 Basic H-bond nomenclature

The nucleus of H-bond nomenclature is the distinction between donor and acceptor. In any D–H···:A bond, D–H is the H-bond donor (and also a Brønsted acid, a Lewis acid, and an electron acceptor) and :A the H-bond acceptor (and also a Brønsted base, a Lewis base, and an electron donor). Donors and acceptors can combine in different ways following the traditional terminology summarized in Chart 2.1 as essentially derived from Jeffrey (1997).

From a chemical point of view, H-bonds can be classified as *homonuclear* and *heteronuclear* (C2.1.a1 and a2), the former being further divided in *homomolecular* and *heteromolecular* (C2.1.a1′ and a1″) according to whether the substituents R_1 and R_2 are equal or less. Normal D–H···A bonds are usually referred to as *two-center* or *monodentate* (C2.1.b1) while the situation of multiple H-bond acceptor is indicated by the terms *three-center*, *bifurcated* or *bidentate* (C2.1.b2 and b2′) and *four-center* or *tridentate* (C2.1.b3). Multiple H-bond donation is indicated as *double* or *chelated* (C2.1.c1) or *triple* (C2.1.c2). Combinations are allowed and C2.1.c1′ is both *double and three-center.*

H-bond connectivity in crystals may be extremely complex and special nomenclature systems have been developed to describe it exactly (Kuleshova and Zorky, 1980; Etter, 1990; Etter *et al.*, 1990; Bernstein *et al.*, 1995; Grell *et al.*, 2002; Bernstein, 2002). A trivial distinction is between *intramolecular* (C2.1.d1) and *intermolecular* (C2.1.d2–d5) bonds, and the most frequent intermolecular patterns are *dimers*, *chains*, *planes*, and *3-dimensional tetrahedral* structures.

Finally, H-bonds widely differ for the geometry of the D–H···A fragment (C2.1.e1–e4) between the two extreme configurations *dissymmetric and bent* (C2.1.e3) and *symmetric and linear* (C2.1.e4) that are the typical situations occurring for weak and very strong bonds, respectively. The symbols commonly used in this book are $d_1 = d(\mathrm{D–H})$, $d = d_2 = d(\mathrm{H\cdots A})$, $D = d(\mathrm{D\cdots A})$ for distances and $\theta = \alpha(\mathrm{D–H–A})$ for angles (C2.1.e1–e2).

2.2 Formal H-bond definitions

It can be said that any important author has given a more or less different definition of what the H-bond should be. All definitions, however, can be traced back to only two main types according to whether they are based on purely empirical

CHART 2.1. HB BASIC NOMENCLATURE

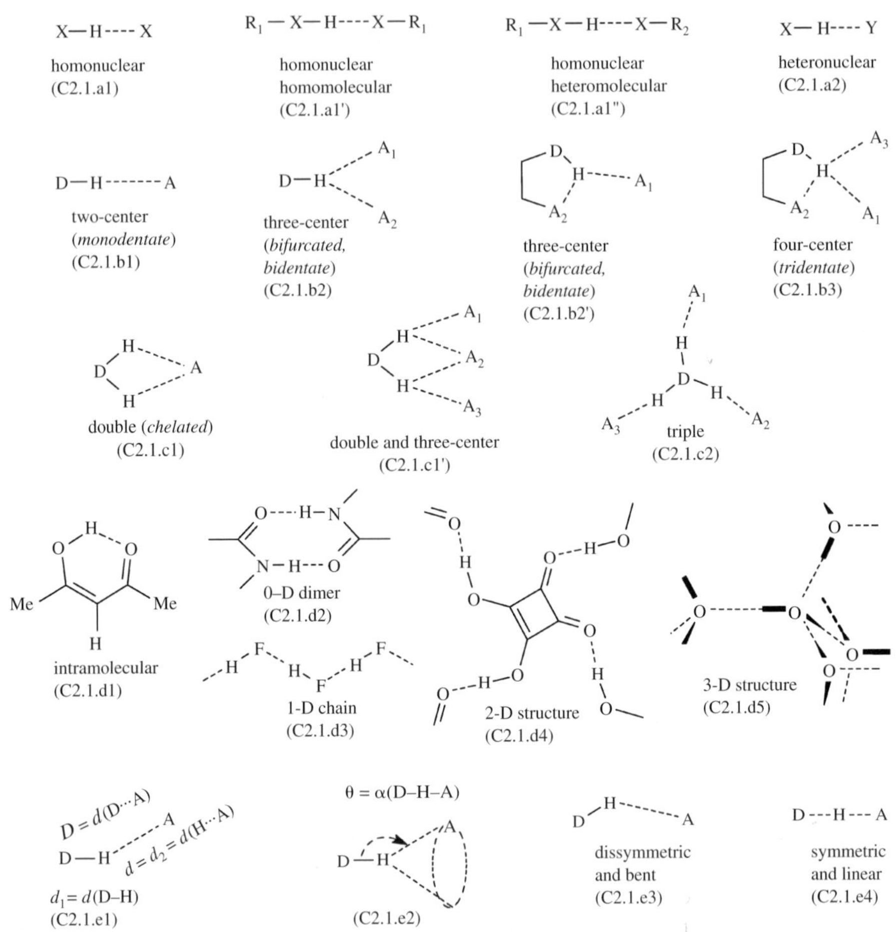

X—H----X
homonuclear
(C2.1.a1)

R_1—X—H----X—R_1
homonuclear
homomolecular
(C2.1.a1')

R_1—X—H----X—R_2
homonuclear
heteromolecular
(C2.1.a1")

X—H----Y
heteronuclear
(C2.1.a2)

D—H-------A
two-center
(*monodentate*)
(C2.1.b1)

three-center
(*bifurcated,*
bidentate)
(C2.1.b2)

three-center
(*bifurcated,*
bidentate)
(C2.1.b2')

four-center
(*tridentate*)
(C2.1.b3)

double (*chelated*)
(C2.1.c1)

double and three-center
(C2.1.c1')

triple
(C2.1.c2)

intramolecular
(C2.1.d1)

0–D dimer
(C2.1.d2)

1-D chain
(C2.1.d3)

2-D structure
(C2.1.d4)

3-D structure
(C2.1.d5)

$D = d(D\cdots A)$
$d = d_2 = d(H\cdots A)$
$d_1 = d(D–H)$
(C2.1.e1)

$\theta = \alpha(D–H–A)$
(C2.1.e2)

dissymmetric
and bent
(C2.1.e3)

symmetric
and linear
(C2.1.e4)

considerations or rather make specific reference to a more general theory of the chemical bond.

Probably the best empirical definition was given by Pimentel and McClellan (1960) in their book *The Hydrogen Bond*:

> A H bond [A–H···B] exists between a functional group A–H and an atom or a group of atoms B in the same or a different molecule when (*a*) there is evidence of bond formation (association or chelation) and (*b*) there is evidence that this new bond linking A–H and B specifically involves the hydrogen atom already bonded to A.

The authors remark that 'It is convenient to separate parts (*a*) and (*b*) because there are many physical measurements that provide information relative to (*a*) but only a few substantiating (*b*). It is specifically criterion (*b*) that differentiates the H bond from other types of associative interaction....' This statement underlies the purely empirical character of the definition. While the evidence for association or chelation is straightforwardly derivable from several physico-chemical properties of the system studied (e.g. altered freezing and boiling points, changes of solubility, dielectric properties and conductivity, deviations from ideal-gas and solution laws), the proof of a direct involvement of the proton requires other specific measurements, the most common being the frequency shift of the IR and Raman ν(A–H) stretching band, the ^{1}H NMR chemical shift of the proton, δ(AH), together with the A···B and AH···B distances determined by X-ray and neutron diffraction becoming far shorter than the sum of their van der Waals radii.

The second type of H-bond definition, originally proposed by Latimer and Rodebush (1920) in the first known paper on the H-bond and successively completed by Vinogradov and Linnell (1971) in their *Hydrogen Bonding* book, makes direct reference to the theory of chemical bonding that Lewis (1923) was developing in those years and that became later known by the name of VB theory (Pauling, 1939, 1940, 1960; Coulson, 1961; McWeeny, 1979). With some minor adjustments it can be summarized as follows:

> Hydrogen bonding occurs between a proton-donor group A–H and a proton-acceptor group B, where A is an electronegative atom, O, N, S, X (F, Cl, Br, I) or C, and the acceptor group is a lone pair of an electronegative atom or a π bond of a multiple bond (unsaturated) system. Generally, a H-bond can be characterized as a proton shared by two lone electron pairs.

These two definitions are not mutually exclusive but, in some way, complement each other. The first is purely empirical and can be called upon a posteriori to assess that what is experimentally observed is most probably due to H-bond formation. The second neglects experiments, is based on the *electronic structure* of the interacting moieties, and can predict H-bond formation (before the experiment) as well as justify it (after the experiment). The two definitions have, however, different scope and discrimination capacity. For the first, all attractive interactions where the proton is securely involved are to be classified as H-bonds, so that the B–H–B bond in boranes is also to be considered as a true H-bond as, on the other hand, explicitly proposed by Pimentel and McClellan (1960). Only the second, entailing the presence of the lone pair on the acceptor atom, can distinguish the three-center–two-electron (3c–2e) bonds occurring in boranes from the three-center–four-electron (3c–4e) interactions to which is normally assigned the status of true H-bonds.

Another important distinction concerns the correct bond polarity that true H-bonds should have. The first definition does not even contemplate the point, while the second includes atomic electronegativity conditions assessing the correct polarity to be $A^{\delta-}$–$H^{\delta+}$···$B^{\delta-}$, so excluding other interactions with reversed

$A^{\delta+}$–$H^{\delta-}\cdots B^{\delta+}$ polarity, such as agostic interactions or the so-called inverse H-bonds (see below). To notice, anyway, that these polarity rules are not devoid of elements of confusion, such as: (*i*) H-bond polarities do not depend on atomic but rather on group electronegativities as modified by all other surrounding atoms, a quantity that can be difficult to appreciate in many practical cases; and (*ii*) extensive analysis of weak H-bonds (Desiraju and Steiner, 1999) seems to indicate that the polarity rule is more strictly obeyed on the donor than on the acceptor side, a fact particularly evident when dealing with π-bond acceptors where the accepting multiple bond can be defined as an *electron-rich* rather than an *electronegative* moiety. These difficulties can be overcome by considering that *the H-bond is not a bond*, in the sense accepted for normal chemical bonds, but rather *a dual (or multiple) bond* made by a single proton with two (or more) electron pairs located on the surrounding ligands. This feature, stressed by Latimer and Rodebush since their 1920 seminal paper ('a H-bond can be characterized as a proton shared by two lone electron pairs'), makes it clear that the true H-bond polarity must be $^{\delta-}A{:}\cdots H^{+}\cdots{:}B^{\delta-}$.

These few considerations seem then to indicate the Vinogradov and Linnell (1971) definition has to be preferred. Accordingly, it is adopted here with only a few changes intended to stress the importance of the 3c–4e bond and to avoid an excessive mixing between H-bond and Brønsted–Lowry acid–base theory, according to the Arnett and Mitchell (1971) criticism. To this aim, the traditional A–H···B notation, where A–H is clearly an acid and B a base, has been changed into the more neutral *D–H···:A* form, intending that *D–H* is the *H-bond donor* and *:A* the *H-bond acceptor* that carries the free-electron pair. With these minor modifications, our final H-bond definition becomes:

> *The H-bond is a three-center–four-electron (3c–4e) shared-proton interaction having the general form R–D–H···:A–R′, where D is the proton donor (an electronegative atom, such as F, O, N, C, S, Cl, Br and I) and :A the proton acceptor or lone-electron-pair carrier (a second electronegative atom or the π-bond of a multiple bond). The H-bond can also be seen as a single proton sharing two lone-electron pairs from two adjacent electronegative atoms or groups: $R{-}D^{-}{:}\cdots H^{+}\cdots{:}A{-}R'$.*

A further simplification adopted all through the text is that, unless specifically stated, the H-bond donor is always *a neutral acid D–H* and *never a protonated base $A{-}H^{+}$* and the H-bond acceptor is always a *neutral base :A* and never a *deprotonated acid $:D^{-}$*. This means, for instance, that in the $R{-}COO^{-}\cdots{}^{+}H{-}NR_3$ bond the donor remains R–COOH (an acid) and the acceptor $:NR_3$ (a base) irrespective of the proton transfer that occurred. This procedure will simplify the relationships between H-bond strength and pK_as of the donor and acceptor treated in Chapter 4.

This H-bond definition is strictly connected to the VB theory by the so-called Coulson formalism (Coulson, 1959; Coulson and Danielsson, 1954a,b) that is illustrated, for the particular case of the O–H···O bond, in Chart 2.2. It regards

CHART 2.2. THE COULSON VB FORMALISM

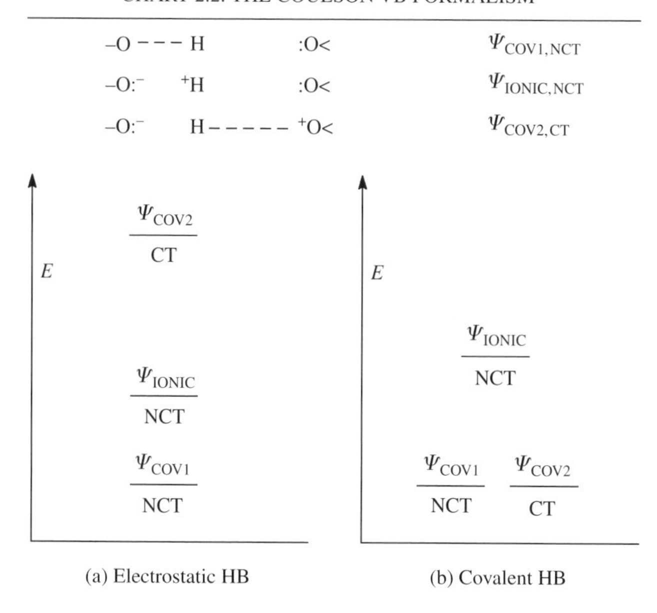

(a) Electrostatic HB (b) Covalent HB

the H-bond as a mixture of three main VB canonical wavefunctions, $\Psi_{\mathrm{COV1,NCT}}$, $\Psi_{\mathrm{IONIC,NCT}}$ and $\Psi_{\mathrm{COV2,CT}}$, which mix with respective coefficients ω_1, ω_2, and ω_3 (CT and NCT mean with or without charge transfer). An essentially electrostatic bond would have $\omega_1 > \omega_2 \gg \omega_3$, while a purely covalent 3c–4e one will be characterized by $\omega_1 = \omega_3 \gg \omega_2$. Since the mixing between two canonical wavefunctions is the larger the smaller their energy difference is, these two extreme types of H-bond can also be represented on a semiquantitative energy scale by schemes (a) and (b) of Chart 2.2, which clearly show that any discussion on whether the H-bond is electrostatic or covalent in nature is purely academic because its true nature is actually *twofold* and can be switched from electrostatic to covalent simply by changing the relative energy order of the three VB resonance forms.

The drawing also stresses the importance of symmetry in H-bond formation because the energy equivalence, $E(\Psi_{\mathrm{COV1,NCT}}) = E(\Psi_{\mathrm{COV2,CT}})$, of the two mixing O–H···O and O···H–O forms requires that they are also geometrically identical, that is symmetry related by a mirror plane perpendicular to the bond. This suggests that very strong H-bonds must be wholly symmetric from both a geometrical (they must be linear and proton-centered) and a chemical (they must be homonuclear and homomolecular) point of view (extension of this concept to the formation of strong heteronuclear H-bonds will be discussed in detail in Section 3.2.3.4). For the moment, it may be remarked that Coulson's model is consistent with the simple concept that the increasing degree of covalency is the factor shifting the H-bond properties between two extremes: (i) the weak, dissymmetric and mostly bent bond of electrostatic nature; and (*ii*) the strong, symmetric and linear interaction classifiable as a true 3c−4e covalent bond. What is needed to really understand the H-bond is then to identify the physical factors that are actually able to determine such a degree of covalency.

2.3 The H-bond as a shared-proton interaction: A chemical classification

The main feature that characterizes the H-bond and makes it different from any other donor−acceptor interaction is the *sharing of the proton* between the H-bond donor and acceptor atoms. In other words, H-bonds can always be seen as *shared-proton interactions*. The converse, however, may not be true and not all the shared-proton interactions necessarily conform to the H-bond definition given above.

Chart 2.3 is an attempt to summarize the types of shared-proton interactions so far described in the chemical and crystallographic literature and to organize them in a more general classification based on purely chemical criteria. This classification owes much to previous systematic investigations carried out by other authors by taking advantage of the newly introduced crystal databases and in particular, to the pre-screening performed by Jeffrey (1997) and Desiraju and Steiner (1999) in their recent H-bond books and, as far as H-bonds formed by metal centers are concerned, by Brammer (2003) in his recent review on this subject. Part of the nomenclature, such as conventional H-bonds or weak H-bond donors and acceptors, is also mutuated from the Desiraju and Steiner book.

The basic distinction is between 3c−4e and 3c−2e interactions, examples of the latter being *boranes* (Group 3.2) (Lipscomb, 1954), *agostic interactions* (Group 3.1) (Crabtree, 1993; Braga *et al.*, 1996a) and *lithium bonds* ($Li^{\delta+}-H^{\delta-}\cdots Li^{\delta+}-H^{\delta-}$; Group 3.3), also called *inverse H-bonds* (Dill *et al.*, 1977; DeFrees *et al.*, 1987) because the hydrogen is negatively instead of positively charged. None of these 3c−2e interactions can be classfied as true H-bonds according to our previous definition but it may be interesting to briefly discuss two of them, agostic interactions and lithium bonds, to see how this exclusion may be actually justified by other chemical-bond considerations.

An agostic interaction is said to occur when an electron-deficient metal makes a close approach to an electron-rich C−H bond giving rise to a nearly T-shaped M···(H−C) contact (Group 3.1) that is normally, but not exclusively, intramolecular. The most frequently involved metals are Ti, Ta, Zr, Ni and Fe for which the reported interaction energies are remarkably high (from 7 to 15 kcal mol^{-1}). The interaction, however, does not bear an evident relationship with normal H-bonds from the structural point of view (H-bonds are tendentiously linear, while agostic interactions are T-shaped) as well as from the electronic one (H-bonds are essentially caused by the :A Lewis basicity, while agostic interactions are due to the metal atom acidity). Also 3c−2e lithium bonds (Group 3.3) can hardly be considered H-bond-like. In this case the interaction occurs between a reversed $D^{\delta+}-H^{\delta-}$ dipole and an electron-deficient center, while

CHART 2.3. TAXONOMY OF SHARED-PROTON INTERACTIONS.
A CHEMICAL CLASSIFICATION OF THE HB

1st Category. 3-Center–4-electon bonds

$$X\bullet\!-\!\bullet H \;----\; :Y \;\longleftrightarrow\; {}^{-}X: \;----\; H\bullet\!-\!\bullet Y^{+}$$

Class 1. Hydrogen Bonds Involving Main-Group Elements

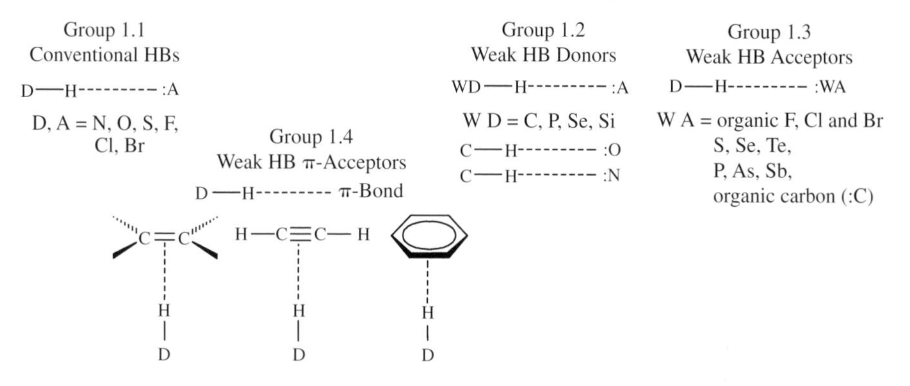

Class 2. Hydrogen Bonds Involving Metal Centers

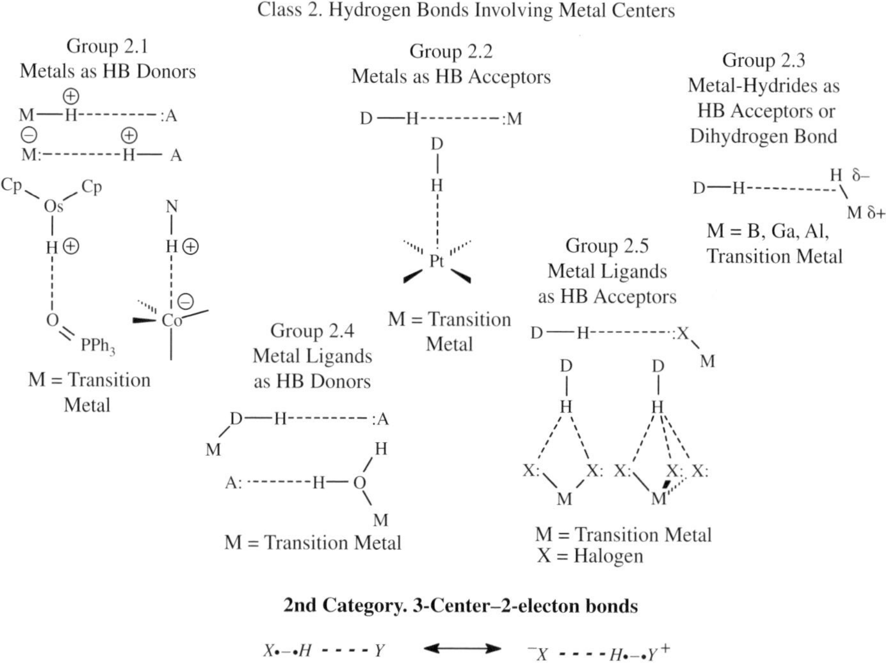

2nd Category. 3-Center–2-electon bonds

$$X\bullet\!-\!\bullet H \;----\; Y \;\longleftrightarrow\; {}^{-}X \;----\; H\bullet\!-\!\bullet Y^{+}$$

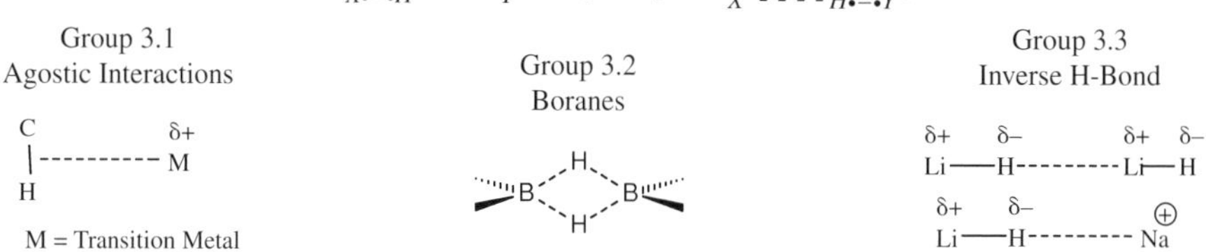

normal H-bonds are interactions of a $D^{\delta-}$—$H^{\delta+}$ dipole with an electron-rich atom. It is clear that, by accepting this inverse H-bond as a true H-bond of another type (which, by the way, would be compatible with the Pimentel and McClellan definition), any relationship between H-bond strength and the electronegativity of the donor and acceptor atoms will be lost, together with any possibility of rationalizing the electron density shifts associated with the D–H···:A three-center interaction that we are used to call H-bond.

Conversely, all the 3c—4e shared-proton interactions of Chart 2.3 appear to be perfectly consistent with the H-bond definition given above and are then to be considered, at least formally, as true H-bonds in every respect. The expression 'at least formally' means that it has still to be verified whether the final distribution of charges in all these different bonds actually corresponds to the $D^{\delta-}$—$H^{\delta+}\cdots A^{\delta-}$ distribution we have considered to be mandatory for all true H-bonds. Even with exclusion of 3c—2e bonds, the number and variety of H-bonds observed in nature remains very large, making it indispensable to organize them into more homogeneous classes. A partition that is easily made is between *H-bonds involving main-group elements (Class 1)* and *H-bonds involving metal centres (Class 2)*, a division that has the further advantage of keeping separate the bonds formed by metals, which represent a rather new field that has been only partially investigated, and whose classification may undergo the greatest changes in the future. The following discussion will show, however, that this division is not purely formal but can be grounded on precise chemical-bonding considerations.

2.4 H-bonds involving main-group elements (Class 1)

The main-group elements that can participate in H-bond formation are those of the upper-right corner of the Periodic Table listed, together with their Pauling's electronegativities, in Table 2.1. Electronegativities of donors and acceptors are known from the earliest H-bond studies to be one of the main determinants of H-bond strength, so that they can be used as a quite natural criterion of further H-bond classification. The division in groups (from 1.1 to 1.4) reported in Chart 2.3 is, in fact, a classification in terms of decreasing electronegativity.

The first subgroup (*Group 1.1: Conventional H-bonds*) includes the most electronegative atoms that, for this reason, can also form the strongest H-bonds, while

TABLE 2.1. Pauling's electronegativities of H-bond donors and acceptors. Atoms more electronegative than hydrogen are marked in italics.

						H	2.20
C	*2.55*	*N*	*3.04*	*O*	*3.44*	*F*	*3.98*
Si	1.90	P	2.19	*S*	*2.58*	*Cl*	*3.16*
Ge	2.01	As	2.18	*Se*	*2.55*	*Br*	*2.96*
		Sb	2.05	Te	2.1	*I*	*2.66*

the others cannot because of the intrinsic weakness (i.e. too low electronegativity) of the donor (*Group 1.2: Weak H-bond donors*) or of the acceptor (*Group 1.3: Weak H-bond acceptors* and *Group 1.4: Weak H-bond π-acceptors*).

2.4.1 *Conventional H-bonds (Group 1.1)*

The name *conventional H-bonds* has entered into use to indicate all the interactions that have been called 'hydrogen bonds' or 'hydrogen bridges' from the early times of H-bond studies and that often play a determinant role in chemistry, biology, and biochemistry. This particular group includes the bonds formed by main-group elements with electronegativities greater than some 3.0 (N, O, F, Cl and Br) that are, therefore, donors and acceptors strong enough to display, according to the circumstances, the full spectrum of H-bond energies and geometries. H-bonds formed by sulphur (S) as donor and acceptor are normally considered to be conventional as well, in spite of the relatively lower electronegativity of the sulphur atom (2.58). Conventional H-bonds include practically all cases of strong H-bonds so far known and, in a sense, all the other groups listed in Chart 2.3 (including metal centers) could be grouped under the common heading of *unconventional H-bonds* or, though less properly, of *weak H-bonds*.

From the point of view of their strengths and geometries, conventional H-bonds represent a quite inhomogeneous family covering the full range from *strong, short, linear and proton-centered bonds of covalent nature* to *weak, long, bent and proton-outcentered ones having electrostatic character*, though weak bonds are so frequent that strong ones can be considered somewhat of an exception. A rich literature has flourished on strong H-bonds since the times of Pauling (Pauling, 1939, 1940, 1960; Speakman, 1972; Emsley, 1980, 1984). Understanding the reasons for this great variability is the main aim of this book and it will be shown in Chapter 3 that the first step in doing that is to subdivide conventional H-bonds in six subgroups, called the six *chemical leitmotifs (CLs)* (Gilli *et al.*, 1994a, 1996a, 2004; Gilli and Gilli, 2000, 2002). This partition is anticipated in Table 2.2, without proof, to facilitate the following discussion and to illustrate from now the important point that really strong H-bonds can only be associated with a very small number of chemical patterns. These six CLs can be summarized as follows:

- CL # 1. *OHB* or *ordinary H-bond* (**2.1.I**). H-bonds that are neither charge-assisted nor π-bond or σ-bond cooperative (see below) and, for this reason, behave as rather weak electrostatic interactions. More than from their strength, OHBs derive their importance from their widespread occurrence in nature, which outnumbers that of all strong H-bonds;
- CL # 2. (±)*CAHB* or *double charge-assisted H-bond.* Strong H-bonds associated with pairs of donor and acceptor molecules having very similar pK_a values, such as the H-bonded complex **2.1.II** between pyridine-*N*-oxide ($pK_a = 0.79$) and trichloroacetic acid ($pK_a = 0.66$);
- CL # 3. (−)*CAHB* or *negative charge-assisted H-bond.* Strong H-bonds corresponding to the general formula $[X{\cdots}H{\cdots}X]^-$ and exemplified by the

TABLE 2.2. Division of conventional H-bonds into the six chemical leitmotifs (CLs)

CL #	Name	Acronym	Strength	Examples
1	Ordinary H-bond	OHB	Weak	2.1.I
2	Double charge-assisted H-bond	(±)CAHB	Strong	2.1.II
3	Negative charge-assisted H-bond	(−)CAHB	Strong	2.1.III
4	Positive charge-assisted H-bond	(+)CAHB	Strong	2.1.IV
5	Resonance-assisted H-bond (π-bond cooperative H-bond)	RAHB	Strong	2.1.V
6	Polarization-assisted H-bond (σ-bond cooperative H-bond)	PAHB	Moderate	2.1.VI

$[F\cdots H\cdots F]^-$ bifluoride anion **2.1.IIIa** or by the acid salts of carboxylic or inorganic acids (**2.1.IIIb**);

CL # 4. *(+)CAHB* or *positive charge-assisted H-bond.* Strong H-bonds with general formula $[X\cdots H\cdots X]^+$ and reducible to the situation where two identical molecules are bridged by a proton donated by a strong acid, such as the H-bonded complex formed by two dimethyl-sulphoxide molecules (**2.1.IV**);

CL # 5. *RAHB* or *resonance-assisted H-bond* (Gilli *et al.*, 1989; Bertolasi *et al.*, 1991) or *π-bond cooperative H-bond* (Jeffrey and Saenger, 1991). A peculiar type of bond where the donor or acceptor atoms are at the two ends of a short π-conjugated molecular fragment as, for instance, in the H-bonded dibenzoylmethane enol of scheme **2.1.V**.

CL # 6. *Polarization-assisted H-bond* or *PAHB* (Gilli and Gilli, 2000). A type of bond first proposed by Jeffrey and Saenger (1991) with the name of *σ-bond cooperative H-bond* to represent the bonds associated with chains of $\cdots O{-}H\cdots O{-}H\cdots$ hydroxyl groups occurring, e.g. in phenol crystals (**2.1.VI**); the bonds formed are only moderately strong and their $O\cdots O$ distances are located on the shorter borderline of ordinary H-bonds (OHBs; CL # 1).

2.4.2 *Weak H-bonds: General properties (Groups 1.2–4)*

H-bonds are known to encompass a wide and continuous scale of bonding energies going from, say, 0.2 to 45 kcal mol^{-1}, the weakest bonds being barely distinguishable from van der Waals interactions and the strongest ones very similar to covalent bonds. Terms such as 'weak' and 'strong' contain an inevitable element of arbitrariness without establishing a precise energy borderline between them. In this book the convention proposed by Jeffrey (1997) is generally adopted, for which all the H-bonds are divided into *strong*, *moderate*, and *weak* by assigning them energy ranges of 15–45, 4–15, and <4 kcal mol^{-1}, respectively (see Table 2.4 below).

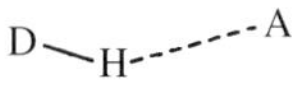

O–H···O inter: 2.70–3.00 Å
O–H···O intra: 2.65–2.90 Å
F–H···F inter: > 2.49 Å

(2.1.I)

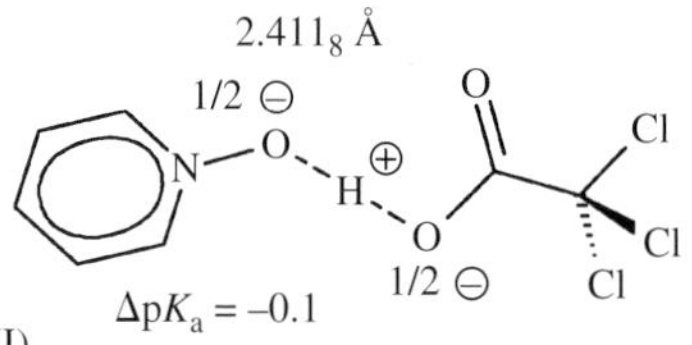

2.411_8 Å

$\Delta pK_a = -0.1$

(2.1.II) Pyridine-*N*-oxide - trichloroacetic acid

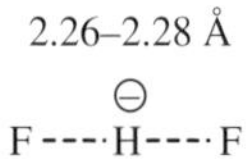

2.26–2.28 Å

Hydrogen difluoride (2.1.IIIa)

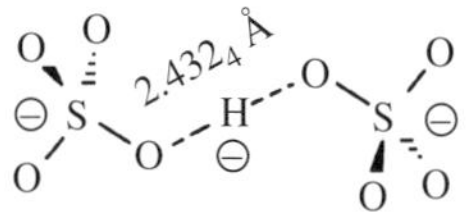

2.432_4 Å

Hydrogen-bis(sulphate) (2.1.IIIb)

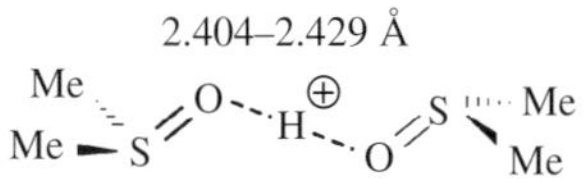

2.404–2.429 Å

H⁺-Bis(dimethyl-sulphoxide) (2.1.IV)

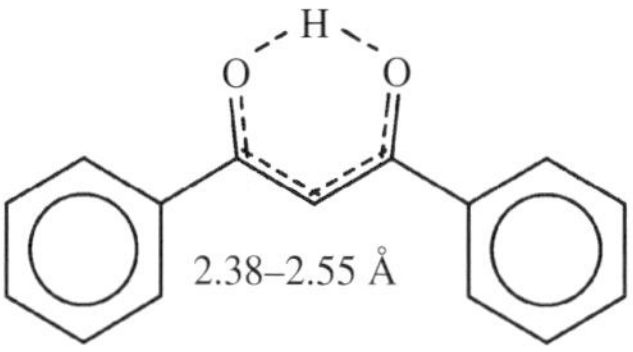

2.38–2.55 Å

Dibenzoylmethane enols (2.1.V)

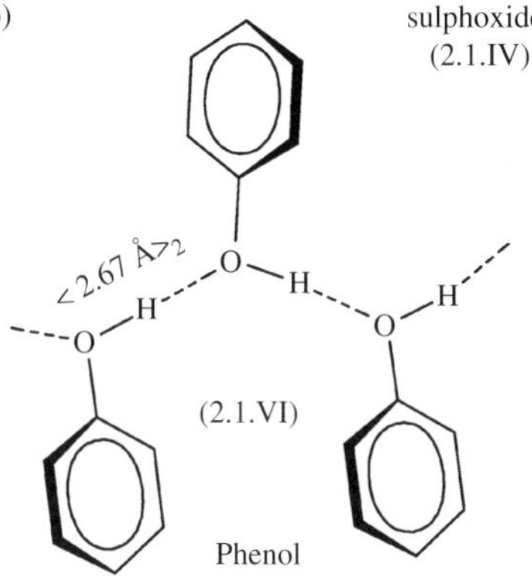

< 2.67 Å > 2

(2.1.VI)

Phenol

Another use of the word 'weak' is in expressions like 'weak H-bond donors' or 'weak H-bond acceptors' used in the classification of Chart 2.3. In this case 'weak' is not used as the opposite of strong but rather to indicate groups of bonds that are systematically weak because of some particular property. In this second case the expression 'weak H-bonds' is a synonym of 'unconventional H-bonds' and *a group of weak H-bonds* is intended to identify chemically similar bonds that are intrinsically weak because of the low electronegativity of the donor, acceptor or both, the expression 'intrinsically weak' specifying that bonds of this group can never become strong. To notice that, according to this definition, O–H···O or N–H···O bonds can never be defined as 'weak H-bonds' because, though able to give weak or very weak bonds in many circumstances, they can also form strong bonds, e.g. charge- or resonance-assisted ones.

Clearly, these classes of weak H-bonds can only be defined in tentative terms that can be challenged by the discovery of new strong H-bonds in the same classes. However, given the great number of structures so far screened by crystal databases or emulated by theoretical calculations, it seems improbable that much is going to change for main-group elements, though some doubts may persist for metal centers whose H-bonds are a relatively new and barely studied topic. In general, both experiments and calculations seem to point to the same scheme where about fifteen nonmetallic elements and a relevant (though presently undefined) number of transition metals can form a variety of 3c–4e shared-proton interactions to be recognized as true H-bonds. The majority of them will consist, however, of weak H-bonds because strong ones seem, at least so far, to be confined to the group of conventional H-bonds.

Anyway, weak H-bonds remain, at least for their great number, a quite important area that deserves to be accurately investigated. The problem has been extensively debated by Desiraju and Steiner (1999) in their book *The weak hydrogen bond*, providing the first efficient and rational classification of these bonds together with a complete analysis of the strategies to be used for facing the very oxymoron of weak H-bonds, that of becoming less and less demonstrable, while becoming more and more weak.

The basic problem is to establish whether these weak H-bonded interactions are to be classified as true (though weak) H-bonds or as van der Waals interactions, being generally recognized that H-bond formation should be characterized by changes in the state of the proton detectable by spectroscopic methods, in particular NMR, IR or Raman techniques. Not by chance, the existence of C–H···O bonds was firstly suggested by the changes of ^{1}H NMR chemical shift of chloroform when mixed with acetone (Huggins *et al.*, 1955). Moreover, H-bond formation heavily affects the stretching band of the D–H bond by red shifting its frequency, ν(D–H), and increasing its bandwidth and intensity. According to Vinogradov and Linnell (1971), there is a nearly linear relationship between frequency shifts, $\Delta\nu$, and H-bond enthalpies, ΔH_{HB}, in weak bonds (the so-called Badger and Bauer relationship) in the respective ranges of $0 \leq \Delta\nu \leq 150$ cm^{-1} and $0 \leq -\Delta H_{HB} \leq 3.0$ kcal mol^{-1}, so that also very weak interactions can be detected. Many weak H-bonds were discovered by this method, which gives clear evidence of the proton involvement together with a reasonable estimate of the interaction energy, though with the not trivial disadvantage of not giving any information on its plausible geometry.

It is not surprising that, with the years, X-ray and neutron diffraction have become the methods of choice for the study of weak H-bonds, though even these methods may have serious drawbacks because of a source of error that, being intrinsic, cannot be eliminated: the weak forces of the crystal field, that are normally unable to perturb strong H-bonds, may become a relevant source of noise for the weak ones making the correlations between H-bond strengths and H···A distances often misleading (Bernstein *et al.*, 1974). A number of different strategies have been developed to circumvent this problem, which have become characteristic of crystal studies on weak H-bonds and are briefly analyzed below (H-bond geometries according to Scheme **2.1.VII**).

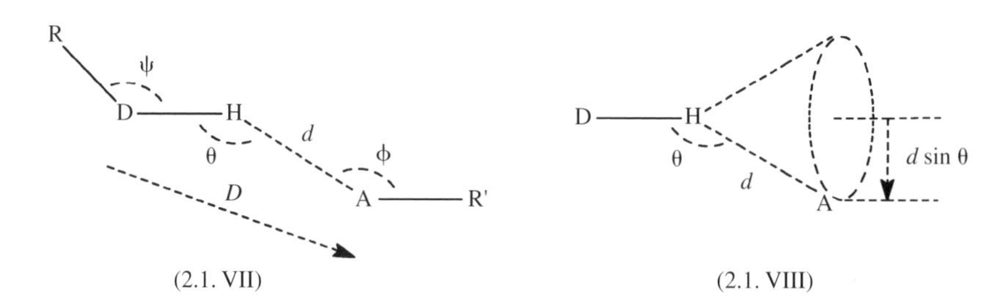

(2.1. VII) (2.1. VIII)

2.4.2.1 *Treatment of H···A contact distances*

In the early times X-ray analysis was unable to localize hydrogens, so that H-bond occurrences were established only from the contact distances, $D = d(\text{D}\cdots\text{A})$, between heavy atoms. To this aim, crystallographers were making use of the Hamilton and Ibers (1968) rule stating that a heavy atom distance less than the sum of their van der Waal radii was perhaps a sufficient, though not necessary, condition for hydrogen bonding, a kind of test working rather well for strong and moderate H-bonds but completely inappropriate for very weak ones. Moreover, since H-bond energies decrease while the D–H–A angle moves away from linearity (Schroeder and Lippincott, 1957), the measure of D alone is also inadequate to evaluate H-bond strengths.

More accurate localization of proton nuclei by neutron crystallography has led to the introduction of the parameter $d = d(\text{H}\cdots\text{A})$ as a much more reliable indicator of H-bond strength (Jeffrey and Saenger, 1991; Jeffrey, 1992) and a new benchmark in H-bond research. To notice that X-ray diffraction methods give X–H bond lengths (X = C, N, O, ...) that are systematically 0.1–0.2 Å shorter than neutron ones because X-rays do not interact with nuclei, as neutrons do, but with electrons. Hence, they do not locate the proton nucleus but the centroid of the electron density surrounding it, which is the more shifted towards X the more the electrons are localized on the internuclear X–H axis, i.e. the weaker the H-bond is. This error affects the proton, which is deprived of core electrons, but does not influence the non-hydrogen atoms (the so-called heavy atoms) whose core electron density outweighs the valence one. The systematic error on X–H bond lengths is particularly inconvenient because X-ray outnumber neutron structures by a factor of nearly 300 and it is then imperative to find a way for correcting it (Jeffrey, 1992, 1997).

An efficient method is the so-called *X–H distance normalization*, which consists in correcting X–H distances shorter than an established D–H standard bond length (Table 2.3) by moving the proton along the X–H vector up to the standard value, while leaving longer distances unchanged. In two different versions of the method, the values of the R–X–H angle are either used as they are or recalculated on the basis of bonding geometry considerations, a method particularly efficient for *sp*- and *sp*^{2}-hybridised carbons. Comparison between selected neutron and X-ray structures shows that the method is accurate enough

TABLE 2.3. Standard H–D bond lengths (Desiraju and Steiner, 1999) and vdW H···A cutoff distances together with the atomic van der Waals radii (in italics) (Bondi, 1964). All distances in Å

				H–H 0.742 *1.20 H* H···H 2.40
H–B 1.19 *≈1.8 B* H···B ≈3.00	H–C 1.083 *1.70 C* H···C 2.90	H–N 1.009 *1.55 N* H···N 2.75	H–O 0.983 *1.52 O* H···O 2.72	H–F 0.917 *1.47 F* H···F 2.67
H–Al 1.59 *– Al*	H–Si 1.50 *2.10 Si* H···Si 3.30	H–P 1.42 *1.80 P* H···P 3.00	H–S 1.338 *1.80 S* H···S 3.00	H–Cl 1.27 *1.75 Cl* H···Cl 2.95
Ga–H 1.62 *– Ga*	H–Ge 1.51 *≈1.9 Ge* H···Ge ≈3.1	H–As 1.52 *1.85 As* H···As 3.05	H–Se 1.46 *1.90 Se* H···Se 3.10	H–Br 1.42 *1.84 Br* H···Br 3.04
	H–Sn 1.71 *– Sn*	H–Sb 1.70 *≈2.1 Sb* H···Sb ≈3.3	H–Te 1.69 *2.06 Te* H···Te 3.26	H–I 1.61 *1.98 I* H···I 3.18

to make a reliable decision on the occurrence of weak H-bonds by direct comparison of the corrected $d = d(\mathrm{H}\cdots\mathrm{A})$ distance with the corresponding van der Waals H···A cutoff (*vdW cutoff*) obtained by summing up the van der Waals radii of H and A. Selected values of standard D–H bond lengths compiled by Desiraju and Steiner (1999), available van der Waals radii according to Bondi (1964) and vdW cutoffs for the main-group elements involved in H-bond formation are collected in Table 2.3.

2.4.2.2 *H-bond directionality*

It is empirically established that H-bonds tend to become increasingly linear with their increasing strength. This phenomenon is preeminent in short and proton-centered H-bonds of covalent type, whose typical angular range is $160 \leq \theta \leq 180°$, but occurs also in longer bonds of electrostatic nature because the $\mathrm{D}^{\delta-}\mathrm{-H}^{\delta+}\cdots\mathrm{A}^{\delta-}$ arrangement must have an energy minimum when the $\mathrm{D}^{\delta-}\mathrm{-H}^{\delta+}$ dipole is collinear with the punctual charge $\mathrm{A}^{\delta-}$. It may then be expected that also weak H-bonds may show a trend to linearity, though with broader angular distributions, and that observation of such a trend can become independent evidence for H-bond formation. In other words, the *donor directionality* associated with the D–H–A angle, θ, can be used, in principle, as an indication of the true H-bond nature of the interaction even when the $d(\mathrm{H}\cdots\mathrm{A})$ distance is fading towards its van der Waals limit.

A typical application of this linearity concept is displayed in Fig. 2.1 that compares the D–H–A distributions (derived by searching the CSD structural database) for two sets of H-bonds of quite different strengths: (*a*) 41 weak C–H···O interactions with $d < 2.4$ Å as determined by neutron diffraction (Taylor and Kennard, 1982); and (*b*) 196 conventional O–H···O bonds in carbohydrate X-ray structures as collected by Kroon and Kanters (1974). Angular distributions cannot be used as they are, but must be precisely corrected for the fact that the probability, $f(\theta)$, of observing any given angle θ is proportional to $\sin(\theta)$ for purely geometrical reasons (scheme **2.1.VIII**). Application of such a *cone correction* is illustrated in Fig. 2.1, where the two histograms on the left

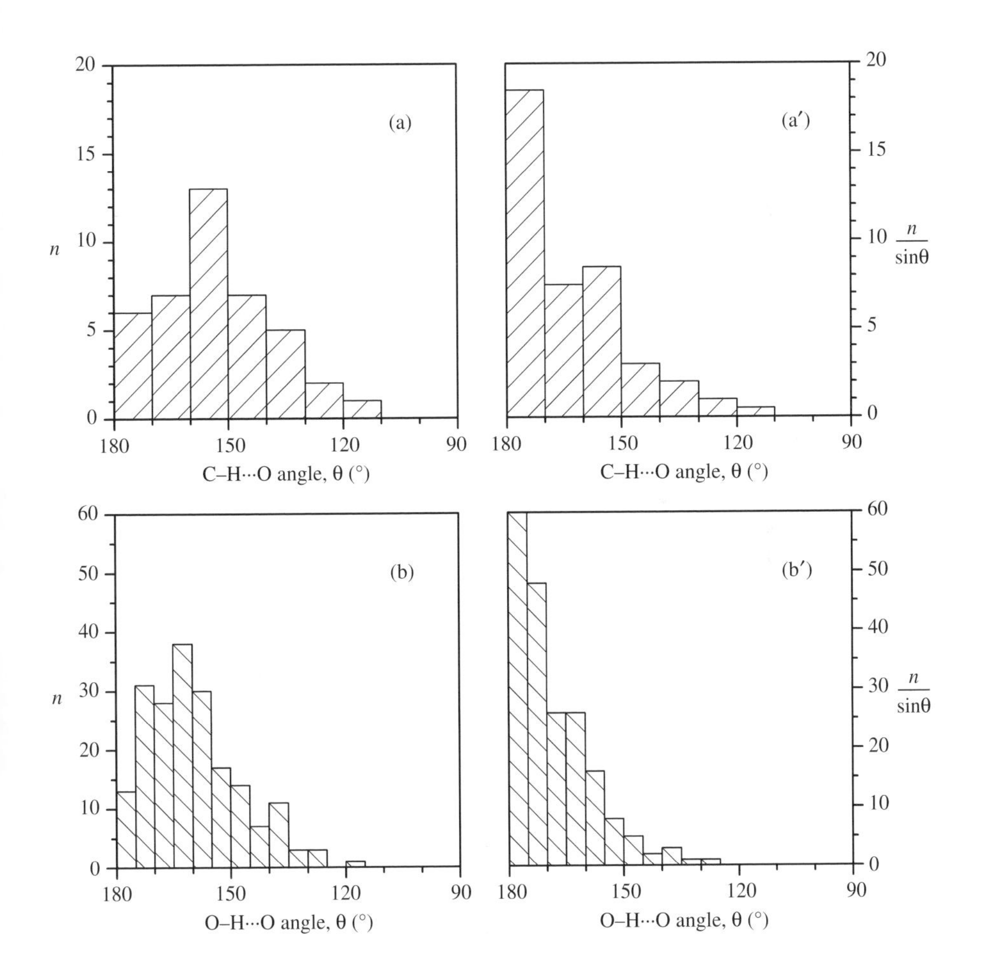

FIG. 2.1. Distribution of D–H···A angles, θ, in two classes of weak and conventional H-bonds. (a) C–H···O bonds in neutron structures with $d < 2.4$ Å. (Data from Taylor and Kennard, 1982). (b) O–H···O bonds in the X-ray structures of carbohydrates. (Data from Kroon and Kanters, 1974).

represent the non-corrected and the two on the right the corrected distributions. Comparison of the corrected histograms indicates that also weak C–H···O interactions conform to a well-defined trend to linearity with angular dispersion only slightly broader than that of the much stronger O–H···O bonds.

A second type of H-bond directionality is associated with the acceptor and measured by the H–A–R′ angle, ϕ (**2.1.VII**). The effect is due to the uneven spatial localization of the lone-pair electron densities around the acceptor and can be observed only when this localization becomes significant as, for instance, around the sp^2-hybridised oxygen of the carbonyl. An extended CSD study on conventional N–H···O and O–H···O bonds is available (Murray-Rust and Glusker, 1984). A beautiful example of lone-pair directionality (Fig. 2.2) has been obtained by Allen *et al.* (1997) by comparing the distribution of the angular values obtained from a CSD search on weak X–H···S=C< and conventional X–H···O=C< bonds (X=N and O). It shows that both types of bonds have a strong directional preference in the plane of the >C=S or >C=O groups with the only quantitative difference that the H···S–C angles are centered somewhat below 110° while the H···O–C ones gather around 120°, in good agreement with the common belief that lone pairs of S and O have greater p and sp^2 character, respectively.

2.4.2.3 *Importance of crystal-packing patterns*

The H···D distance is often too long for assessing without doubt the presence of a weak H-bond. In these cases, the fact that these weak interactions can be shown to play a substantial role in determining the crystal packing can become independent evidence of their attractive nature. This type of reasoning has been often used to assess the nature of the intermolecular forces that drive the packing, for instance the $>C{=}O{:}\rightarrow C({=}O)R_2$ donor–acceptor interactions that bind carbonyls to themselves in crystals. In the field of H-bonding, this approach has been employed to confirm the importance of weak C–H···X (X = O and N) bonds in crystals, as pioneered by Bernstein *et al.* (1974) in their paper on the structural chemistry of quinones. Figure 2.3 shows, as an example, the structure of 1,4-benzoquinone (Trotter, 1960) that represents one of the first cases of crystal packing shown to be completely determined by binding C–H···O interactions.

2.4.3 *Weak H-bond donors (Group 1.2)*

In theory, *weak H-bond donors* should include the hydrides of the main-group elements C–H, P–H, Si–H, As–H, Se–H, and also S–H, though S has been previously classified as a conventional H-bond donor. Very few of these weak H-bond donors, however, can be seriously documented by at least one reasonably accurate crystal structure, so that the group is practically reduced to C–H, P–H and S–H donors only. At present, it is impossible to say whether this result is produced by real chemical factors (e.g. the too low electronegativity of the atoms involved) or depends on the lack of systematic studies in this field.

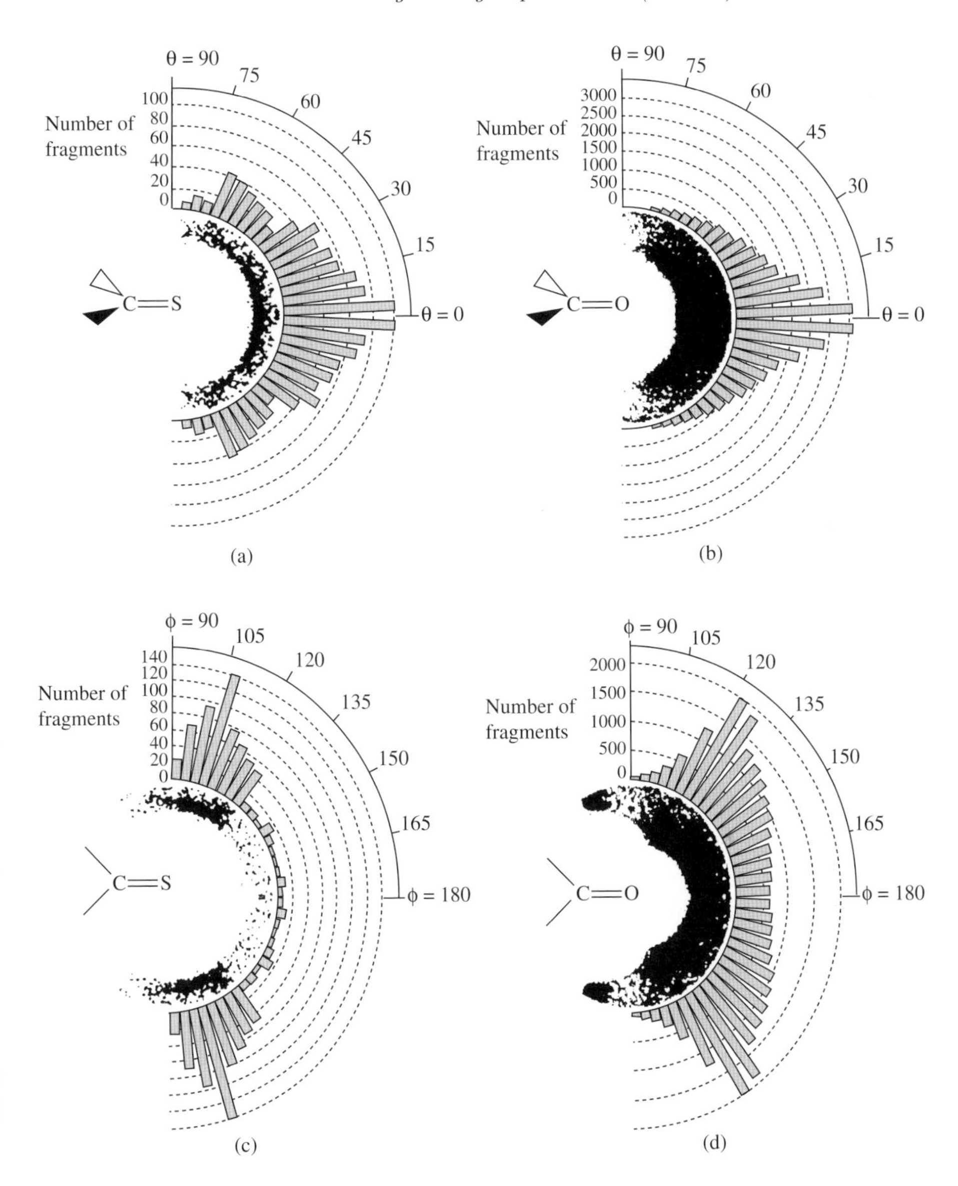

FIG. 2.2. Lone-pair directionality in the plane and perpendicularly to the plane of the H-bond acceptor fragments $R_2C{=}S$ (left) and $R_2C{=}O$ (right). (Reproduced by permission from Allen *et al.*, 1997).

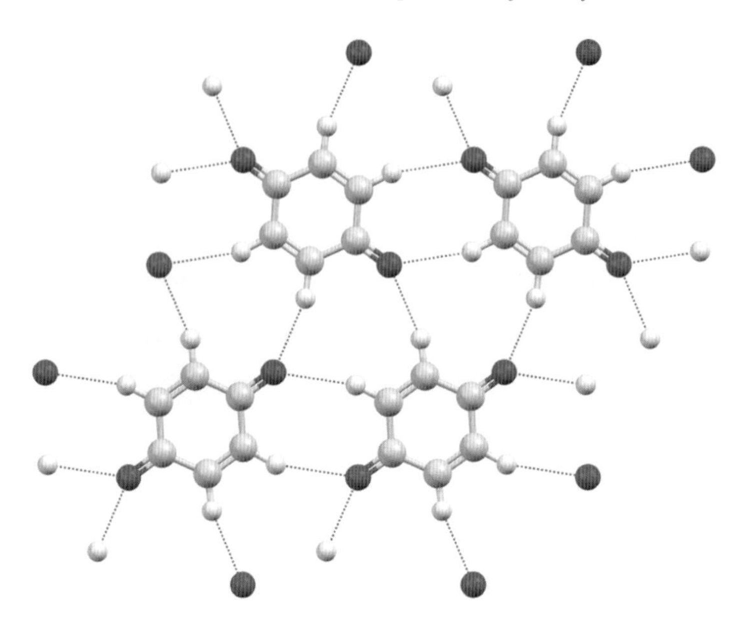

FIG. 2.3. The crystal structure of 1,4-benzoquinone (Trotter, 1960) is one of the first examples of crystal packing determined by C–H···O interactions.

2.4.3.1 *C–H··· bonds*

The earliest indications that the C–H donor may give origin to C–H···A (A = O and N) H-bonds date back to the 1930s when relatively large red shifts of the IR C–H stretching frequency were observed in consequence of the mixing of strong C–H acids with potential H-bond acceptors. The first crystal structure containing this type of bond is probably that of H–C≡N, where formation of infinite ···H–C≡N···H–C≡N··· chains with D = 3.18 Å was observed by Dulmage and Lipscomb (1951), while the first systematic study on C–H···O bonds in crystals is due to Sutor (1962, 1963) who observed that a number of short H···O contacts (d = 2.20–2.27 Å) were determinant for the packing of a number of purines and pyrimidines and that their lengths were far below the then assumed vdW cutoff of 2.60 Å (and then even smaller than the presently accepted cutoff of 2.72 Å reported in Table 2.3). Unfortunately, Sutor's assessment that these short C–H···O contacts should be considered as true H-bonds was forcefully questioned by Donohue (1968) on the wrong assumption that the true vdW cutoff was 2.20–2.40 and not 2.60 Å and, because of that, the idea of C–H···O H-bonds was challenged, or at least considered with deep suspicion, by most crystallographers for many years. It was the paper by Taylor and Kennard (1982), a rather sophisticated piece of work for the time making use of the newly developed crystallographic databases and exclusive results of neutron crystallography, to conclude the controversy and to provide conclusive evidence of the existence of true C–H···O H-bonds in crystals.

From then on, C–H···A (A = strong acceptor) interactions have been considered quite normal H-bonds, though rather long and weak. The normally accepted ranges of contact distances are 3.0–3.8 for D and 2.0–2.8 Å for d, though also slightly longer bonds can be sometimes considered to be attractive (and then weak H-bonds) due to their evident ability to control the crystal packing. The main factor that appears to control both lengths and energies is the acidity of the C–H group and theoretical calculations indicate bond energies of some 2.0 or 0.5–1.0 kcal mol^{-1} for C–H···O/N bonds formed by remarkably or slightly acidic C–H donors, respectively (see below). Recently, increasing attention has been also given to the C–H interactions with weak π-acceptors, in particular triple bonds and aromatic rings. Also the H-bond nature of these last interactions seems to be ascertained, at least for the most acidic donors, but the weak binding energies involved make the subject too complex to be treated here. The reader is addressed to more detailed expositions (Desiraju, 1991; Desiraju and Steiner, 1999; Scheiner, 1997; Gu *et al.*, 1999).

In spite of their weakness, C–H···O/N interactions exert an important action in the organisation of the natural world. This derives from their overwhelming occurrence and from the often neglected fact that the external surfaces of organic molecules are wholly covered by a layer of C–H groups with the H atoms pointing outwards, a layer that is interrupted only in a few points by the emerging hetheroatoms of the functional groups (essentially O, N, S and halogens atoms). The consequences are particularly evident in molecular crystals, where C–H groups are seen to saturate all possible H-bond acceptors not otherwise engaged in stronger H-bonds for lack of more efficient H-bond donors. The role played by this external C–H mantle in determining the properties of organic molecules in a biological environment, that is in water, has not been yet investigated in any detail, but it seems rather probable that C–H···O_W interactions may play a role in the formation of the water cages that embed hydrophobic surfaces, giving rise to the class of phenomena going under the name of hydrophobic interactions (see Chapter 7). The importance of C–H···O_W interactions is well witnessed by a statistical analysis carried out on 101 water molecules occurring in 46 very accurate neutron structures of molecular crystals (Steiner and Saenger, 1993) that has shown that over 50 per cent of these water molecules accept C–H···O_W bonds with $d < 3.0$ Å.

Another singular feature of C–H···A bonds is that the acidity of the C–H group can sweep over a wide range of pK_a values prompted by changes of the hybridization state of the carbon (C–H acidity is increasing in the $sp^3 < sp^2 \ll sp$ order) and for effect of electron-withdrawing substituents. As an extreme example, the pK_as in water of CH_4, $H_2C{=}CH_2$, $H{-}C{\equiv}C{-}H$ and $(N{\equiv}C)_3C{-}H$ decrease in the order 48, 44, 26 and −5. This imparts to C–H···A bonds a capacity of being modulated in strength and length by chemical factors that is exceeded only by conventional H-bonds. The effects of C–H acidity on the geometry of the C–H···O bonds have been studied by Pedireddi and Desiraju (1992) on the basis of a CSD analysis extended to around twenty categories of C–H acids covering a wide range of pK_a values and C···O contact distances ($3.0 \leq D \leq 4.0$ Å). Distances are found to correlate reasonably well with pK_a values, shorter bonds being systematically associated with more acidic C–H groups, and vice versa. The role played by C–H acidity is confirmed by the structure of the C–H···:O adduct of trinitromethane (pK_a = 0.0) with dioxane (p$K_a \approx 2.1$) reporting the shortest C···O distance (2.936 Å) ever observed (Bock *et al.*, 1993).

2.4.3.2 *S–H··· bonds*

S displays non-dissimilar behaviour from O in spite of its much lower electronegativity (2.58 against 3.44). Accordingly, it can be considered a conventional H-bond donor and is discussed here for reasons of convenience.

S–H can donate to O, N, S and even to π-acceptors and can simultaneously act as donor and acceptor in analogy with the O–H group. The bonds formed are, however, rather weaker, as shown by the fact that mean S–H···O distances are $d = 2.1 - 2.3$ and $D = 3.3 - 3.6$ Å with a contraction with respect to the H···O vdW cutoff (2.72 Å; Table 2.3) of nearly 19%, while the corresponding distances in ordinary O–H···O bonds are $d = 1.7 - 2.0$ and $D = 2.7 - 3.0$ Å with a vdW cutoff contraction of more than 32%.

For C(sp^3)–S–H···S bonds, d and D values around 2.8 and 4.0 Å have been reported, which amounts to a shortening of only 7% with respect to the H···S vdW cutoff of 3.00 Å (Table 2.3). These rather long S–H···S bonds are considerably shortened when involved in resonance-assisted (RAHB) or charge-assisted (CAHB) H-bonds. Two examples of RAHB are shown in Schemes **2.1.IX** and **X**. The formation of the resonant dimer in ethylphenyldithiophosphinic acid (**2.1.IX**; Krebs, 1983) reduces d to 2.51 and D to 3.81 Å and that of the resonant chain ···S=P–S–H···S=P–S–H··· in diphenyldithiophosphinic acid (**2.1.X**; Krebs and Henkel, 1981) to $d = 2.47$ and $D = 3.79$ Å, with a vdW cutoff shortening of 16 and 18%, respectively. Even more significantly, the formation of the charge-assisted [S–H···S]$^-$ bond in the dithiolate salt (**2.1.XI**: Boorman *et al.*, 1992) gives rise to the shortest intermolecular S–H···S bond ever reported with a distance d of only 2.12 Å, amounting to a vdW cutoff shortening as high as 29%. Similar shortening is associated with the charge-assisted arrangement S^+–H···Hal$^-$, for instance in Me_2S^+–H···Cl^- (Mootz and Deeg, 1992) having $d = 2.14$ Å and a vdW cutoff contraction of 29%.

2.4.3.3 *P–H··· bonds*

P and H have practically identical electronegativities so that it is not expected that P–H be a sufficiently good H-bond donor unless the P atom is bound to other strongly electron-attracting groups. This is verified by the structures of the very few compounds forming neutral P–H···O bonds, such as **2.1.XII**, where the molecules are linked in resonant ···O=P–H···O=P–H··· chains having a d(H···O) of 2.23 Å and a vdW cutoff contraction of 18% (Saenger and

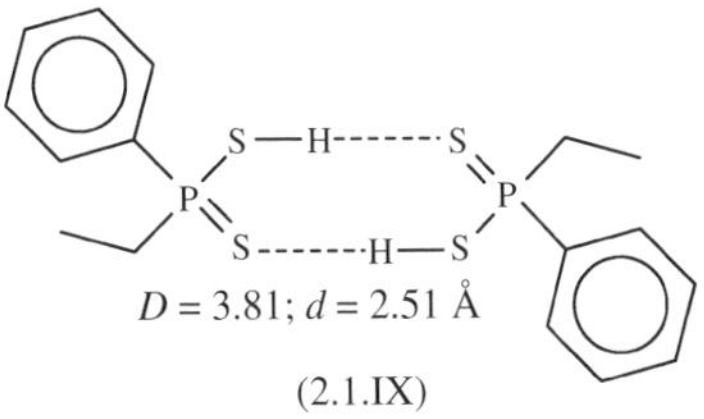

$D = 3.81$; $d = 2.51$ Å

(2.1.IX)

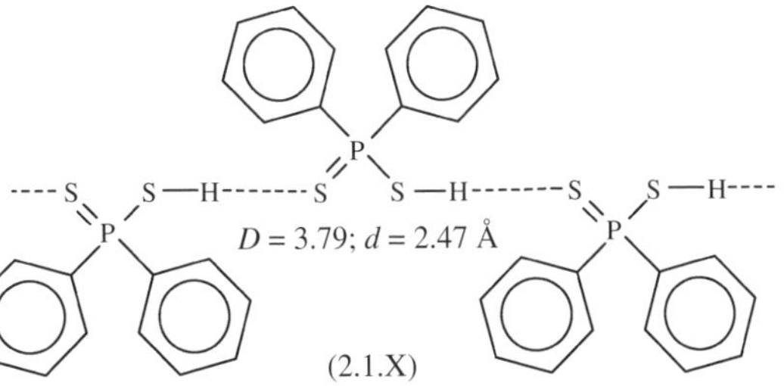

$D = 3.79$; $d = 2.47$ Å

(2.1.X)

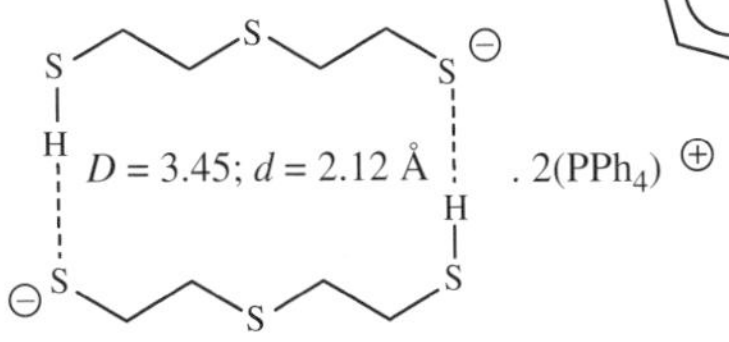

$D = 3.45$; $d = 2.12$ Å . $2(PPh_4)^{\oplus}$

(2.1.XI)

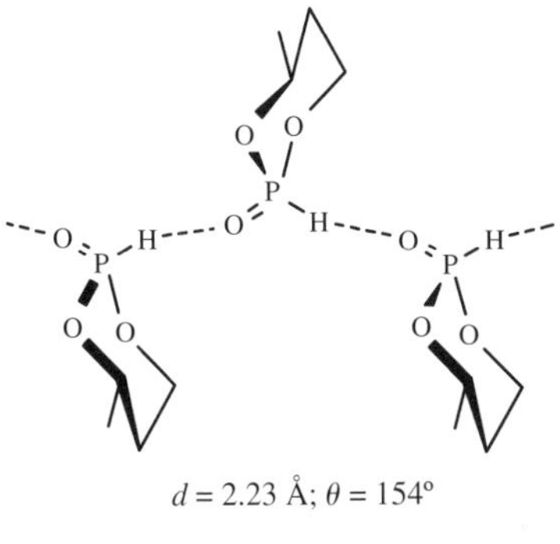

$d = 2.23$ Å; $\theta = 154°$

(2.1.XII)

$D = 3.81$; $d = 2.43$ Å

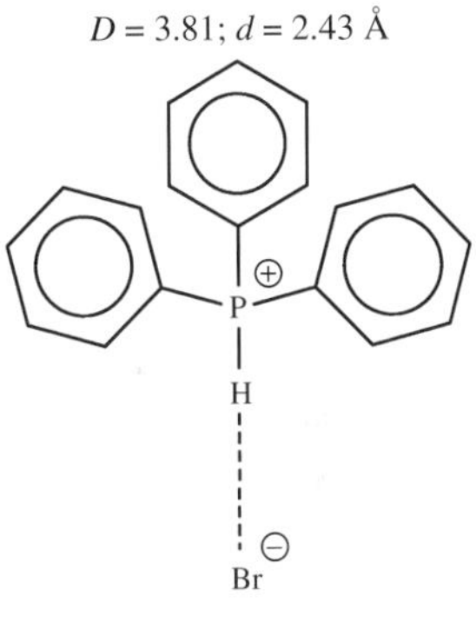

(2.1.XIII)

Mikolajczyk, 1973). Larger contractions of 22–26% for d(H···Br) distances of 2.43–2.37 Å are also observed in the two triphenylphosphonium bromide derivatives **2.1.XIII** (Bricklebank *et al.*, 1993; Schmutzler *et al.*, 1984) that can be imputed to the formation of a charge-assisted type of bond.

2.4.3.4 *Si–H, As–H and Se–H··· bonds*

No convincing proof of unambiguous H-bonds involving these donors has been produced so far, at least in the solid state, a fact particularly surprising since the Se atom is nearly as electronegative as S. It may be supposed that the synthesis of new compounds where the acidity of the D–H donor is enhanced by addition of proper electron-attracting substituents could help, in analogy with what was already done for P, to make up for such a shortcoming.

2.4.4 *Weak H-bond acceptors (Group 1.3)*

In this section we will discuss main-group acceptors other than N, O, F, Cl and Br. While only few elements are weak H-bond donors (mostly C, P, and S), a much longer list of them (C, P, As, Sb, S, Se, Te) can be shown to behave as weak acceptors, suggesting that electronegativity effects may be less pronounced for acceptors than for donors. This list must be extended to the so-called 'organic halogens', i.e. F, Cl and Br atoms covalently bound to organic C, which are traditionally classified as weak acceptors.

2.4.4.1 *C–Hal (Hal = F, Cl, Br) as acceptors*

Halogens are known to form quite conventional H-bonds, except when covalently bound to organic carbons. The observation that conventional O–H and N–H donors seldom bind to organic F has prompted a number of database analyses (Murray-Rust *et al.*, 1983; Shimoni and Glusker, 1994; Howard *et al.*, 1996). In the most recent (Dunitz and Taylor, 1997) 5974 C–F bonds of molecules endowed with at least one potential O–H or N–H donor group were retrieved, out of which only 37 (0.6%) were found to be involved in O/N–H···F–C bonds with $d < 2.6$, so showing without doubt how weak an acceptor the organic F may be. Further studies have shown that also organic C–Cl (Aullón *et al.*, 1998) and C–Br groups (Desiraju and Steiner, 1999) are very poor H-bond acceptors, though slightly better than C–F. These findings apparently contrast with the intuitive idea that highly electronegative atoms, such as halogens, should be good H-bond acceptors, but can be accounted for in terms of the quite low proton affinities of halogens when bound by a strong heteronuclear bond to carbon, as can be easily verified from the following comparison of PA values (Hunter and Lias, 2005) of some common H-bond acceptors, i.e. MeF : MeCl : MeBr : $MeNH_2$: pyridine : Me_3N = 143.1 : 154.7 : 165.7 : 214.1 : 220.5 : 233.2 kcal mol^{-1}.

2.4.4.2 *S, Se and Te as acceptors*

Pauling's electronegativities of these atoms (Table 2.1) decrease in the order 2.58, 2.55 and 2.10. The characteristics of S–H···S bonds and lone-pair directionality of the >C=S group have already been discussed in the previous sections. The geometrical features of the O–H···S bonds have been reviewed by Allen *et al.* (1997) who have shown that thioketones, >C=S, are much more efficient acceptors than thioesters, >S, average d and D distances and vdW cutoff contractions being 2.41, 3.32 Å and 20% for the former against 2.67, 3.39 Å and 11% for the latter. The corresponding N–H···S bonds are, on average, some 10 per cent longer than the O–H···S ones.

In general, S is a weaker H-bond acceptor than O, as can be shown by comparing the d and D values and vdW cutoff contractions of the thioketones above with those of ordinary O–H···O=C< bonds (1.884, 2.790 Å and 31%). However, the O–H···S=C< bond can be considerably strengthened by RAHB

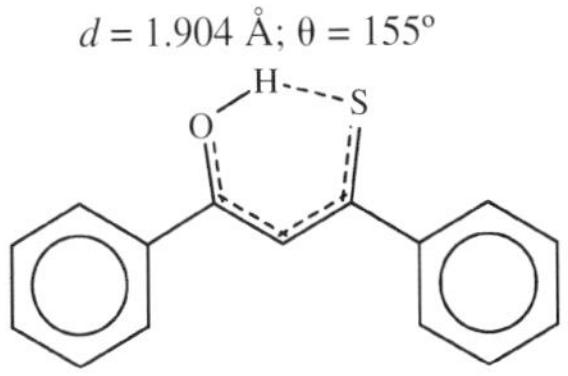

$d = 1.904$ Å; $\theta = 155°$

(2.1.XIV)

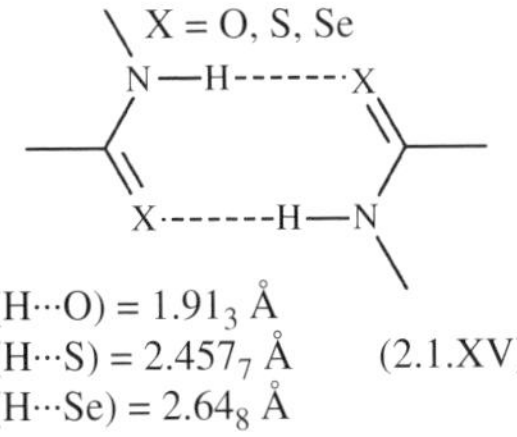

X = O, S, Se

$d(\mathrm{H}\cdots\mathrm{O}) = 1.91_3$ Å
$d(\mathrm{H}\cdots\mathrm{S}) = 2.457_7$ Å
$d(\mathrm{H}\cdots\mathrm{Se}) = 2.64_8$ Å

(2.1.XV)

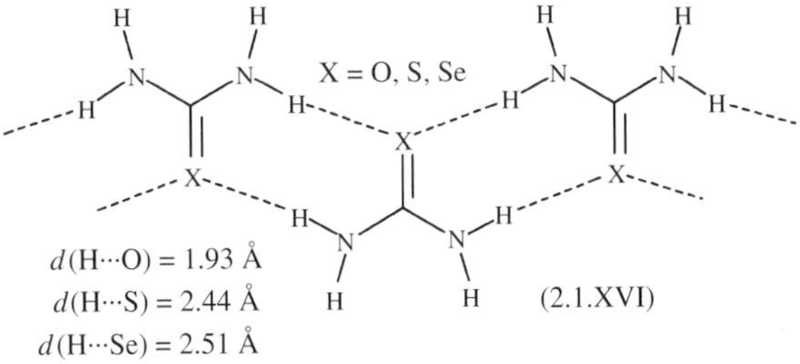

X = O, S, Se

$d(\mathrm{H}\cdots\mathrm{O}) = 1.93$ Å
$d(\mathrm{H}\cdots\mathrm{S}) = 2.44$ Å
$d(\mathrm{H}\cdots\mathrm{Se}) = 2.51$ Å

(2.1.XVI)

formation, as exemplified by the neutron structure of the β-thioketone enol **2.1.XIV**, where $d(\mathrm{H}\cdots\mathrm{S})$ is as short as 1.904 Å and the resonant nature of the bond is verified by the almost complete π-delocalization of the interleaving O–C=C–C=S fragment (Steiner, 1998a, b). The ability of S of forming resonance- as well as charge-assisted $[\mathrm{S{-}H\cdots S}]^-$ H-bonds (see Section 2.4.3.2) is another indication that the H-bonds formed by S are better classified as conventional H-bonds.

While Te does not seem able to accept H-bonds, a limited number of O–H···Se and N–H···Se bonds with >C=Se and >Se acceptors are reasonably well documented. An interesting parallel among the N–H···X bonds with X = O, S, Se has been drawn out by Desiraju and Steiner (1999) for both amide dimers (**2.1.XV**) and urea ribbons (**2.1.XVI** and Fig. 2.4) compared with their S and Se analogues. Comparison of the average $d(\mathrm{H}\cdots\mathrm{O})$, $d(\mathrm{H}\cdots\mathrm{S})$ and $d(\mathrm{H}\cdots\mathrm{Se})$ contact distances with their vdW cutoffs (2.72, 3.00 and 3.10 Å) leads to respective contractions of 29, 18–19 and 16–18%, showing that the S and Se are much weaker acceptors than O but strictly comparable among themselves, as anticipated by their very similar electronegativities.

2.4.4.3 *P, As and Sb as acceptors*

These elements can behave as H-bond acceptors when carrying a free-electron pair, which almost exclusively happens for $\mathrm{R_3X}$: trisubstituted phosphines, arsines and stibines. However, only very few crystal structures containing weak O/N–H···P bonds have been so far reported, so that most of what we know on acceptor potentialities of P, As and (much less) Sb is based on their complexes in

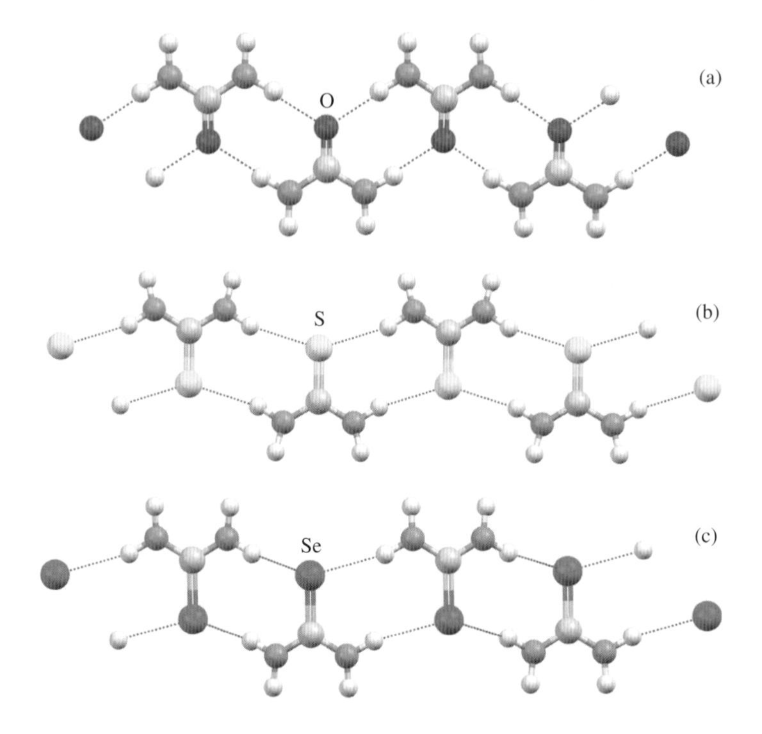

FIG. 2.4. Isostructural molecular ribbons formed by (b) thiourea in its crystal (Truter, 1967); (a and c) urea (Li *et al.*, 1995) and selenourea (Wright and Meyers, 1980) in two lattice inclusion compounds.

the gas phase, solid argon matrices or dilute solutions of strong donors. A review on the matter is available (Sennikov, 1994).

2.4.4.4 *C as acceptor*

Carbon can act as H-bond acceptor in two ways: (*i*) as a π-acceptor (as described in the next paragraph); and (*ii*) when the C atom is endowed with a free-electron pair. According to Desiraju and Steiner (1999) this happens in four classes of compounds: isonitriles (or isocyanides), carbenes, carbanions, and *N*- or *P*-ylides, as depicted in Schemes **2.1.XVII–XXI**. Their H-bonds are not necessarily weak but can certainly be considered unconventional.

Isonitriles are lone-pair donors because of the predominant contribution of the $R{-}N^{+}{\equiv}C{:}^{-}$ form to their ground state and, because of the *sp* hybridisation of the carbon, are expected to behave as H-bond acceptors along the main symmetry axis of the molecule. The O–H···:C≡N–R H-bonds were firstly identified from the red shifts of the IR ν(O–H) stretching frequency of some 150–250 cm^{-1} observed in alcoholic solutions of isonitriles (Ferstanding, 1962; Schleyer and

Isonitriles (2.1.XVII) Carbenes (2.1.XVIII) Carbanions (2.1.XIX)

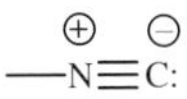

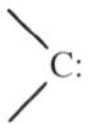

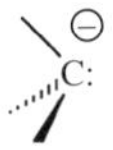

N-ylides (2.1.XX) *P*-ylides (2.1.XXI) 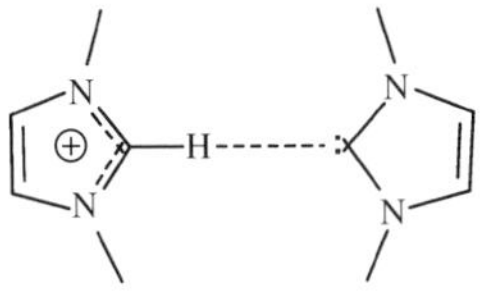 (2.1.XXII)

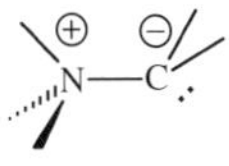

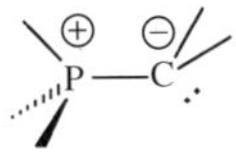

Allerhand, 1962). Very few bonds of this type have been found in crystals and a last-moment CSD search performed on the O–H···:C≡N–R fragment has identified only four cases with mean values of d = 2.12(2) and D = 3.02(2) Å. Assuming a vdW cutoff of 2.90 Å for the H···C contact, the observed distances correspond to a contraction of some 27%, which is comparable with ordinary O–H···O bonds.

Carbenes are not normally observed in crystals because of their high reactivity. The only known example (Arduengo *et al.*, 1995) concerns the H^+-*bis*(imidazole) derivative **2.1.XXII** (d = 2.10, D = 3.18 Å, θ = 173°) which represents a rare case of strong $[C\cdots H\cdots C]^+$ positive charge-assisted H-bond with an important 28% of vdW cutoff contraction.

H-bonds donated to carbanions were postulated by Ferstanding (1962). At present, they are mostly a theoretical possibility, though some attempts to stabilize the lone pair in *N*- and *P*-ylides have been reported. Examples of *P*-ylides forming weak C–H···:C bonds in the crystalline state have been reported by Batsanov *et al.* (1996).

2.4.5 *Weak π-acceptors (Group 1.4)*

Electron-rich π-electron systems, such as aromatic rings and C=C and C≡C multiple bonds, can behave as acceptors in respect to most H-bond donors, giving origin to a rich typology of intra- and intermolecular interactions (**2.1.XXIII**) to be classified as weak D–H···π bonds produced by the electrostatic attraction between the $D^{\delta-}$–$H^{\delta+}$ polarised donor and the negatively charged π-electron cloud.

The first evidence for intramolecular O–H···π interactions dates back to the work of Wulf, Liddel and Hendricks (1936) who studied a number of *o*-substituted phenols (such as *o*-phenyl, *o*-ethynyl and *o*-allylphenol) by IR spectroscopy detecting ν(O–H) red shifts of 40–120 cm^{-1}. The intermolecular O–H···π bond was later confirmed by similar IR experiments on diluted CCl_4 solutions of phenol and π-acceptors (Yoshida and Osawa, 1966), a result soon extended to the N–H donor. The first examples of C–H···π(phenyl) bond are probably due to Reeves and Schneider (1957), who found, by means of the newly developed NMR techniques, that the interaction of chloroform with benzene and olefines was a type of H-bond, and to Nakagawa and Fujiwara (1961) and Hatton and Richards (1961) who observed a weak ^{1}H NMR downfield shift in R–C≡C–H derivatives when dissolved in aromatic liquids. By the mid-1960s, many D–H···π interactions were reasonably well known thanks to the work of IR and NMR spectroscopists and, in particular, it was already known that all these O/N–H···π bonds are weak (say, less than 4 kcal mol^{-1}) and C–H···π ones even weaker (normally less than 0.6 kcal mol^{-1}) and at the borderline with van der Waals interactions.

More recently, there has been a renewed interest for these weak D–H···π complexes and their geometries have been intensively studied by MW rotational spectroscopy in the gas phase, particularly at the very low temperatures achievable by supersonic expansion techniques. MWs provide a measure of the equilibrium distance, D(D···π), between D and the mean plane crossing the π-system that, corrected for the D–H bond length (Table 2.3), gives an estimate of the H···π distance, d(H···π). This last distance is weakly affected by π-system changes (e.g. ethylene, acetylene or benzene) and much more by the different acidities of the D–H donors. Representative d values are: 2.18–2.25 Å for F–H···π, 2.35–2.44 Å for Cl–H···π, and 2.73–2.74 Å for H–C≡C–H···π, which correspond to H···C vdW cutoff reductions of 23, 17, and 6%, respectively. Both techniques and results have been surveyed (Legon and Millen, 1987a,b). Typical gas-phase dimers studied by MW rotational spectroscopy are displayed in **2.1.XXIII**.

A particularity of the C–H···π bonds is their displaying *anti-H-bond behaviour* (Hobza *et al.*, 1998; Hobza and Havlas, 2000; Cubero *et al.*, 1999; Barnes, 2004). At variance with conventional H-bonds, which are characterized by a lengthening of the C–H bond and consequent red shift of its IR stretching frequency, anti-H-bonds display a small contraction of the C–H length and then a *blue shift* of the vibrational frequency. Blue-shifting effects have been studied by high-level theoretical calculations and found to be caused by a different redistribution of the electron density occurring only in the weakest H-bonds (Gu *et al.*, 1999).

Despite the convincing spectroscopic evidence, structural crystallography took little interest in H-bonds with π-acceptors till the end of the 1980s, though the recovery of the last fifteen years has been so good that crystal data on D–H···π H-bonds must be dealt with in specialized publications. The problems connected with the study of these weak interactions in the solid state and the considerable differences with the parallel spectroscopic evidence can be illustrated by an example. The gas-phase MW structures of complexes such as benzene···HCl (Read *et al.*, 1983) or acetylene···HCl (Legon *et al.*, 1981) have provided highly symmetrical geometries with the axis of the Cl–H donor perpendicular to the π-system. Much less regular pictures are obtained from crystal structures and a

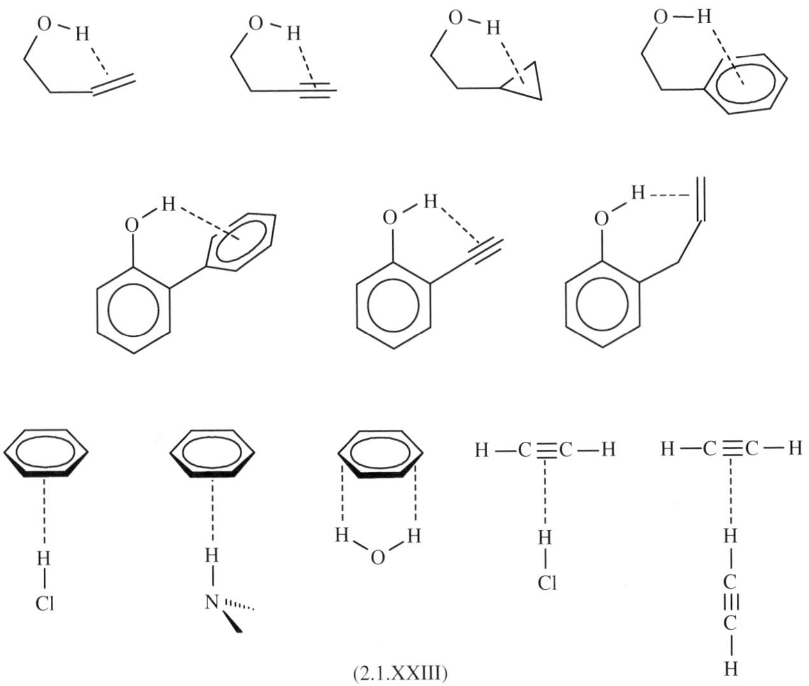

(2.1.XXIII)

comprehensive CSD analysis of D–H···π(phenyl) bonds (Malone *et al.*, 1997) has observed such a large variability of the D–H···phenyl interaction geometries that it has been necessary to organize them in six different categories where the donor is seen to aim at different points of the phenyl surface making quite different impact angles with it. This is most probably the consequence of the too weak binding energies that are unable to contrast efficiently the perturbations induced by the crystal field.

2.5 H-bonds involving metal centers (Class 2)

This is a relatively new field dating back to the beginning of the 1990s when the first evidence of the involvement of transition metals in true H-bonds started to be accumulated by IR and NMR spectroscopic methods in solution (Belkova *et al.*, 2005; Epstein and Shubina, 2002; Kazarian *et al.*, 1993) and neutron diffraction in the crystal state (Brammer *et al.*, 1995; Brammer, 2003). It is now generally accepted that the H-bond is not confined to the organic and biological domain but must be extended to include a complex typology of bonds occurring in coordination and organometallic compounds.

H-bonds formed with metals can be divided into two large subgroups according to whether they directly involve the metal center (Chart 2.3 – Groups 2.1–2.3) or the ligands of its first coordination sphere (Groups 2.4 and 2.5), the latter being more or less conventional H-bonds whose donor–acceptor properties are modulated by the metal bond and that are discussed in this section for reasons of convenience. The former are the real *metal H-bonds (MHBs)*, i.e. unconventional interactions based on three possible mechanisms: (*i*) metal hydrides with the dipolar structure $M^{\delta-}-H^{\delta+}$ can become H-bond donors forming M–H···:A bonds where :A is a conventional H-bond acceptor (Group 2.1); (*ii*) metal complexes of the late transition series endowed with filled *d*-orbitals can become H-bond acceptors from conventional D–H donors and form D–H···:M bonds (Group 2.2); (*iii*) the negatively charged hydrogen of metal hydrides having the dipolar structure $M^{\delta+}-H^{\delta-}$ can become acceptor of a H-bond donated by a conventional D–H donor forming the $D^{\delta-}-H^{\delta+}\cdots:H^{\delta-}-M^{\delta+}$ dihydrogen bond (Group 2.3).

The clue to MHB formation are then metal hydrides, a complex class of compounds where the M–H bond may equally display ionic, partially covalent, or covalent properties (Cotton and Wilkinson, 1988). The *ionic metal hydrides* formed by alkalis and alkaline earths are the only ones to really contain the hydride ion, H^-, and, as a consequence, cannot form H-bonds but only salt-like ionic crystals. The first interactions that can be classified as H-bonds appear in *partially covalent main-group hydrides*, such BeH_2, BH_3, AlH_3 and GaH_3, whose M–H bond, however, cannot act as a H-bond donor because of its strong $M^{\delta+}-H^{\delta-}$ polarization but rather as H-bond acceptor in what has become known as a *dihydrogen bond* (Group 2.3). Finally, *partially covalent transition-metal hydrides* share the very interesting property that their M–H bonds can be modulated by metal electronegativity and electronic properties of the other ligands in such a way as to impart to the hydrogen atom a character changing continuously from protonic (acidic) to hydridic (basic) through fully covalent, so allowing the formation of bonds where the M–H group is either a H-bond donor (Group 2.1) or acceptor (dihydrogen bond; Group 2.3).

2.5.1 *Metals as H-bond donors (Group 2.1)*

The most studied M–H donors are the metal carbonyl hydride complexes, such as $HCo(CO)_4$, $H_2Fe(CO)_4$ or $HMn(CO)_5$, and their analogues where one or more carbonyls are substituted by other π-acceptor ligands [PPh_3, $P(OPh)_3$, dppe = bis(diphenylphosphino)ethane, Cp = C_5H_5, Cp* = C_5Me_5, etc.]. Substitution of the carbonyls by weaker π-acceptors allows tuning of the acid–base properties of the M–H acid, as exemplified by the series $HCo(CO)_4$, $HCo(CO)_3P(OPh)_3$, $HCo(CO)_3PPh_3$, and $HCo(CO)_2(PPh_3)_2$, whose $pK_a(H_2O)$ values are 0.9, 4.9, 7.0 and >14, respectively.

Since some of these neutral hydrides are rather strong acids, they can be supposed to form neutral $M^{\delta-}-H^{\delta+}\cdots:A^{\delta-}$ H-bonds with uncharged acceptors. However, IR and Raman studies carried out by Norton and coworkers (Kristjansdottir *et al.*, 1991) failed to detect H-bond formation between $HCo(CO)_4$, a strong acid, and a number of neutral nitrogen and oxygen bases. Later, it has become increasingly evident that the metal can act as H-bond donor only in charge-assisted H-bonds of the types $[M-H^{+}\cdots:A][anion]^{-}$ or $[M^{-}\cdots H-A^{+}]$, probably because only these arrangements can actually produce the required $M^{\delta-}\cdots H^{\delta+}$ polarization.

A limited number of $[M-H^{+}\cdots:A]$ bonds have been confirmed by the spectroscopic and/or crystallographic evidence. The formation of the $Os-H^{+}\cdots:O{=}PR_3$ bond in the adduct **2.2.I** between $[Os(\eta^5\text{-}C_5Me_5)_2H]PF_6$ and Ph_3PO in CH_2Cl_2 solutions was suggested on the basis of the small red shifts observed for the Os–H (20 cm^{-1}) and P=O (26 cm^{-1}) IR stretching vibrations (Epstein *et al.*, 1993). Similar findings were reported by Peris and Crabtree (1995) for cationic iridium hydrides as potential H-bond donors. The crystal structure of $WH_3(dppe)_2[OC(Me){=}O]$ (**2.2.II**) has detected the formation of an intramolecular $W-H^{+}\cdots{}^{-}OOCMe$ interaction with an H···O distance of 2.33(6) Å (Fairhurst *et al.*, 1995) that was later confirmed by ^{1}H NMR spectroscopy (Steiner, 2002). A systematic CSD investigation carried out by Braga *et al.* (1996a,b) led to the identification of a number of intermolecular M–H···O H-bonds between cationic hydrides and the oxygen side of coordinated carbonyls, an example of which is the complex $[Cp_2Mo(H)CO]^{+}$ $[CpMo(CO)_3]^{-}$ shown in **2.2.III**, where the H···O distance amounts to 2.51 Å. The authors conclude that these rare interactions have metrics and, presumably, strengths comparable with C–H···O H-bonds.

$HCo(CO)_4$ and its $HCo(CO)_3L$ analogues (L = triarylphosphine) have been extensively studied for their ability to produce, in the presence of strong nitrogen bases, a complete proton transfer from the donor to the acceptor with formation of remarkably strong (short) charge-assisted $N-H^{+}\cdots:Co^{-}$ H-bonds (sometimes misinterpreted as H-bonds accepted by the metal anion, Co^{-}). An example is the neutron structure of $Et_3N-H^{+}\cdots Co^{-}(CO)_4$ (**2.2.IV**) at 15 K, which shows a perfectly linear N–H···Co bond with a slightly elongated N–H distance [N–H 1.054(1), H···Co 2.613(2), N–H···Co 180.0°] (Brammer *et al.*, 1992). The acid–base properties of these $N-H^{+}\cdots Co^{-}(CO)_3L$ bonds have been investigated by comparing the structures of a series of complexes where the nitrogen base and/or the metal hydride had been slightly modified. The complete series is discussed by Brammer *et al.* (2000), while Scheme **2.2.V** illustrates the particular case of three of these complexes obtained by reacting the same base (1,4-diaza[2.2.2]bicyclooctane or DABCO; pK_a of 8.82) with three different acids: $HCo(CO)_4$ (pK_a 0.9), $HCo(CO)_3P(p\text{-tol})_3$ (pK_a unknown), and $HCo(CO)_3PPh_3$ (pK_a 7.0). A full analysis is impossible because the second pK_a is missing, but the comparison of the other two is probably sufficient to understand what is happening. Their $\Delta pK_a = pK_a(\text{acid}) - pK_a(\text{base})$ values amount to −7.9 and −1.8 for H···Co distances of 2.392 and 2.247 Å, respectively, supporting the idea that the H-bond is strengthened by the pK_a matching of the H-bond donor and acceptor moieties (see Chapter 4 for a full discussion). Various IR spectroscopic studies in solution have confirmed that these N–H···Co bonds occurring in the crystal state are also retained in non-polar solvents and that their

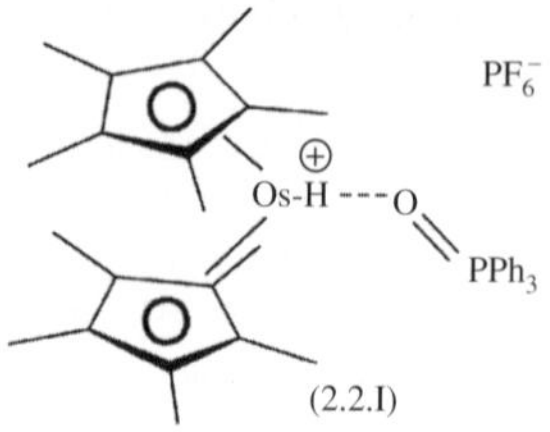

(2.2.I)

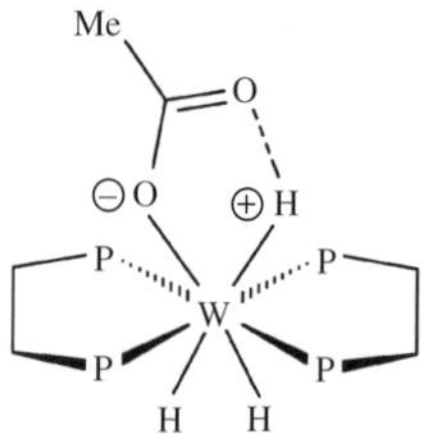

d(H···O) = 2.33 Å

(2.2.II)

d(Mo–H) = 1.80 Å

d(H···O) = 2.51 Å

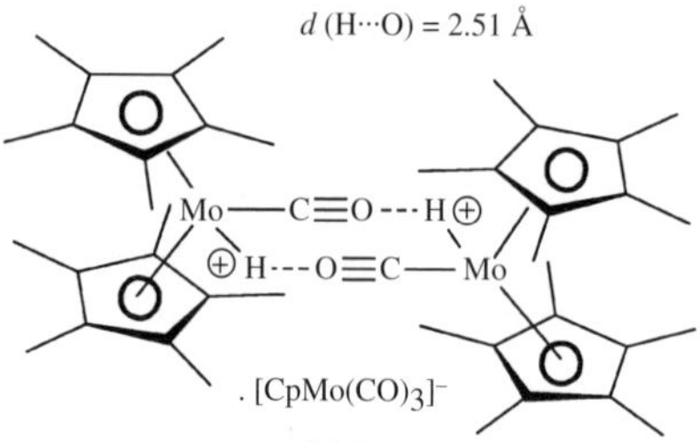

. $[CpMo(CO)_3]^-$

(2.2.III)

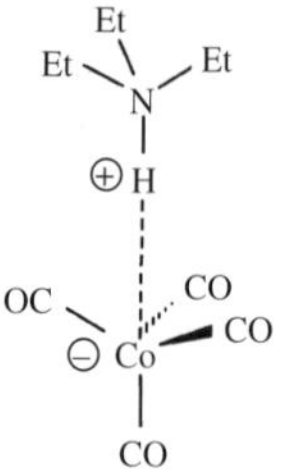

d(H···Co) = 2.613 Å

(2.2.IV)

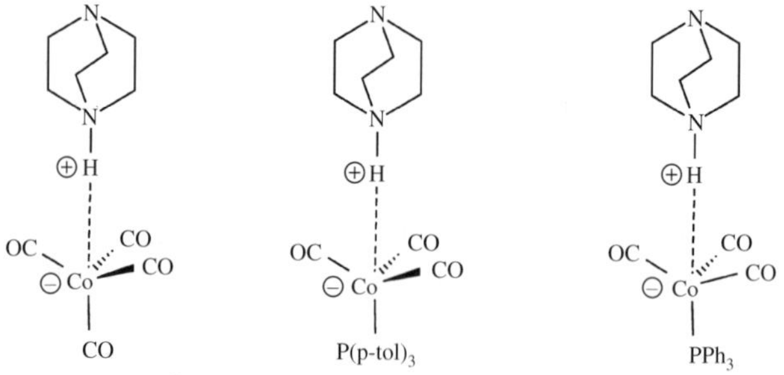

d(H···Co) = 2.392 Å

d(H···Co) = 2.322 Å

d(H···Co) = 2.247 Å

(2.2.V)

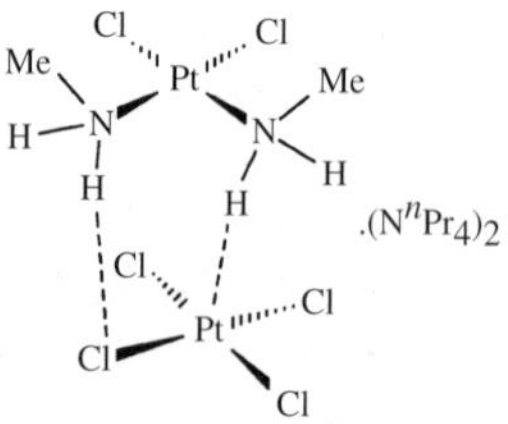

d(H···Pt) = 2.26 Å

(2.2.VI)

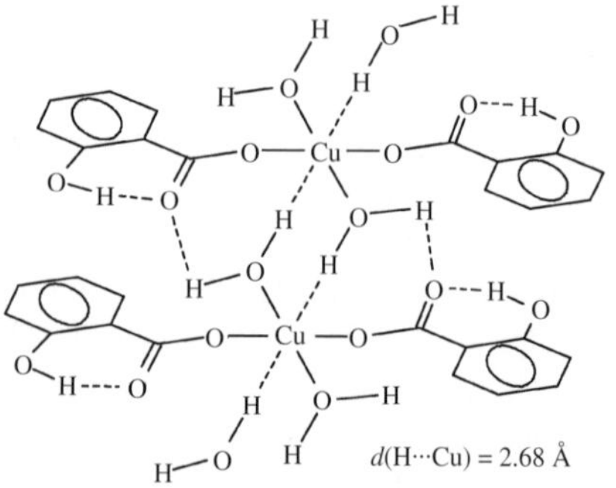

d(H···Cu) = 2.68 Å

(2.2.VII)

energies, as evaluated by thermodynamic methods, are in the medium-strength range ($-\Delta H_{HB}$ = 4–7.6 kcal mol^{-1}) (Belkova *et al.*, 2005).

2.5.2 *Metals as H-bond acceptors (Group 2.2)*

As introduced above, D–H···M interactions may have quite different properties according to whether electron-deficient or electron-rich metal centers are involved. The former give rise to the family of agostic interactions (Group 3.1) that have already been excluded from the list of the true H-bonds because of their 3c–2e nature, the latter to true 3c–4e D–H···:M H-bonds, provided the electron-rich metal is endowed with at least a doubly filled *d*-orbital in suitable orientation to accept the bond. This situation is typically associated with low-oxidation-state metal complexes with d^6, d^8 and d^{10} electronic configurations and a limited number of coordination geometries. The general aspects of these compounds have been recently reviewed (Brammer *et al.*, 1995; Brammer, 2003; Epstein and Shubina, 2002; Yao *et al.*, 1997; Braga *et al.*, 1997).

IR spectroscopy in non-polar or low-polar solvents was used to study the association of conventional H-bond donors with sandwich and half-sandwich metallocenes (Cp_2M, Cp^*_2M, M = Fe, Ru, Os; Cp^*ML_2, M = Co, Rh, Ir; L = CO, PR_3) as well as $ML_2(dppe)_2$ complexes (Shubina *et al.*, 1993; Epstein *et al.*, 1994). An accurate IR study in supercritical medium was also performed by Kazarian and Poliakoff (Kazarian *et al.*, 1993) on the H-bonds formed by fluorinated alcohols (R^FOH) with Cp*MLCO (M = Co, Rh, Ir; L = CO, PPh_3, C_2H_2, N_2) complexes.

Structural data for the interactions between N–H and C–H donors and d^8 square planar metals complexes along the axial direction have been reviewed by Yao, Eisenstein and Crabtree (1997). They identified a number of nearly linear N–H···M interactions with H···M distances of 2.11–2.81 Å that are shorter than the sum of van der Waals radii. A less demanding linear preference was found in weak C–H···M bonds displaying intermediate structure between H-bonds and 3c–2e agostic interactions. A good example of short intermolecular N–H···:Pt interaction in a d^8 square-planar complex comes from the neutron structure of the compound $[N^nPr_4]_2[PtCl_4]$.*cis*-$[PtCl_2(NH_2Me)_2]$ shown in **2.2.VI** (Brammer *et al.*, 1991), where one N–H of the NH_2Me ligand was found to bind to the Pt at the center of the nearly perpendicular $[PtCl_4]^{2-}$ anion with an essentially linear N–H···Pt angle [167.1(9)°] and H···Pt distance of 2.26(1) Å. The distribution of the X–H···M geometries (X = C, N, O) in crystals has also been investigated by Braga *et al.* (1997) and an example of intermolecular O–H···Cu bond found in this study is shown in **2.2.VII**.

2.5.3 *Metal hydrides as H-bond acceptors or dihydrogen bond (DHB) (Group 2.3)*

Recently, considerable attention has been given to the interactions between conventional H-bond donors and M–H σ-bonds, with M less electronegative than H, as acceptors (Shubina *et al.*, 1997; Crabtree *et al.*, 1996, 1998; Crabtree, 1998a,b;

Alkorta *et al.*, 1998; Custelcean and Jackson, 2001). These protonic-to-hydridic interactions are more commonly called *dihydrogen bonds (DHBs)* and have been shown to be strong enough to exert significant effects on molecular structure and reactivity in solution and in the solid state, with particular concern for organometallic catalysis. Two classes of DHBs have been described: (*i*) DHB to main-group hydrides, and (*ii*) DHB to transition-metal hydrides.

2.5.3.1 *DHB to main-group hydrides*

The compounds involved are the partially covalent hydrides of Group 3 elements: B, Al, and Ga. The first observations date back to the late 1960s when Brown and Heseltine (1968) studied the IR spectra in CCl_4 of some boron donor–acceptor complexes [$L{\cdot}BH_3$ and $L{\cdot}BH_2X$ ($L = Me_3N, Et_3N$, Py, Et_3P; X = Cl, Br, I)] in the presence of MeOH, PhOH, or *p*-F-PhOH and suggested the formation of a new type of H-bonded interactions having energies around 1.7–3.5 kcal mol^{-1} and where *B–H was the proton acceptor* despite its lack of lone pairs or π electrons. These findings were overlooked for years and only recent IR and NMR studies by Epstein and coworkers (Epstein *et al.*, 1998) have confirmed the ability of boron hydride complexes to act as H-bond acceptors with respect to alcoholic donors. The authors could conclude that these unconventional O–H···H–B bonds were comparable in energy with ordinary H-bonds and that their energy was strongly affected by their charge, being 1.1–3.7 kcal mol^{-1} for neutral (e.g. $Et_3N{\cdot}BH_3$ or $(EtO)_3N{\cdot}BH_3$) and 2.3–6.5 kcal mol^{-1} for ionic ones (e.g. $Bu_4N^+{\cdot}BH_4^-$).

Solution studies could not suggest any plausible geometry for these bonds. The recently determined neutron structure of $NaBH_4{\cdot}H_2O$ (**2.2.VIII**) shows the formation of three different O–H···H–B bonds with H···H contact distances of 1.79, 1.86 and 1.94 Å, all substantially shorter than the H···H vdW cutoff of 2.4 Å (Custelcean and Jackson, 2001). However, the O–H vector is found to point towards the middle of the B–H bond, suggesting an association with the σ-bond electrons rather than with the hydrogen atom. Similar conclusions come from a systematic CSD search carried out by Crabtree and coworkers on boron–nitrogen complexes (Richardson *et al.*, 1995). A total of 26 intermolecular N–H···H–B bonds with H···H contacts in the range 1.7–2.2 Å were found, all showing a strong preference for a bent geometry with typical NH–H–B angles of 95–120° and much more linear N–H–HB angles of 150–170°. These same geometrical features [H···H 1.82 Å; NH–H–B 98.8 and N–H–HB 158.7°] were reproduced by DFT emulation of the $H_3N{\cdot}BH_3$ dimer of C_2 symmetry (**2.2.IX**) for which a dimerization energy of 12.1 kcal mol^{-1} was calculated.

The hydrides of heavier Group 3 elements are also capable of forming DHBs. In 1994 Raston and coworkers (Atwood *et al.*, 1994) provided X-ray crystallographic evidence for intramolecular N–H···H–Al bonding in an adduct of alane with tetramethylpiperidine. Computational MP2 simulation (Cramer and Gladfelter, 1997) on $H_3N{\cdot}AlH_3$ dimer of C_2 symmetry and isostructural with **2.2.IX** has given H···H 1.781 Å, NH–H–Al 119.4 and N–H–HAl 172.0° with a dimerization energy of 11.8 kcal mol^{-1}. Finally, the neutron structure of cyclotrigallazane

$d(\mathrm{H}\cdots\mathrm{H})$= 1.79, 1.86, 1.94 Å

(2.2. VIII)

$d(\mathrm{H}\cdots\mathrm{H})$= 1.82 Å

(2.2. IX)

$d(\mathrm{H}\cdots\mathrm{H})$= 1.97 Å

(2.2. X)

SbF_6^-

X = PPh_3

$d(\mathrm{H}\cdots\mathrm{H})$= 1.80 Å

(2.2. XI)

$d(\mathrm{H}\cdots\mathrm{H})$= 1.8–1.9 Å

M = Os, Ru, Ir
X = $P(i\text{-}Pr)_3$

(2.2.XII)

$d(\mathrm{H}\cdots\mathrm{H})$= 1.7–2.2 Å

M = Al, B, Ga
Ir, Mo, Mn, Os,
Re, Ru, W

X = F, O, N, C

$\phi(\mathrm{H}\cdots\mathrm{H{-}M})$ = 90 – 135°

(2.2.XIII)

L = PMe_3, PEt_3, $P(i\text{-}PrO)_3$, PPh_3
R = Ph, $CH(CF_3)_2$, $C(CF_3)_3$

(2.2.XIV)

(**2.2.X**) provided a good example of intermolecular N–H···H–Ga DHB with parameters: H···H 1.97 Å, NH–H–Ga 131° and N–H–HGa 145° (Campbell *et al.*, 1998).

2.5.3.2 *DHB to transition-metal hydrides*

The first unequivocal evidence of DHB involving transition metal hydrides is due to Crabtree (Lee *et al.*, 1994, Peris *et al.*, 1995) and Morris (Lough *et al.*, 1994; Park *et al.*, 1994). For example, Crabtree's X-ray crystal structure of the iridium hydride (**2.2.XI**) shows unexpected tautomerization of the amidic ligand to its iminol form, a fact that was interpreted by an attractive O–H···H–Ir interaction between the acidic $O^{\delta-}$–$H^{\delta+}$ proton and the hydridic $Ir^{\delta+}$–$H^{\delta-}$ hydrogen. In the following years a considerable number of other intramolecular as well as intermolecular M–H···H–X DHBs were identified, suggesting that these interactions could be taken advantage of in applications of molecular recognition and crystal engineering. A beautiful example of intermolecular linkage is displayed in Scheme **2.2.XII** (Abdur-Rashid *et al.*, 1998, 2000). For a more detailed analysis on the subject the reader is addressed to the comprehensive review by Custelcean and Jackson (2001).

The geometries of transition-metal DHBs are not significantly different from those of Group 3 elements, so that they are displayed together in Scheme **2.2.XIII**. The rather short H···H distances, which are in the range 1.7–2.2 Å, are to be compared with the H···H vdW cutoff of 2.40 Å. The most salient feature is the strongly bent XH–H–M angle, typically 90–135°, which indicates that the X–H bond does not interact with a hypothetical lone pair located on the hydridic hydrogen, as would happen in conventional D-H···:A bonds, but rather with the bonding electron pair of the M–H σ-bond. This clearly suggests that DHBs are to be classified as completely unconventional H-bonds.

As for DHB interaction energies, a thorough IR and NMR analysis carried out by Epstein and coworkers (Shubina *et al.*, 1996) on tungsten hydride–alcohol complexes (**2.2.XIV**) in solution has shown exclusive formation of O–H···H–W interactions, whose strengths increase with the donor ability of the L ligand (PMe_3 > PEt_3 > $P(i\text{-}PrO)_3$ > PPh_3) and with the increasing acidity of the proton-donating alcohol ($(CF_3)_3COH$ > $(CF_3)_2CHOH$ > PhOH). Association energies fall in the range 4.1–6.9 kcal mol^{-1} as evaluated by the shifts of the IR ν(O–H) stretching frequency and by the variations of the binding constants with temperature.

2.5.4 *Metal ligands as H-bond donors or acceptors (Groups 2.4–5)*

Conventional H-bond donors and acceptors coordinated to metal centers persist in forming H-bonds that are still conventional except for the fact that the acidity of the donor or the basicity of the acceptor may be substantially affected by the linkage with the metal. A well-known case of metal ligands acting as H-bond donors concerns metal-coordinated waters that normally display greater acidity and better H-bond ability than non-coordinated ones. This suggests the possibility of tuning the H-bond donor properties of water by coordination with metals having different electron-withdrawing abilities. The effect is well documented in ionic crystals and minerals and, at the molecular level, in organometallic and coordination compounds, though no systematic study on this topic is yet available.

A case of metal-coordinated H-bond acceptors extensively studied by CSD methods (Braga *et al.*, 1995; Braga and Grepioni, 1997) concerns carbonyl ligands, particularly in respect to the prevalent formation of C–H···O≡C–M bonds. It seems of interest that the H-bond strengths, evaluated by the H···O distances, increase in the order M(CO) < M_2(μ-CO) < M_3(μ_3-CO) (Scheme **2.2.XV**), which is also the predictable order of increasing basicity of the carbonyl oxygen as determined by π-back donation from the metal(s) to the CO π^* orbital.

By far the most studied metal-coordinated H-bond acceptors are the halide ions that, for their property of forming quite strong H-bonds with conventional donors, are having increasing applications in crystal engineering as hinges between coordination polyhedra or connections between these polyhedra and variously designed organic moieties (Mareque Rivas and Brammer, 1998; Brammer *et al.*, 2002). In two extensive CSD investigations (Brammer *et al.*, 1999; Aullón *et al.*, 1998), the geometries of many thousands of D–H···X–M (D = N, O; X = F, Cl, Br, I) have been examined and compared with those

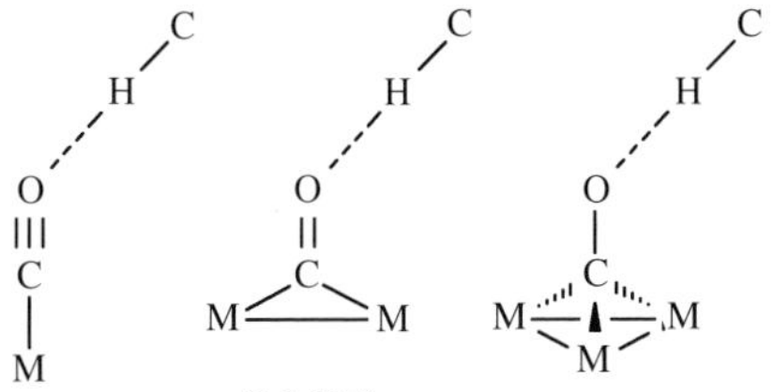

(2.2.XV)

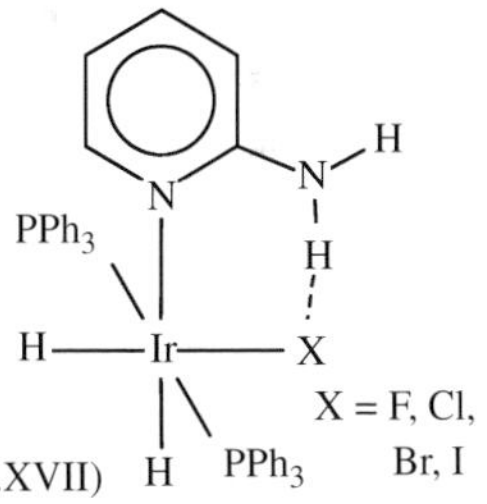

(2.2.XVII) X = F, Cl, Br, I

(a) O—H···Cl—M (b) O—H···Cl⊖ (c) O—H···Cl—C

(d) N—H···Cl—M (e) N—H···Cl⊖ (f) N—H···Cl—C

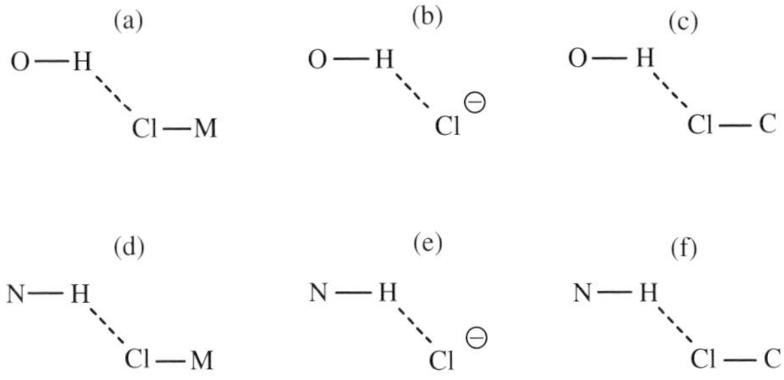

(2.2.XVI)

of the corresponding D–H···X–C and D–H···X⁻ bonds. The results obtained for chlorine are summarized in Fig. 2.5 which shows the histograms of the H···Cl distances for the six classes of H-bonds depicted in Scheme **2.2.XVI**. The :Cl–M acceptor is seen to form rather short H-bonds that are comparable with strong charge-assisted X–H···Cl⁻ conventional ones and much stronger than those formed by :Cl–C organic acceptors, a class of bonds reported above under the heading of weak H-bond acceptors. This well illustrates the influence of the metal in enhancing the H-bond acceptor ability of the halogen in consequence of the increased $M^{\delta+}$–$Cl^{\delta-}$ bond polarization. Moreover, comparison among different halogens leads to the conclusion that the relative strengths of the D–H···X–M bond decrease in the order D–H···F–M > D–H···Cl–M > D–H···Br–M > D–H···I–M, in agreement with expectation.

A not dissimilar trend was observed by Cratbee, Eisenstein and coworkers (Peris *et al.*, 1995) who made use of combined NMR spectroscopy and *ab initio* calculations to monitor the strength of the intramolecular N–H···X–Ir bond (X = F, Cl, Br, I) in the Ir complex displayed in **2.2.XVII**.

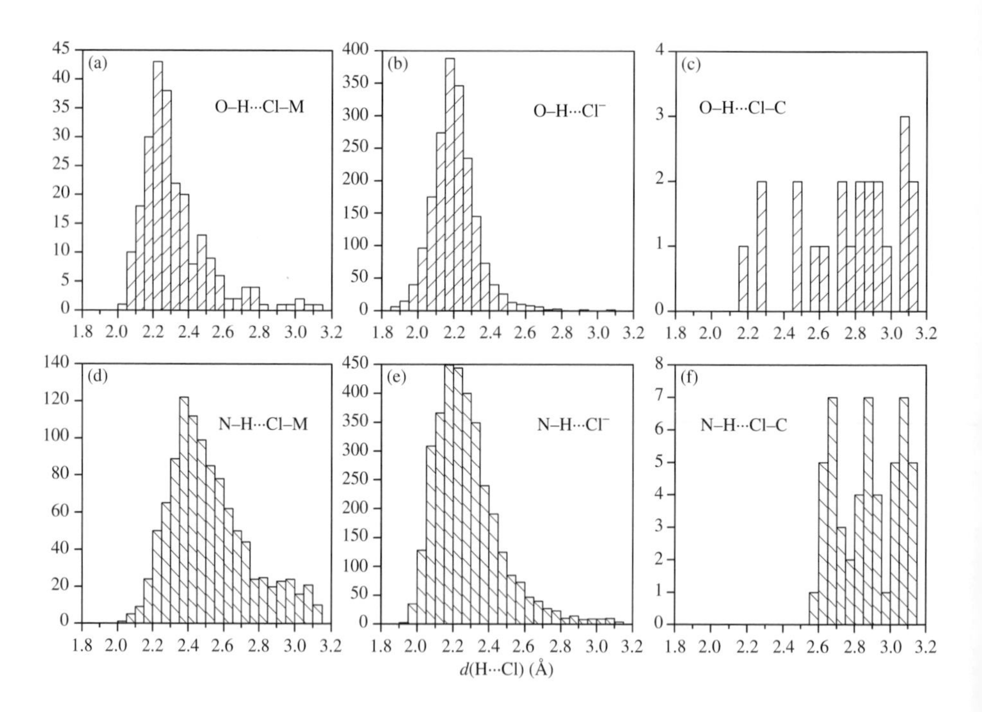

Fig. 2.5. Histograms of the H···Cl distances (Å) for the six classes of H-bonds depicted in Scheme **2.2.XVI**.: (a) O–H···Cl–M; (b) O–H···Cl⁻; (c) O–H···Cl–C; (d) N–H···Cl–M; (e) N–H···Cl⁻; (f) N–H···Cl–C. (Data from Aullón *et al.*, 1998)

2.6 H-bond classification by physical properties: Weak, moderate, and strong H-bonds

The chemical classification of the H-bond given above is not the only one possible. H-bonds can also be usefully classified in terms of their physicochemical properties according to the scheme of Table 2.4, originally proposed by Jeffrey (1997) following the Pimentel and McClellan (1960) empirical treatment that identifies the H-bond itself with the set of its phenomenological manifestations, i.e. the set of physicochemical quantities that can be experimentally measured or theoretically calculated.

The classification is limited to the H-bonds formed by main-group elements (Chart 2.3 – Class 1) because, at present, too little is known about bonds involving metal centers (Class 2). The physical quantities considered are the most currently used in H-bond studies and for which the largest set of accurate experimental results is normally available, and include: (*i*) H-bond geometries, as derived from X-ray or neutron crystallography, MW rotational spectroscopy, and gas electron diffraction; (*ii*) H-bond energies, E_{HB}, as derived from thermodynamic studies of binding equilibria in gas phase or non-polar solvents; (*iii*) red shifts of the ν(D–H) stretching frequency of the D–H bond, as derived from IR or Raman vibrational spectroscopy; (*iv*) downfield shifts of the δ(DH) chemical shift of the H-bonded proton, as derived from ^{1}H NMR spectroscopy.

In principle, H-bond geometries, energies, vibrational frequencies, and NMR chemical shifts can also be obtained by *ab initio* or DFT quantum-mechanical emulation. At present, emulated data are far fewer than experimental data, particularly those collected from crystallographic and thermodynamic databases. Greater drawbacks are, moreover, that they are difficult to access because they are not organized in systematic databases and difficult to compare among themselves because of the quite large changes that have occured in computing methods in the last thirty years. For these reasons, these data are not systematically used here, but only taken advantage of in particular cases.

Conversely, computational methods are *de rigueur* in assessing the shapes of PT pathways of the H-bonds of different strengths. This topic will be treated in greater detail in Chapter 6 but some preliminary definitions are indispensable here for understanding the following discussion. To this aim, Schemes **2.3.Ia–d** depict the somewhat idealized shapes that PT profiles are expected to assume for weak (**2.3.Ia**), moderate (**2.3.Ib**), strong (**2.3.Ic**), and very strong H-bonds (**2.6.Id**). The nomenclature is taken from Gilli *et al.* (2004) and makes use of the following symbols for the description of the shapes of PT profiles and of the heights of the corresponding PT barriers: (*i*) *s* = *symmetric* and *a* = *asymmetric*; (*ii*) *SW* = *single-well* and *DW* = *double-well*; (*iii*) *HB* = *high-barrier*, *MB* = *medium-barrier*, *LB* = *low-barrier*, and *NB* = *no-barrier*. The two extremes are represented by weak (*aSW-HB* = asymmetric single-well/high-barrier) and very strong H-bonds (*sSW-NB* = symmetric single-well/no-barrier), while H-bonds of intermediate strengths are endowed with double-well (*DW*) PT profiles of

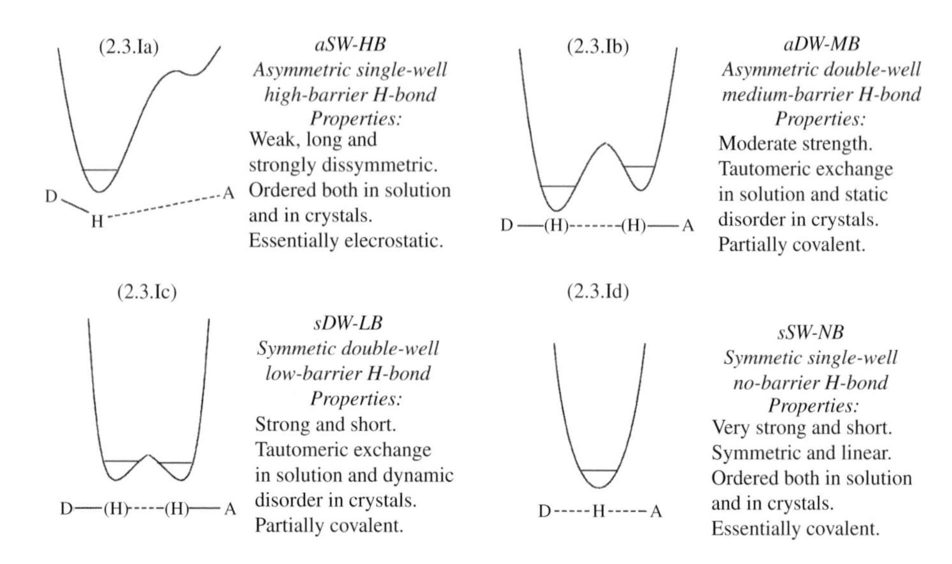

variable symmetry and barrier height (*aDW-MB* = asymmetric double-well/ medium-barrier and *sDW-LB* = symmetric double-well/low-barrier).

(2.3.Ia)
D
H
A
aSW-HB
Asymmetric single-well
high-barrier H-bond
Properties:
Weak, long and
strongly dissymmetric.
Ordered both in solution
and in crystals.
Essentially elecrostatic.
(2.3.Ib)
D—(H)-------(H)—A
aDW-MB
Asymmetric double-well
medium-barrier H-bond
Properties:
Moderate strength.
Tautomeric exchange
in solution and static
disorder in crystals.
Partially covalent.
(2.3.Ic)
D—(H)----(H)—A
sDW-LB
Symmetic double-well
low-barrier H-bond
Properties:
Strong and short.
Tautomeric exchange
in solution and dynamic
disorder in crystals.
Partially covalent.
(2.3.Id)
D-----H-----A
sSW-NB
Symmetic single-well
no-barrier H-bond
Properties:
Very strong and short.
Symmetric and linear.
Ordered both in solution
and in crystals.
Essentially covalent.

Table 2.4 shows that the properties considered (geometry, energy, IR red shift, ^{1}H downfield shift, and PT profile) provide a criterion for partitioning all H-bonds in three different groups called, according to Jeffrey (1997), *weak, moderate,* and *strong H-bonds* and to which arbitrary E_{HB} ranges of 1–4, 4–15, and 15–45 kcal mol^{-1} have been assigned. In parallel to energy, each group is characterised by well-defined ranges of all other properties that monotonically increase, or decrease, from the weakest to the strongest bond. The extremely wide range encompassed by all parameters considered well illustrates the extraordinary wide latitude spanned by the H-bond phenomenon. The table is completed in its lower part by a compilation of the most common donor and acceptor groups and of the H-bonds they may form, that can be considered more typical of each class of strength. The results obtained reproduce well the distinction between weak and conventional H-bonds of the chemical classification illustrated above and the partitioning of conventional bonds in the six chemical leitmotifs listed in Table 2.2. It is confirmed, in particular, that strong H-bonds are associated with very specific atomic arrangements not dissimilar from true chemical species, out of which three are charge-assisted (leitmotifs # 2, 3, and 4) and one resonance-assisted (leitmotif # 5). The subject will be resumed in Chapter 3.

2.7 Correlation among physical descriptors: The problem of the driving variable

The discussion above has clearly shown that the H-bond is by no means a simple and well-delimited subject, but rather a complex phenomenon where a large

TABLE 2.4. Physicochemical properties of weak, moderate and strong H-bonds

H-bond	Weak H-bond	Moderate H-bond	Strong H-bond
D–H···:A bond	electrostatic	electrostatic–covalent	mostly covalent
Bond lengths	D–H $\ll$ H···A	D–H $<$ H···A	D–H $\approx$ H···A
H···A (Å)	3.2–2.2	2.2–1.5	1.5–1.2
D···A (Å)	4.0–3.2	3.2–2.5	2.5–2.2
D–H–A angle (°)	90–150	130–180	165–180
H···A vdW cutoff contraction (%)	<20	20–40	40–55
Bond energy, E_{HB} (kcal mol^{-1})	1–4	4–15	15–45
PT profile and height of the PT barrier	*aSW-HB*	*aSW-HB* *aDW-MB*	*sDW-LB* *sSW-NB*
Decrease of the IR ν(D–H) stretching (%)	<10	10–20	20–80
^{1}H δ(DH) chemical shift (ppm)	–	<14	14–22
Typical donors	C–H, P–H, S–H, Se–H	–O–H, =N–H	[=O–H]$^{+}$, [≡N–H]$^{+}$
Typical acceptors	C–Hal:, =S:, =Se:, π-bond	=O:, ≡N:, P=O:	[Hal]$^{-}$, [–O]$^{-}$, [P–O]$^{-}$
Typical H-bonds	C–H···O, C–H···N C/O/N–H···Hal–C C/O/N–H···π-bond	–O–H···O=, –O–H···N≡ –N–H···O=, –N–H···N≡ *Polarization-assisted or σ-cooperative H-bond:* ···Ò–H···Ò–H··· *Resonance-assisted or π-cooperative H-bond:* ···O=C–C=C–OH··· ···O=C–C=C–N(R)H···	*Charge-assisted H-bond:* $R_3N^{½-}$···H^{+}···$^{½-}$OOCR [F···H···F]$^{-}$ [RCOO···H···OOCR]$^{-}$ [O_3SO···H···OSO_3]$^{-}$ [H_2O···H···OH_2]$^{+}$, proton sponges [≡N···H···N≡]$^{+}$ *Resonance-assisted or π-cooperative H-bond:* ···O=C–C=C–OH··· ···O=C–C=C–N(R)H····

number of chemical elements, acting as H-bond donors, acceptors, or both, combine among themselves to produce an exceptionally wide spectrum of possible interactions. In such a situation, an efficient classification becomes mandatory and in the previous paragraphs we have summarized two of these classifications, respectively based on *purely chemical criteria* and on the *strengths of the H-bonds formed.* The chemical classification, in particular, has proved to be able to assemble together, under a unique H-bond definition, the full spectrum of 3c–4e interactions derived from both main-group and transition elements and to account reasonably well for the electronegativity effects of the H-bond donor and acceptor atoms.

Neither of the two classifications, however, can account for what we like to call the *H-bond puzzle* (Gilli and Gilli, 2000), i.e. the unique feature of the H-bond by which bonds made by the same donor–acceptor pair may display an extremely wide range of energies and geometries. This extreme variability is well represented by the R_1–O–H···:O–R_2 bonds that, according to the choice of R_1 and R_2, can span d(O···O) values from 2.38 to 3.00 Å and d(H···O) ones from 1.20 to 2.00 Å, while E_{HB} collapses from 25–30 to less than 1.0 kcal mol^{-1}. Even more impressive are the effects induced by the changing acid–base properties of the environment. The H-bond in the HO–H···OH_2 neutral water dimer has an energy somewhat less than 5 kcal mol^{-1} for D and d distances of about 2.75 and 1.75 Å. In acid or basic environments, however, the dimer switches to its protonated $[H_2O\cdots H\cdots OH_2]^+$ or deprotonated $[H–O\cdots H\cdots O–H]^-$ forms, both of which have E_{HB}s up to 27–32 kcal mol^{-1} and D and d distances down to 2.38–2.40 and 1.20–1.24 Å. A similar effect is well known to occur for the ···F–H···F–H··· bond [$E_{HB} \simeq 6$ kcal mol^{-1}, d(F···F) $\simeq$ 2.49 Å] that changes in a basic environment into its deprotonated $[F\cdots H\cdots F]^-$ form that is considered to be the strongest H-bond occurring in nature [$E_{HB} \simeq$ 40–45 kcal mol^{-1}, d(F···F) $\simeq$ 2.21–2.28 Å].

What is observed for O–H···O and F–H···F bonds can be easily shown to occur for all conventional H-bonds, leading to the surprising conclusion that: *It cannot be generally stated that there are strong or weak H-bonds in themselves because any given D–H···:A system may form H-bonds in a wide range of strengths and distances, whose two extremes are (i) the weak, bent, and proton-out-centered bond of purely electrostatic nature; and (ii) the strong, short, linear, and proton-centered one classifiable as a true 3c–4e covalent bond.*

This completely changes the outlook of the H-bond problem. The position of the donor and acceptor atoms in the Periodic Table remains fundamental in determining electronegativity effects but we see now that there are *other changes occurring within each box of the Periodic Table* that may supersede the effects of periodicity and need to be accounted for if we want to achieve a sufficient degree of understanding of the H-bond phenomenon. The new challenge is then: *Which chemical factor, or factors, can induce these marked variations of H-bond strengths?*

The problem can be considered from a different point of view. The H-bond was discovered in 1920 and thousands of papers have been dedicated to measuring or

TABLE 2.5. A selection of the H-bond physicochemical quantities of more widespread use in the description of R–D–H···:A–R′ H-bonded systems

Gas-phase and solution thermodynamics
H-bond binding equilibrium:
$K_{eq,HB}$; ΔH_{HB}; ΔS_{HB}; ΔG_{HB}
Proton Affinities (PA):
PA(R–D$^-$); PA(R′–A)
ΔPA = PA(R–D$^-$) – PA(R′–A)
Acid–base constants (pK_a):
pK_a(R–D–H); pK_a(R′–A–H$^+$)
ΔpK_a = pK_a(R–D–H) – pK_a(R′–A–H$^+$)
Computational and theoretical chemistry
H-bond energies: ΔE_{HB}; ΔH_{HB}; ΔG_{HB}
Optimized H-bond geometries
Vibrational normal-mode parameters
PT pathways; PT barriers
Electron densities; electrostatic potentials
Dipole moments
Deuteron nuclear quadrupole resonance
Deuteron quadrupole coupling constant (e^2Qq/h)
X-ray and neutron crystallography
Solid-state H-bond geometries:
d(D–H); d(H···A); d(D···A); θ(D–H–A)
Proton populations in double-well H-bonds
Electron densities
Microwave (MW) rotational spectroscopy and gas electron diffraction (g.e.d.)
Gas-phase H-bond geometries
IR and Raman vibrational spectroscopy and neutron inelastic scattering
Stretching: ν(D–H); ν(H···A)
In-plane bending: δ(D–H); δ(H···A)
Out-of-plane bending: γ(D–H); γ(H···A)
Band intensity; deuterium isotope effect
Solution and solid-state NMR spectroscopy
^{1}H chemical shift: δ(DH)
^{13}C, ^{14}N, ^{15}N, ^{19}F chemical shifts
Deuterium isotope effect
Deuterium fractional factor

to calculating any possible physicochemical quantity associated with its formation (a list of the most important ones is displayed in Table 2.5). All together, these quantities constitute a set of *H-bond physicochemical descriptors* that cover the most salient aspects of the H-bond and can be shown to be highly intercorrelated, another fact that, over the years, has been the object of a large number of scientific publications. This mutual correlation, in itself, is a positive fact because it indicates that we are actually studying a unique scientific problem (the H-bond) from different points of view. Less positive is that such an intercorrelation among variables does not tell us which is the cause and which the effect, a consideration that redefines the problem we have to find a solution to in a different way, that is: *Which H-bond physicochemical descriptor, among the many intercorrelated ones, is the independent variable that drives the H-bond strength and, in turn, all other dependent properties?*

The following chapters (3 and 4) will try to answer this basic question by a systematic exploration of the structural and thermodynamic H-bond data accessible from the presently existing crystallographic and thermodynamic databases. Chapter 5 will summarize the results obtained, that constitute the empirical laws governing the H-bond and will show that the driving variable sought has to be identified in the acid–base indicators (proton affinities in the gas phase and related acid–base dissociation constants in solution) of the H-bond donor and acceptor.

3

MODELLING THE H-BOND BY CRYSTALLOGRAPHIC METHODS

3.1 Crystallographic databases and structural correlations

3.1.1 *A survey of structural databases*

At the end of the nineteenth century *molecular structure* was a purely topological (or stereochemical) concept intended to define the types and numbers of atoms constituting the molecules and the ways these atoms were mutually connected by chemical links. The first time that this concept of structure was expanded to include *molecular metrics*, that is the full set of interatomic distances and angles, dates back to the determination of the crystal structure of diamond and alkaline halogenides carried out by Bragg and Bragg (Bragg, 1913; Bragg and Bragg, 1913) making use of the newly discovered technique of X-ray diffraction. Immediately, the crystal structures of elements and simple covalent and ionic compounds started to accumulate until Pauling (1939, 1940, 1960) used them to lay down the foundations of a new branch of chemistry, structural chemistry. From there on, X-ray (and, later, neutron) crystallography became the standard technique for assessing the chemical structure of matter and, in a crescendo of scientific and technological advances, extended its application range from relatively small organic and organometallic molecules to proteins, nucleic acids, and viruses up to ribosomes. These impressing scientific achievements were parallelled by an amazing expansion of the number of crystal structure determinations that, at present, amount all together to something like 400 000 items. The growth of the structural information for organic and organometallic compounds, by far the most numerous group of crystal structures, is illustrated in Fig. 3.1.

Since such an amount of data could hardly be handled by ordinary bibliographic methods, they have been arranged in *structural databases*, that is in continuously updated libraries of structural data associated with a set of computer programs suited to search them, to retrieve the single items and to compare the data retrieved for large sets of structures having specified chemical similarities. This idea has been so successful that, today, all structural data are organised in structural databases (Table 3.1) particularly addressed to organic and organometallic compounds (CDS; Allen *et al.*, 1979; Allen, 2002), inorganic salts and minerals (ICSD; Bergerhoff *et al.*, 1983), metals and intermetallic compounds (CRYSTMET; Calvert, 1981; Calvert and Rodgers, 1984), and biological macromolecules (PDB; Bernstein *et al.*, 1977). This crystallographic information

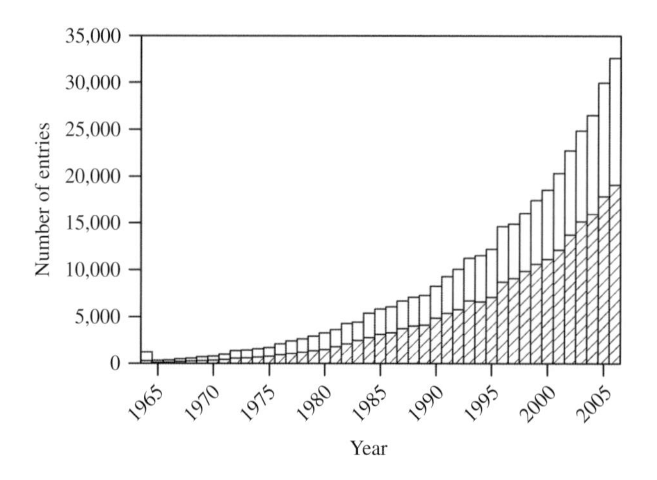

FIG. 3.1. Number of new crystal structures registered each year in the Cambridge Structural Database (CSD) (Allen *et al.*, 1979; Allen, 2002) from 1964 to 2006. Structures of organometallic and organic compounds are in dark and light, respectively.

is complemented by the molecular structures determined in the gas phase by microwave spectroscopy and gas electron diffraction (MOGADOC; Vogt and Vogt, 2004), while a new section of the PDB collects the structures of relatively small biological macromolecules in solution as determined be the combined use of Overhauser-enhanced NMR spectroscopy and molecular dynamics techniques.

All together, databases constitute the greatest library of interatomic distances ever conceived by man that, for our aims, can be divided into three distinct subsets according to whether they are shorter than the sum of the bond radii (the intramolecular range), longer than the sum of the van der Waals radii (the non-bonded or dispersive range) or encompassed in the intermediate interval of distances characteristic of those molecular interactions (multipolar interactions, charge transfer and strong H-bonds) that are attractive enough to produce significant relaxation of the adjacent bond and contact distances. This intermediate range of distances constitutes a complete set of experimental data that, when properly collected and organised in a rational way, can lead to a full interpretation of the H-bond phenomenon in terms of chemical bond theory. The basic problem is to determine such a rational way of dealing with the data and the experimentation of the last few decades has shown that the most efficient solution comes by combining systematic CSD search with *crystal-structure correlation (CSC)* methods.

Though each database has a crucial importance in its specific field (small molecules, biomacromolecules, minerals, and metals), CSD has certainly had the most wide applications in studies of chemical bonding, in general, and H-bond,

TABLE 3.1. Structural databases of molecules and macromolecules

Database name	Content	Experimentals	Number of Entries*	References
CSD. Cambridge Structural Database	Organic and organometallic compounds	Crystal phase X-ray and neutron diffraction	368481	Allen *et al.*, 1979; Allen, 2002
ICSD. Inorganic Crystal Structure Database	Inorganic salts and minerals	Crystal phase X-ray and neutron diffraction	89 384	Bergerhoff *et al.*, 1983
CRYSTMET. Crystallographic Metals Data File	Metals and intermetallic compounds	Crystal phase X-ray and neutron diffraction	81 117	Calvert, 1981; Calvert and Rodgers, 1984
MOGADOC. Molecular Gas Phase Documentation	Molecules and molecular complexes	Gas phase microwave spectroscopy	5900	Vogt and Vogt, 2004
	Molecules and molecular complexes	Gas phase electron diffraction		
PDB. Protein Data Bank	Biological macromolecules	Crystal phase X-ray and neutron diffraction	28 807	Bernstein *et al.*, 1977
	Small biological macromolecules	Solution Overhauser-enhanced NMR	5124	

*Number of entries updated at the end of year 2006.

in particular. The reasons for this success can be traced back to three main factors:

(*i*) for the nature of the compounds considered and the large number of entries (some 370 000 at present) CSD provides a practically complete coverage of all types of H-bonds occurring in the condensed phase, whether derived from main-group or metallic elements;

(*ii*) sufficiently accurate H-bond studies need crystal structures with a remarkably high-quality standard, which means C–C bond standard deviations not exceeding 0.005–0.007 Å and reliable localization of all the hydrogen atoms within the molecule. Results of this quality can be achieved only for CSD molecular crystals by accurate X-ray or neutron diffraction experiments, preferably performed at low temperatures. Conversely, macromolecular PDB structures, even at their highest level of resolution, cannot provide heavy-atom distances with standard deviations smaller than some 0.2–0.3 Å and little or no information on hydrogen locations. While data of this quality may be of great value in assessing the complex morphology of the H-bonded nets occurring in macromolecular environments, they cannot be taken advantage of in studies of chemical bonding;

(*iii*) even more importantly, the data needed for the solution of any specific bonding problem must be straightforwardly retrievable by the use of efficient algorithms cast within the database itself. At present, CSD is the only database that fulfils completely this condition because its structural data are completely coded in terms of *interatomic connectivities*, making it possible to search the database not only for specific molecules but also for specific *molecular fragments* that are included in these molecules. This particular search method provides the basis for the correlation methods discussed below.

3.1.2 *Crystal–structure correlation (CSC) methods*

The analysis of crystallographic data presents the problem that, while the amount of data is large and often of great accuracy, the datum provided (the molecular geometry) does not identify any precise physical quantity (or quantum-mechanical observable). This is because nature is governed by the principle of energy conservation and, therefore, only energy-related quantities, such as binding enthalpies, entropies, and free enthalpies, can provide a correct description of the real strength and stability of the H-bonded complex. Unfortunately, there is no parallel principle of geometry conservation: bond distances do not bear any direct connection with bond energies and, therefore, cannot be used to establish the stability of the bond itself.

Because of these difficulties, molecular geometries find application in H-bond studies in two quite different ways: (*i*) *H-bond shortness* is used as a simple indicator of H-bond strength leaving undetermined its relationships with energy; to this aim, $d(\mathrm{D}\cdots\mathrm{A})$ values are often used as coarse indicators of H-bond strength, though other indices, such as $d(\mathrm{H}\cdots\mathrm{A})$ or $d'(\mathrm{D}\cdots\mathrm{A}) = d(\mathrm{D}\cdots\mathrm{H}) + d(\mathrm{H}\cdots\mathrm{A})$, may be more appropriate because they partially account for the dependence of H-bond strengths on the D–H–A angle (Section 3.1.3.3); (*ii*) *energy–geometry relationships* can be systematically investigated in the frame of the traditional VB theory; this method is much more scientifically grounded but with the evident drawback that it requires a complex preliminary analysis of large sets of intercorrelated contact distances and bond distances and angles intended to assess the functional form assumed by these relationships.

To realise this second approach we therefore need a method able to consider the geometries associated with a given H-bonded system not as isolated points but as part of the same family of intercorrelated data. In this way, we enter the field of *crystal–structure correlations* (*CSCs*), a method of treating crystal data that was proposed some forty years ago (Bent, 1968; Bürgi, 1973; Bürgi *et al.*, 1973) and whose interpretation has been the object of several papers (Bürgi, 1975; Murray-Rust *et al.*, 1975; Dunitz, 1979; Bürgi and Dunitz, 1983; Bürgi, 1992; Ferretti *et al.*, 1992) and a few general reviews (Bürgi and Dunitz, 1994; Ferretti *et al.*, 1996).

The CSC method can be thought of as an attempt to obtain information on the dynamic behavior of molecules from the inevitably rather static crystal data and relies on the basic concept that any chemical reaction can be described as a low-energy pathway on a *potential-energy surface* (*PES*) of the reaction itself, which is the function describing the total potential energy of the system in terms of the relative positions of all the atomic nuclei (Eyring and Polanyi, 1931). The *reaction pathway* is that trajectory (*reaction coordinate*) that connects the points of relative minimum on the surface and leads from the reagents to the products through the higher-energy *transition* or *activated state.* This process is well represented by the reactions $H + H_2 \rightleftharpoons H_2 + H$ or $H_2 + D \rightleftharpoons H + HD$ that can be found in all elementary textbooks of physical chemistry.

If it were possible to take *snapshots* of the molecules while moving along the reaction pathway, the complete pattern of the atomic positions along the course of the reaction would become known. The basic CSC idea consists of assuming that the instantaneous displacements of the nuclei during the reaction can be represented by the images of the reaction center collected from a great number of crystal structures where the center itself happens to be accidentally distorted by the local inter- and intramolecular perturbations. In this hypothesis, the different images need only to be ordered in a rational sequence to give the progressive deformations of the reaction center along the reaction coordinate.

The practical tool for extracting the data from the CSD is the search of the database by *chemical substructure* or *molecular fragment (MF)* (Allen, Kennard, and Watson, 1994) where 'fragment' is any chemical moiety that can be given a stereochemical meaning. The individual representations of the fragment retrieved from the CSD are thus the *fragment images* (or snapshots or *frozen-in pictures*), which illustrate the different forms assumed by the fragment when perturbed by the intra- and intermolecular force field present in any actual crystal structure. In general, a fragment of N atoms will have ($3N$–6) *degrees of vibrational freedom* and then its geometry can be described by ($3N$–6) *internal coordinates*, q_i, (bond distances, angles and torsion angles, or their combinations) that define the *configuration* (*or parameter*) *space* of the fragment. In this space the individual fragment images are represented by single points (the *representative points*) and *crystal–structure correlations* are only statistically significant correlations connecting these representative points.

An interesting feature of many of these correlations is that, following the evolution of the fragment image in E^3 (the physical Euclidean space) along the route defined by the correlation itself, atoms appear to move in a coordinated

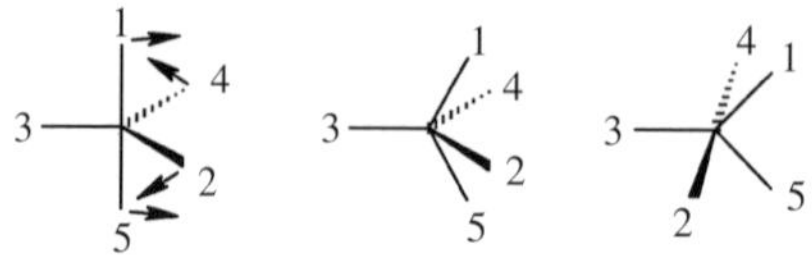

X = O, S, N–, C=

(3.1.I) Berry pseudorotaiton

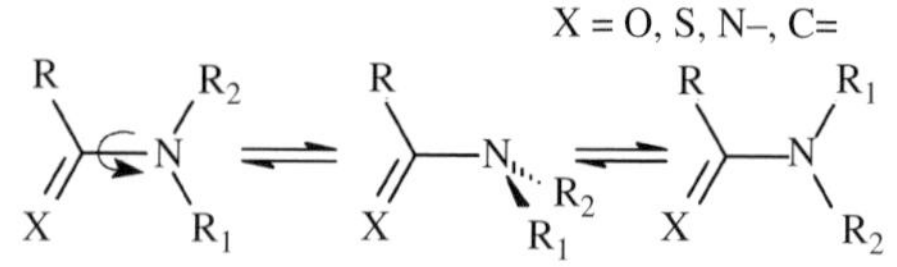

(3.1.II) *Cis-trans* isomerization around the C(sp^2)–N(sp^3) bond

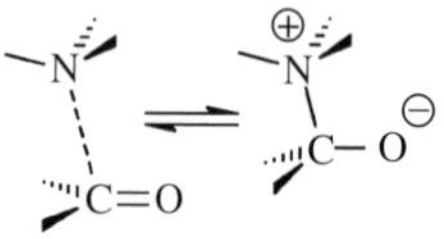

(3.1.III) Nucleophilic addition to carbonyl

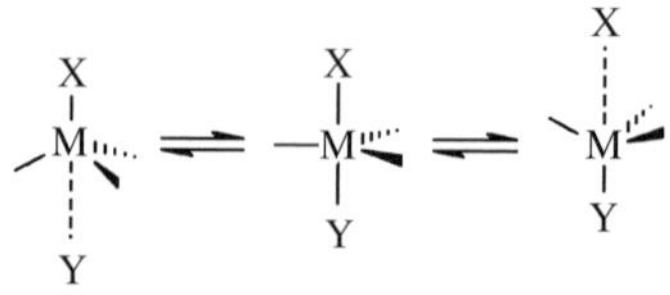

(3.1.IV) Nucleophilic substitution to a metal center

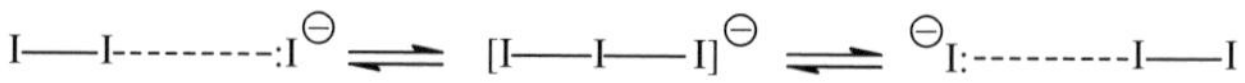

(3.1.V) Tautomerism in the triiodide anion

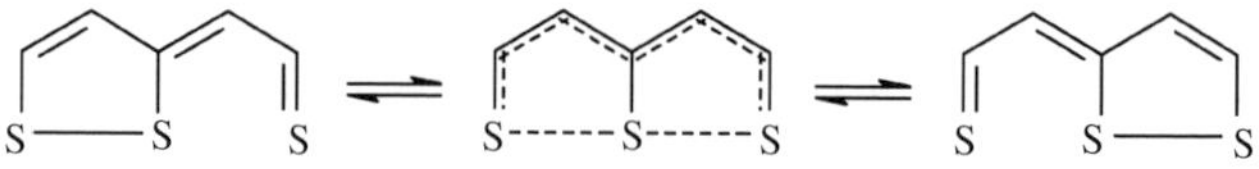

(3.1.VI) Tautomerism in thiathiophthenes

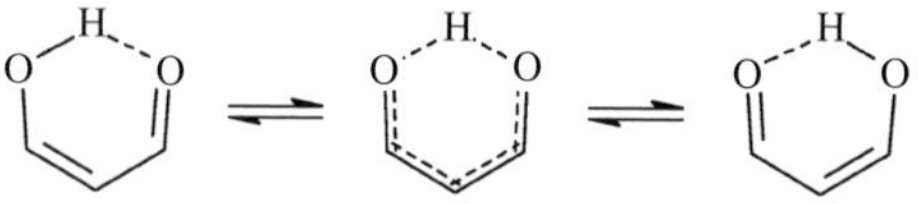

(3.1.VII) Prototropic tautomerism in β-diketone enols

fashion that amazingly resembles what we believe can happen in the course of molecular vibrations or along pathways of chemical reactions or conformational rearrangements. This coordinated motion in E^3 can be called a *relaxation pathway* because the modified fragment geometry is produced by the relaxing of the reference fragment under the effects of the inter- or intramolecular perturbation field. The assumption that CSCs and relaxation pathways actually map low-energy pathways on the PES of the molecular fragment considered is often indicated as the *structure correlation principle* (Bürgi, 1975; Murray-Rust *et al.*, 1975; Dunitz, 1979).

Some schematic examples of chemical reactions that have been studied by correlation methods are reported in Schemes **3.1.I–VII**. The first two are conformational rearrangements. Scheme **3.1.I** represents the well-known Berry pseudorotation process exchanging axial and equatorial ligands in PF_5 (Berry, 1960) and other pentacoordinate phosphorous compounds (Holmes, 1972, 1975, 1979), as well as in bipyramidal-trigonal d^8 metal complexes (Auf der Heyde and Bürgi, 1989a–c), while Scheme **3.1.II** maps the pathway of *cis–trans*

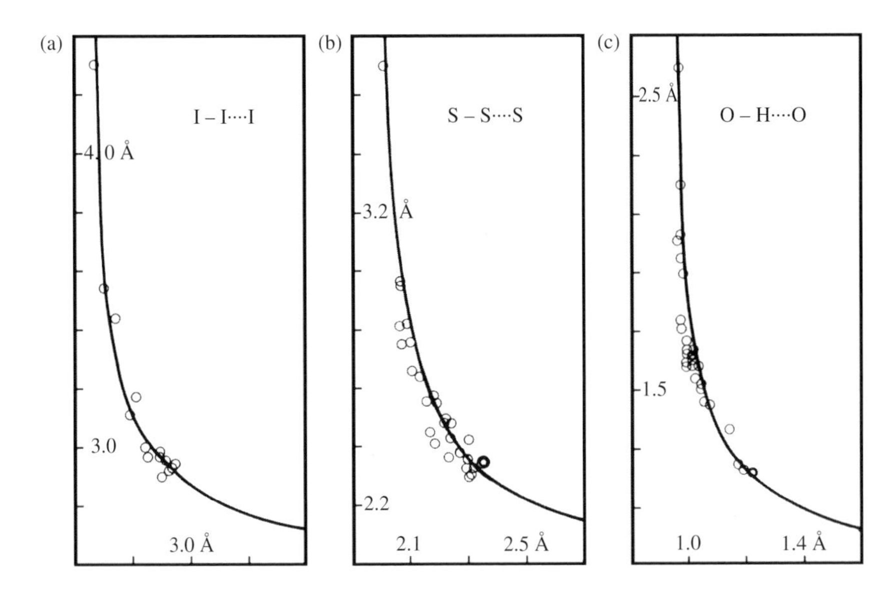

FIG. 3.2. Structure correlations between interatomic distances (in Å) for three linear tautomeric systems: (a) triiodide anions, (b) thiathiophthenes, and (c) linear O–H···O hydrogen bonds. (Reproduced by permission from Bürgi, 1975)

isomerization around the C(sp^2)–N(sp^3) bond occurring in amides and related compounds (Dunitz, 1979; Gilli *et al.*, 1986; Ferretti *et al.*, 1993). The next two examples map true chemical reactions pathways of nucleophilic addition to carbonyl (**3.1.III**; Bürgi *et al.*, 1973, 1974) and to a metal center (**3.1.IV**; Bürgi, 1973). The last three schemes are examples of tautomeric equilibria (Bent, 1968; Bürgi, 1975), respectively, in triodide anions (**3.1.V**), thiathiophthenes (**3.1.VI**), and O–H···O H-bonds (**3.1.VII**). These last three examples are the most interesting from the point of view of molecular interactions because they represent typical 3c–4e processes, known from the seminal papers by Bent (1968) and Bürgi (1975) to be associated with charge-transfer reactions (**3.1.V, VI**) or H-bond formation (**3.1.VII**). Figure 3.2 shows the characteristic shape of the structure correlations between the bond distances I_1–I_2 and I_2–I_3, S_1–S_2 and S_2–S_3, and O_1–O_2 and O_2–O_3. The representative points of each plot are seen to lie on continuous hyperboloids whose functional form is discussed in the next section.

3.1.3 *Bond lengths, bond energies and bond-number conservation rule*

3.1.3.1 *Bond lengths, energies and numbers*

CSC methods give quite useful quantitative relationships among bond distances as well as other geometrical parameters of the fragment studied but have the great drawback that the correlations found do not entertain any direct connection with the bond energies. The problem does not admit a general theoretical solution but can be treated in the framework of the traditional VB theory (Pauling, 1939, 1940, 1960; Coulson, 1961; McWeeny, 1979) by making use of the same empirical functions that have been developed for interpreting the vibrational spectra of simple molecules. So, bond length–bond energy relationships are most frequently derived from the empirical *Morse equation* for diatomic molecules (Morse, 1929)

$$D(d-d_o)=D_o\{1-\exp[-a(d-d_o)]\}^2, \tag{3.1}$$

where $(d-d_o) = \Delta d$ is the displacement of the actual distance d from the equilibrium internuclear distance d_o, D_o is the depth of the potential energy minimum, k the force constant of the stretching vibration, and $a = (k/2D_o)^{1/2}$. A number of alternative Morse-like empirical functions are known that were surveyed by Steele, Lippincott and Vanderslice (1962), the most common being that proposed by Lippincott (1955) to be used in the well-known Lippincott and Schroeder H-bond model (see below).

A further improvement comes from considering the *Pauling relationship* (1947)

$$\Delta d=d(n)-d(1)=-c\log_{10}n, \tag{3.2}$$

where n is called the *bond number* (sometimes *bond order*), $d(1)$ and $d(n)$ are the distances of the simple and n-ple bond, respectively, and c a constant to be determined. It has been shown (Dunitz, 1979; Bürgi and Dunitz, 1987) that this equation can be straightforwardly derived from the Morse relation 3.1, together with the so-called *bond energy–bond order relationship* (*BEBO*: Johnston and Parr, 1963)

$$D(n)=D(1)n^p, \tag{3.3}$$

where $D(1)$ and $D(n)$ are the dissociation energies for the single and n-ple bond, respectively, and p is an empirical parameter ranging from 0.8 to 1.1 according to the type of bond considered. These three quantities, bond energy, D, bond length, d, and bond number, n, are then related by a triangular relationship where eqn 3.1 represents $D = f(d)$, eqn 3.2 $d = f(n)$ and eqn 3.3 $D = f(n)$, so leading to the important *proportionality relationship* $E(\text{bond}) = D \propto n \propto \exp(-d)$ (Ferretti *et al.*, 1996).

3.1.3.2 *Bond-number conservation rule*

The bond number, n, is particularly useful in connection with the so-called *bond-number conservation rule* (Johnston and Parr, 1963; Bürgi, 1973, 1975)

$$n_1+n_2=\text{constant}, \tag{3.4}$$

which finds its main application in nearly linear three-center bonds such as those displayed in Schemes **3.1.V–VII**. Here n_1 and n_2 are the bond numbers of the two bonds connecting the three atoms and the value of the constant depends on the system treated being, for instance, $n_1 + n_2 = 3$ for –C=C–C= ↔ =C–C=C–, and $n_1 + n_2 = 1$ for I–I + I⁻ ↔ I⁻ + I–I or –O–H···O= ↔ =O···H–O–. Substitution of eqn 3.2 into eqn 3.4 for a generic A_1–B···A_2 ↔ A_1···B–A_2 system gives

$$n_1 + n_2 = 10^{-\Delta d1/c1} + 10^{-\Delta d2/c2}, \tag{3.5}$$

where $\Delta d_1 = d_1 - d_1^\circ$ and $\Delta d_2 = d_2 - d_2^\circ$ are the changes of the A_1–B and B–A_2 distances induced by the perturbation (charge-transfer, H-bond or redistribution of single and double bonds in π-conjugated systems) with respect to the unperturbed d_1° and d_2° distances, and c_1 and c_2 are empirical constants typical of each bond. When A_1 and A_2 become identical the equation reduces to

$$n_1 + n_2 = 10^{-\Delta d1/c} + 10^{-\Delta d2/c}, \tag{3.6}$$

with $c = \Delta d'/\log 2$, where $\Delta d'$ is the bond length increment for the symmetrical point having $\Delta d_1 = d_1 - d^\circ$ equal to $\Delta d_2 = d_2 - d^\circ$. This equation can also be written in the more comfortable form

$$d_1 = d^\circ - c\,\log_{10}\left(n_1 + n_2 - 10^{-(d2-d^\circ)/c}\right). \tag{3.6$'$}$$

Its application to the three cases of Fig. 3.2 gives parameters (*a*) $d(1) = 2.67$ Å, $\Delta d' = 0.26$ Å, $c = 0.85$; (*b*) $d(1) = 2.02$ Å, $\Delta d' = 0.31$ Å, $c = 1.03$; and (*c*) $d(1) = 0.96$ Å, $\Delta d' = 0.26$ Å, $c = 0.85$ and the resulting equations, drawn as continuous lines on the figure, are seen to match well the experimental points.

A more complete application of the bond-number conservation rule comes from the analysis of the O–H···O system recently performed (Gilli *et al.*, 1994a; Gilli and Gilli, 2000) by making use of the neutron diffraction data (Jeffrey and Saenger, 1991) collected for nearly linear H-bonds (O–H–O angle $\geq 165°$) occurring in organic compounds and inorganic acid salts.

Calling $r = d$(O–H), $r^* = d$(H···O), and $R = d$(O···O), eqn 3.6′ becomes

$$r^* = r^\circ - c\,\log_{10}\left(1 - 10^{-(r-r^\circ)/c}\right), \tag{3.7}$$

$$R = r + r^*, \tag{3.7$'$}$$

where r° is the O–H distance in the absence of H-bond and c is a constant that can be calculated as $c = [R_{min}/2 - r^\circ]/\log_{10}2$, R_{min} being the shortest possible O···O distance occurring when the proton is perfectly centered. Only two parameters, r° and R_{min}, are then seen to control the functional form of both r versus r^* and r versus R relationships, whose scatter plots are shown in Fig. 3.3(a) and 3.3(b), respectively. Data fitting by eqns 3.7 (continuous curves in the figures) gives values of 2.40 and 0.925 Å for R_{min} and r°, respectively. While $R_{min} = 2.40$ Å is in reasonable agreement with all experimental data so far collected, the value of 0.925 Å for r° is significatively shorter than that measured for the free molecule by gas electron diffraction, which is 0.95–0.96 Å (Vajda and Hargittai, 1992).

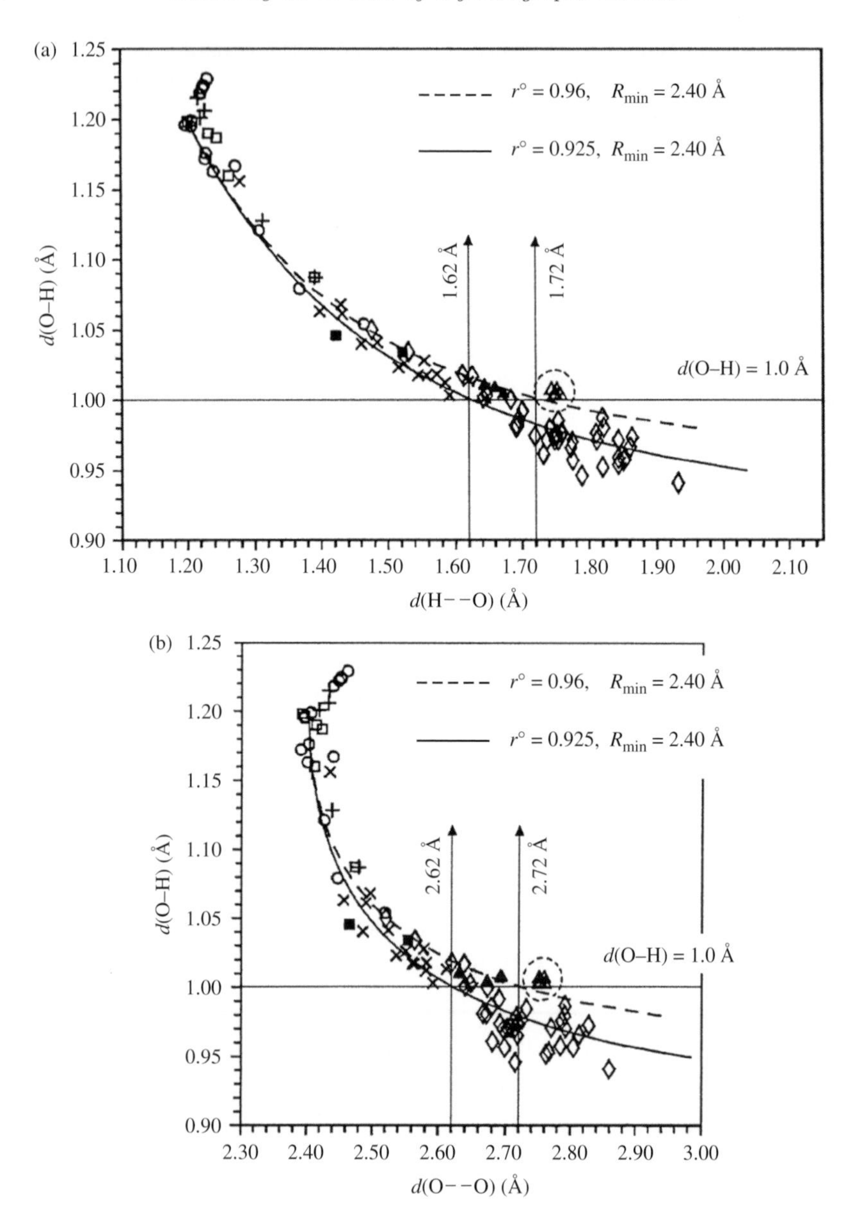

FIG. 3.3. Scatter plots of $r = d(\mathrm{O{-}H})$ as a function of $r^* = d(\mathrm{H{\cdots}O})$ (a) and $r = d(\mathrm{O{-}H})$ as a function of $R = d(\mathrm{O{\cdots}O})$ (b) for selected neutron diffraction data of nearly linear O–H···O bonds with O–H–O angle ≥165°. (Reproduced by permission from Gilli *et al.*, 1994a)

On the other hand, a fitting carried out with the values R_{min}= 2.40 and r° = 0.96 Å (dashed curves) does not appear to fit the experimental data.

This apparent inconsistency has been considered a consequence of the twofold (ionic and covalent) nature of the H-bond (Gilli *et al.*, 1994a). The two dashed curves can be interpreted as the application of the Pauling equation (eqn 3.2) to a *purely covalent* 3c–4e bond and their interpolation for an O–H···O bond having, say, r = 1.0 Å gives values of r^*= 1.72 (Fig. 3.3(a)) and R = 2.72 Å (Fig. 3.3(b)), which would be the correct H···O and O···O distances only if the H-bond were totally covalent without any additional electrostatic contribution. The effect of such an electrostatic term will be that of further reducing r^* and R to the values of 1.62 and 2.62 Å that are located on the underlying continuous curves. This example can be generalized by saying that, for any specific r = d(O–H) value, the two dashed curves give the contribution of the covalent term to the H-bond geometry, while the horizontal shift between the dashed and the continuous curves is a measure of the electrostatic contribution, which causes a further shortening of both H···O and O···O distances. Such a horizontal shift is large for long H-bonds, which are essentially electrostatic, but continuously decreases while the bond shortens, indicating a continuous decrease of the electrostatic component. For the shortest bonds (say, d(O···O) $\leq$ 2.45–2.50 Å) the dashed and continuous curves tend to coincide, showing that the H-bond has become a purely covalent interaction. A similar treatment, based on accurate neutron structures, has been reported by Steiner (1998a, 2002) for the O–H···N/N–H···O system.

A general application of the bond-number conservation rule to the linear O–H···O bond is shown in Fig. 3.4 that reports the plots of the O–H, H···O and O···O distances versus the n(O–H) bond number as calculated by eqns 3.7 with parameters R_{min} = 2.40 and r° = 0.925 Å or 0.960 Å. It can be seen that a single parameter, n(O–H) = 1–n(H···O), controls the full H-bond geometry within the limits of the parametric choice done.

3.1.3.3 *The Lippincott and Schroeder H-bond model (LS-HB)*

The LS-HB model is a semiempirical method for calculating binding energies and IR stretching constants and frequencies of the H-bond starting from its geometry (Lippincott and Schroeder, 1955; Schroeder and Lippincott, 1957). It is based on the Morse-like potential function proposed by Lippincott (1955) that, for an ordinary D–H chemical bond, can be written in the form

$$V=D_0\left\{1-\exp\left[-n(r-r_0)^2/2r\right]\right\},\qquad n=k_0r_0/D_0 \tag{3.8}$$

where D_0 is the bond dissociation energy and k_0 and r_0 are, respectively, the stretching force constant and the D–H bond distance at the equilibrium. This equation gives results strictly comparable with those obtained by the classical Morse formula (eqn 3.1) but, being much simpler to treat mathematically,

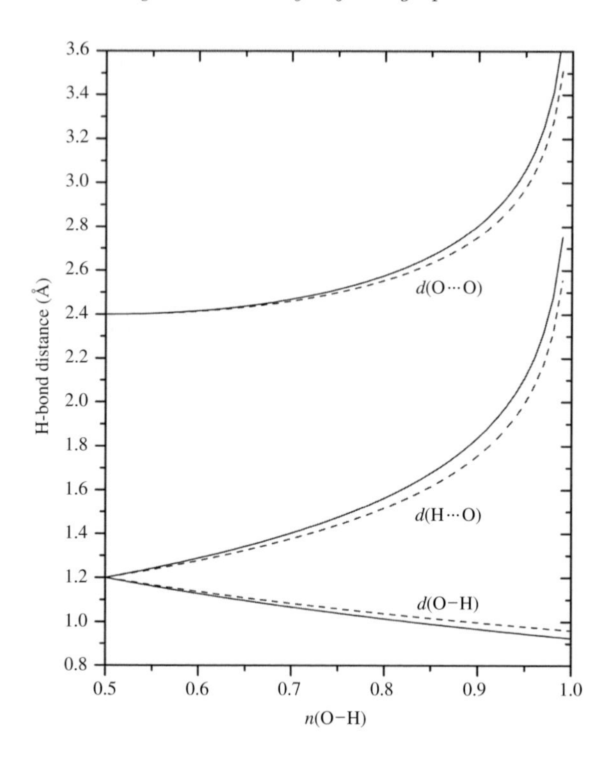

FIG. 3.4. Values of the O–H, H···O and O···O distances for the linear O–H···O bond as calculated by eqns 3.7 as a function of the n(O–H) bond number. Parameters used are $R_{\text{min}} = 2.40$ and $r^{\circ} = 0.925$ Å for the continuous line and $R_{\text{min}} = 2.40$ and $r^{\circ} = 0.960$ Å for the dashed one.

is more suited to build up a model for the neutral D–H···:A bond based on the overlapping of two such Morse-like potentials describing, according to the Coulson formalism of Chart 2.2, the resonance mixing of the two covalent –D–H···:A< and –D$^-$:···H–A$^+$< forms.

Accordingly, the LS-HB model evaluates the total potential energy of the H-bonded system, V, as a sum of the terms

$$V = V_1 + V_2 + V_3 + V_4, \tag{3.9}$$

where

$$\begin{aligned} V_1 &= D_0\{1-\exp(-\alpha)\} \\ V_2 &= D_0^*\{1-\exp(-\beta)\}-D_0^* \\ V_3 &= A\exp(-bR) \\ V_4 &= -B/R^m \end{aligned} \tag{3.10}$$

and where the exponential arguments α and β assume the form

$$\begin{aligned} \alpha &= n(r - r_0)^2 / 2r \\ \beta &= n^*(r^* - r_0{}^*)^2 / 2r^* \end{aligned} \tag{3.11}$$

for the linear bonds defined according to the coordinate system of Fig. 3.5(a). This treatment has been successively extended to bent H-bonds, corresponding to the coordinate system of Fig. 3.5(b), by taking advantage of the modified potential proposed by Moulton and Kromhout (1956). In this case, the arguments α and β assume the slightly more complex form

$$\begin{aligned} \alpha' &= n(r'/\cos\vartheta' - r_0)^2 / (2r'/\cos\vartheta') \\ \beta' &= n^*(r'^*/\cos\vartheta'^* - r_0{}^*)^2 / (2r'^*/\cos\vartheta'^*) \\ r'\tan\vartheta' &= r'^*\tan\vartheta'^* \end{aligned} \tag{3.12}$$

where the angle used is θ' = H–D–A instead of the normally employed angle θ = D–H–A, the approximate relation between them being $\theta \cong 180° - 1.5\,\theta'$.

Figure 3.6 shows a representative example of superposition between the two V_1 and V_2 potential curves for a selected value of the d(O···O) = R distance. V_1 represents the Morse-like potential of a O–H bond whose shape is completely determined by the parameters D_0, the O–H bond dissociation energy, and $n = k_0 r_0 / D_0$, where $r_0 = r_0$(O–H) and $k_0 = k_0$(O–H) are the tabulated equilibrium values of the O–H bond length and of its stretching force constant, respectively. V_2 is the parallel potential function for the highly stretched H–O$^+$< bond that, being out of equilibrium, will be substantially weaker than the O–H one. To take into account this fact, the authors did not attempt to modify k_0 and r_0 because no experimental information was available but, after taking $k_0{}^* = k_0$ and $r_0{}^* = r_0$, more simply downscaled the bond dissociation energy giving it the value $D_0{}^* = D_0/g = cD_0$, while upscaling the constant n to $n^* = k_0 r_0 / D_0{}^* = gn = n/c$. The effect exerted by a change of the constant $c = 1/g$ on the shape of the V_2 potential is shown in the lower part of Fig. 3.6. A value of $g = 1.45$ ($c = 0.69$) suitable for the O–H···O system was derived by the authors by a procedure of best-fitting between computed and experimental data. This value was successively shown to be transferable to other neutral H-bonds, such as N–H···N or N–H···O. So, the degree of superposition between V_1 and V_2 is only dependent on the donor–acceptor distance, R, and all the PT profiles, from strongly asymmetric SW to nearly symmetric DW or SW, are obtainable by a progressive shortening of such an R distance, though the fully symmetric DW or SW profiles will never be achieved because of the intrinsic difference of the V_1 and V_2 potentials used.

The potential terms V_3 and V_4 in eqn 3.10 represent, respectively, the van der Waals repulsion between two oxygen atoms placed at a distance R, expressed in

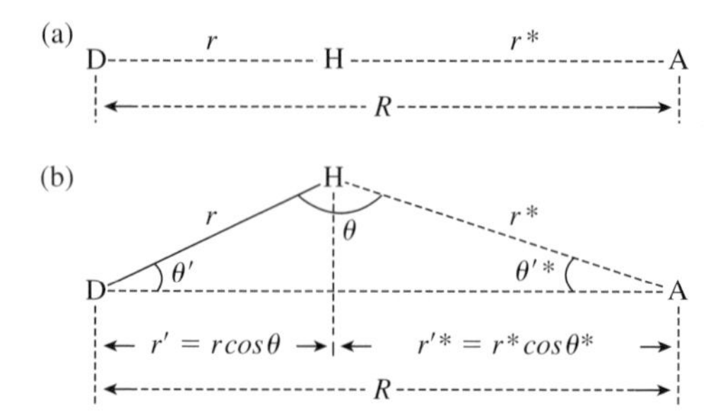

FIG. 3.5. Coordinate system used to describe the linear (a) and bent (b) H-bond.

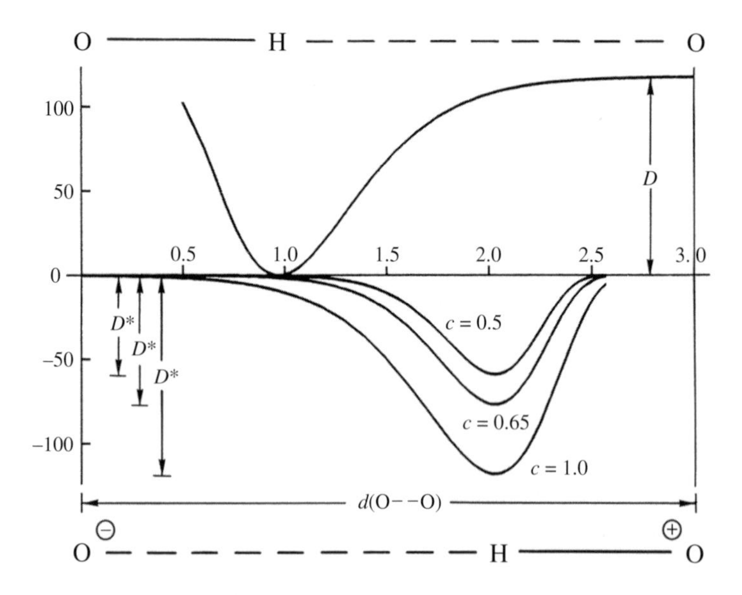

FIG. 3.6. The O–H···O Morse-like potentials V_1 (upper curve) and V_2 (lower curves) (eqn 3.10) drawn for a fixed value of $d(\mathrm{O}\cdots\mathrm{O}) = R$. Three V_2 curves are drawn for three different values of $D^* = cD = (1/g)D$. (Reproduced by permission from Gilli *et al.*, 1994a)

the exponential form $A\ \exp(-bR)$, and the electrostatic attraction between the opposite charges ($\pm B$) established on the two oxygens because of $-\mathrm{D}^-{:}\cdots\mathrm{H}{-}\mathrm{A}^+<$ bond formation and represented in the form $-B/R^m$. In principle, the exponent m allows the selection of the type of electrostatic interaction preferred, being m equal to 1, 2 or 3 for monopole–monopole, monopole–dipole or dipole–dipole interactions, respectively. In practice, the authors have systematically used $m = 1$, having shown that other values of the constant do not significantly affect the final result.

The present version of the LS-HB model can deal with both linear and bent bonds and is parametrized for O–H and N–H as donors and O, N and Cl as acceptors. Parameters provided for each atom (Schroeder and Lippincott, 1957) are D_0, r_0, k_0, ν_0, b, and g, where ν_0 is the IR stretching frequency of the unperturbed D–H bond, b the exponential repulsion constant and $g = 1/c$ the rescaling constant discussed above. All other quantities are successively calculated. After chosing an allowed D–H···A bond, input data consist of (R, θ') couples of D···A distances and H–D–A angles for which computations must be performed, while the output provides the optimized values of $r = d$(D–H) and $r^* = d$(H···A) (Å), k(D–H) (dynes cm^{-1}), ν(D–H) (cm^{-1}), B (fractional charge) and k(D···A) (dynes cm^{-1}) together with the values of the V_1–V_4 potentials and of the total H-bond energy, V, in kcal mol^{-1}. Calculations are complex because the formula giving r as a function of R is recursive and must be solved by a method of successive approximations. The same happens for the quantities A and B appearing in the expressions for V_3 and V_4. For this reason, computations can only be performed by a proper computer program that is available in FORTRAN version on request (Gilli and Gilli, 1992).

Since its formulation, the LS-HB model has been one of the most popular tools for predicting geometrical and vibrational H-bond properties and even today it has not lost its utility as a fast way of screening such properties on a wide range of different situations. Two typical applications are shown in Fig. 3.7. Figure 3.7(a) shows the $r = d$(O–H) and $r^* = d$(H···O) versus $R = d$(O···O) plots (dotted continuous lines) as evaluated by the LS-HB method for linear O–H···O bonds in comparison with the same curves (dashed lines) computed from the empirical eqns 3.7 with parameters $r^0 = 0.925$ and $R_{\mathrm{min}} = 2.40$ Å, an expression already used in Fig. 3.3 to interpolate experimental neutron diffraction data of nearly linear bonds. The agreement is quite good beyond 2.60 Å, that is for the range from weak to moderately strong O–H···O bonds that contains the greatest part of bonds of practical interest. Deviations become significant in the range 2.40–2.60 Å, which is typical of strong charge-assisted or resonance-assisted bonds, for reasons to be discussed below. Figure 3.7(b) illustrates the dependence of the H-bond energy, $E_{\mathrm{HB}} = V$, on the distance $R = d$(O···O) for selected values of the angle $\theta' =$ H–O–O from 0 to 40° in steps of 5°(nearly corresponding to $\theta =$ O–H–O decreasing from 180 to 120° in steps of 7.5°). It is seen that H-bond strengths rapidly decrease while the bond bends, a decrease that is greater for shorter O···O distances, a fact to be carefully considered when comparing intramolecular H-bonds, which are necessarily bent, with the intermolecular ones, which are often not far from linearity.

These calculations may be a basis for an analytical formulation of the H-bond laws. The decay of V for any fixed angle θ' has a clear negative exponential form that can be expressed as $V = V_0(\theta', R_0) \exp[-b(\theta')(R - R_0)]$, where $V_0(\theta', R_0)$

is the value of V for $R = R_0 = 2.40$ Å at any fixed θ' value and the exponential coefficient $b(\theta')$, optimized by least-squares, takes values of 5.8–4.7, while θ' changes from 0 to 40 degrees. Hence, H-bond energies display the same exponential dependence on R, E(H-bond) $\propto \exp(-R)$, than that previously found (p. 72) for normal covalent bonds of length d, E(bond) $\propto \exp(-d)$.

The dependence of $V_0(\theta', R_0)$ on θ' cannot be expressed in analytical form. For $R_0 = 2.40$ Å it is a flat sigmoid with linear regression equation $V_0(\theta', 2.40 \text{ Å}) = 23.9 - 0.56\,\theta'$ ($n = 9$, $r = -0.990$) that does not change significantly for other values of R. The energies of any O–H···O bond in Fig. 3.7(b) can therefore be approximated by the equation

$$E_{\mathrm{HB}} = V \cong (23.9 - 0.56\,\theta')\exp[-5.1(R - 2.40)], \qquad (3.13a)$$

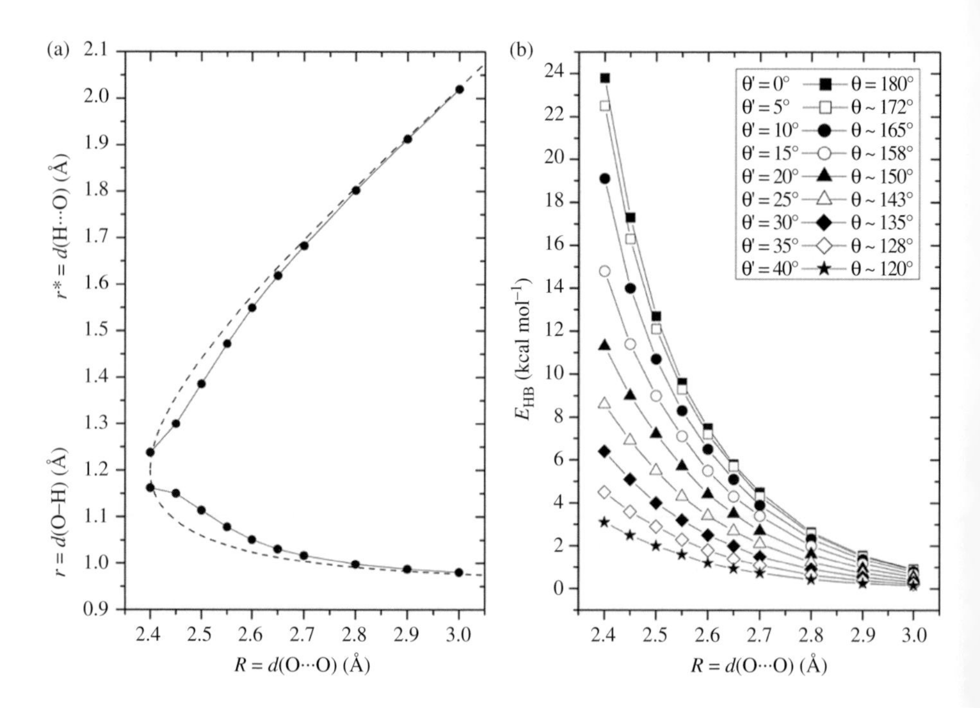

FIG. 3.7. (a) Curves $r = d$(O–H) and $r^* = d$(H···O) versus $R = d$(O···O) (continuous dotted lines) as evaluated by the LS-HB method for linear O–H···O bonds compared with the same curves (dashed lines) as interpolated from the corresponding neutron diffraction data; (b) Dependence of the H-bond energy, V, on the $R = d$(O···O) distance in the range $2.40 \leq R \leq 3.00$ Å for selected values of the angle θ' = H–O–O from 0 to 40° in steps of 5° (nearly corresponding to θ = O–H–O angles decreasing from 180 to 120° in steps of 7.5°).

with $R = d(\mathrm{O}\cdots\mathrm{O})$ in Å, $\theta' = $ H–O–O in degrees and E_{HB} in kcal mol^{-1}, which becomes

$$E_{\mathrm{HB}} = V \cong (-43.8 + 0.38\,\theta)\exp[-5.1(R-2.40)], \tag{3.13b}$$

when using the angle θ = O–H–O in degrees. These equations can be used to approximate E_{HB} for any O–H···O bond of known geometry.

The applicability of the LS-HB model depends on its correct parametrization and, in turn, on the accuracy of the experimental data available. These are particularly good for the O–H···O bonds, which are also the most generally tested, while parameters for the other types of H-bond certainly need some improvement. The lack of good parameters is not, however, the main drawback of the method. As firstly remarked by Reid (1959), the LS-HB treatment is substantially correct only for neutral D–H···:A bonds because only for them is the H–A$^+$< bond certainly *weaker* than the D–H one, so justifying the assumption of the same g factor of 1.45 for bonds of all kinds. This procedure cannot cope, however, with fully symmetric and proton-centered situations associated with very strong charge-assisted H-bonds, such as $[\mathrm{R{-}COO\cdots H\cdots OOC{-}R}]^-$ or $[\mathrm{R_2C{=}O\cdots H\cdots O{=}CR_2}]^+$, for which only a value of $g = 1/c = 1$ would be formally appropriate and could lead to a perfectly symmetric H-bond.

It might be concluded that the original LS-HB model can be used with confidence for all neutral H-bonds, including the resonance-assisted ones whose energies, on the other hand, rarely exceed 20 kcal mol^{-1}. Conversely, it has to be used with some care when dealing with very strong and symmetric bonds, particularly of the charge-assisted type. In these cases Reid (1959) suggests to use values of $g = 1/c$ more nearly similar to one, though this procedure cannot be considered to be strictly correct without a substantial restyling of the entire force field.

3.2 A new class of H-bonds: The resonance-assisted H-bond (RAHB)

3.2.1 *Cooperative H-bonds: An introduction*

The synergism for which a particular combination of chemical bonds is more energetic than the sum of the individual bond energies is known as *non-additivity* or *cooperativity*. Though the most common examples are the 'resonance' effects occurring in π-conjugated chains or rings of alternate single and double bonds, cooperativity can also occur in other circumstances and has been supposed to be 'particularly important in hydrogen bonding because of the diffuse nature and of the high polarizability of the hydrogen and lone-pair electron densities', as remarked by Jeffrey (1997) who also named σ-bond cooperativity and π-bond cooperativity the two forms by which cooperativity can strengthen the H-bond itself (Jeffrey and Saenger, 1991).

The main feature of any cooperative system is its periodicity, which finds expression in the formation of rings or infinite chains. For instance, *σ-bond* *cooperativity* occurs in the homodromic chains **3.2.I** that, for R = alkyl, aryl or hydrogen, describe the polymeric arrangements of alcohols, phenols and waters so frequently observed in crystals. This type of cooperativity is known to induce a moderate increase of H-bond strength (Table 2.4) that can be qualitatively explained in two ways: (*i*) as due to H-bond-induced polarization of the O–H σ-bond, which has the correct sign to further strengthen the H-bond itself (**3.2.Ia**); or (*ii*) as due to the partial VB mixing between the two pseudotautomeric resonance forms **3.2.Ia** $\leftrightarrow$ **3.2.Ib**.

A second type of H-bond cooperativity is π-*bond cooperativity* that derives from the positive synergism between H-bond strengthening and the polarization of a π-conjugated system occurring in periodic chains or rings. The shortest resonant unit is represented by the Y=C–X–H fragment that occurs in carboxylic acids, amides, or amidines and can easily form homomolecular chains (**3.2.II**) as well as homomolecular (**3.2.III**; **3.2.IVa,b**) and heteromolecular (**3.2.IVc**) dimers. Cooperativity can also be associated with longer π-conjugated units, for instance β-diketone enols (**3.2.V–VII**). Since π-bonds are considerably more polarizable than σ ones, it is expected that the H-bond strengthening produced by π-cooperativity is more important, particularly when longer resonant fragments are involved. Qualitative interpretations of π-bond cooperative effects can then be based, in analogy with the σ one, either (*i*) on the setting up of partial charges having the correct sign for tightening the H-bond formed (**3.2.IIa**; **3.2.IIIa**; **3.2.VIIIa**); or (*ii*) on the VB mixing of two pseudotautomeric resonance forms, as illustrated in Schemes **3.2.IIa** $\leftrightarrow$ **3.2.IIb**, **3.2.IIIa** $\leftrightarrow$ **3.2.IIIb** and **3.2.VIIIa** $\leftrightarrow$ **3.2.VIIIb**.

The concept of resonance between tautomers was used by Hunter (1944, 1945, 1947, 1954) to formulate his H-bond theory of the *mesohydric tautomerism*, a theory with several merits that, nevertheless, 'has not won general acceptance' (Pimentel and McClellan, 1960) because, at the time, it was inconceivable that tautomeric equilibrium and resonance mixing of the same forms could not be mutually exclusive (see below). In the period 1934–1957 the concept of *resonance* was specifically proposed prior and subsequent to Hunter's work by a number of authors including Sidgwick, Sherman, Coates, Wirtz, Vand, and Coulson. In his review '50 Years of hydrogen bond theory', Huggins (1971) dedicates a large section to 'coupled hydrogen bonds' including both σ- and π-cooperativity and interpreting them in terms of 'electron resonance between the two oppositely polarised structures'. In particular, he gave π-resonance the particularly effective name of *synchronised oscillation resonance* and suggested that it might play an important role in the H-bonded coupling of adenine and thymine.

Important compounds in the history of π-cooperativity have been β-diketones (**3.2.Va**), most of the work done on them prior to 1984 having been collected in an excellent review by Emsley (1984). These molecules are in tautomeric equilibrium with their β-diketone enol (or β-enolone) forms (**3.2.Vb**) that, in turn, undergo internal ketoenol–enolketone tautomerism in both intramolecular (**3.2.VI**) and intermolecular (**3.2.VII**) arrangements. A variety of physical studies (IR, NMR, UV spectroscopy; magnetic properties; X-ray diffraction) carried

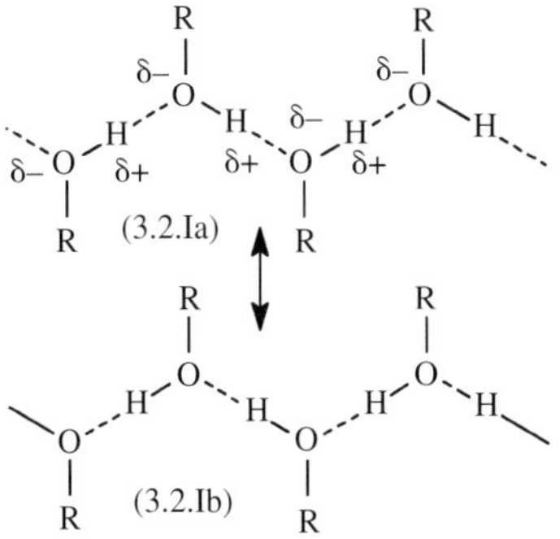

(3.2.Ia) (3.2.IIa)

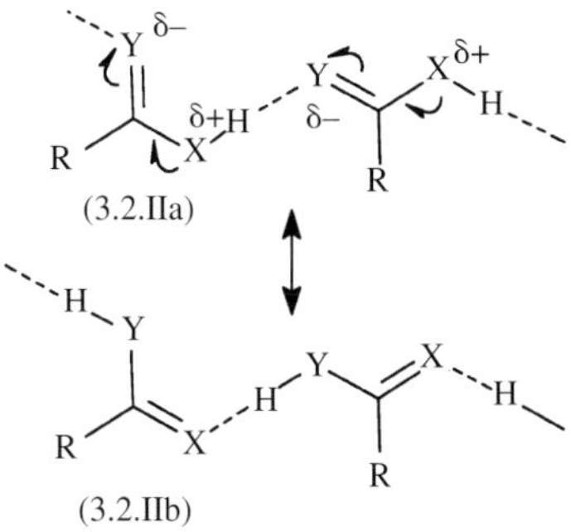

(3.2.Ib) (3.2.IIb)

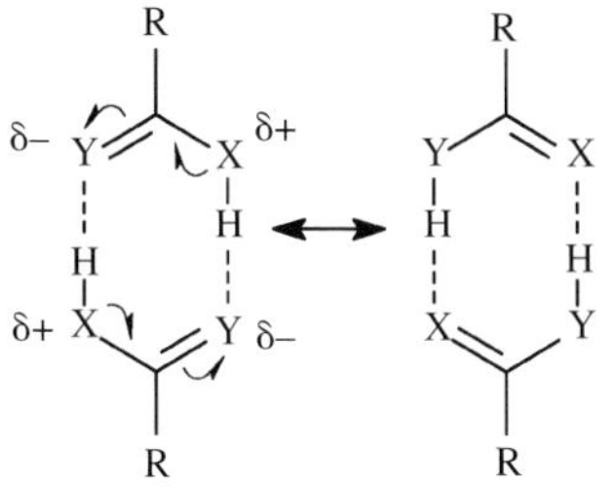

(3.2.IIIa) (3.2.IIIb) 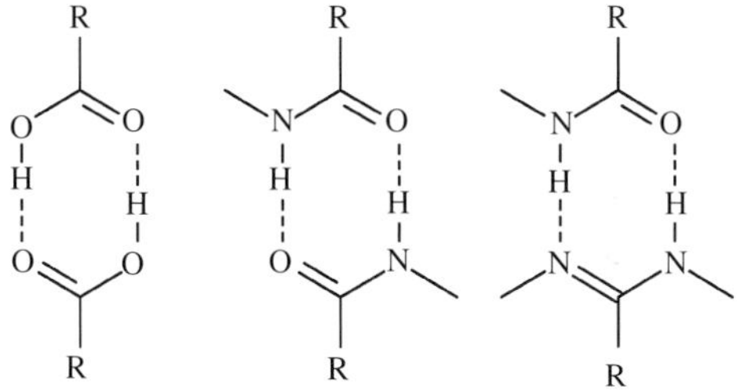 (3.2.IVa) (3.2.IVb) (3.2.IVc)

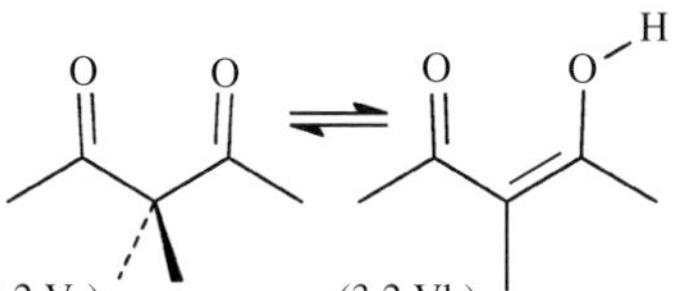

(3.2.Va) (3.2.Vb) 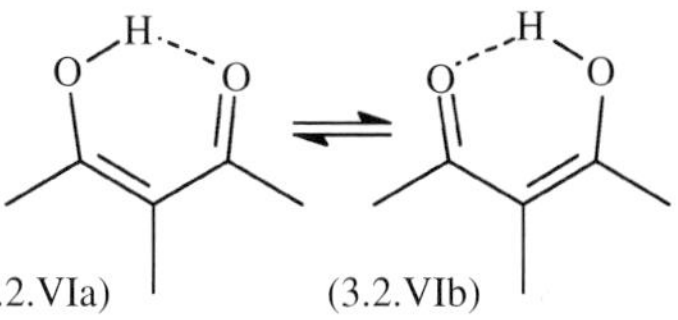 (3.2.VIa) (3.2.VIb)

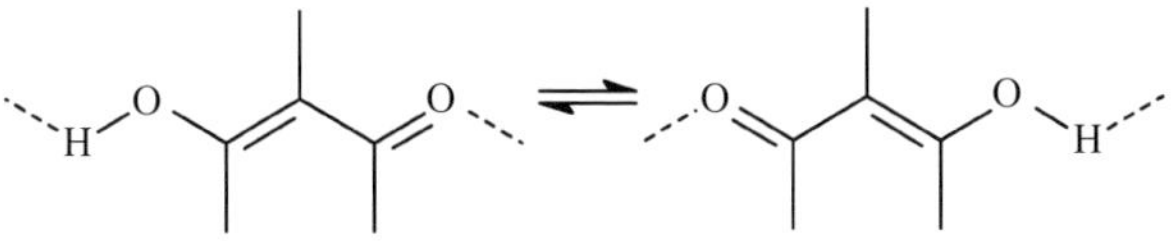

(3.2.VIIa) (3.2.VIIb)

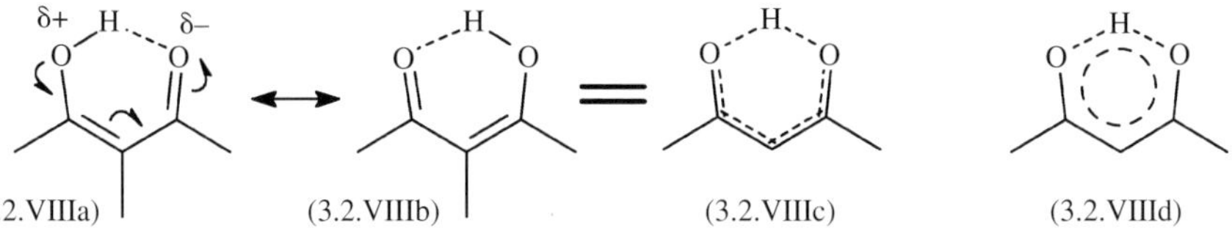

(3.2.VIIIa) (3.2.VIIIb) (3.2.VIIIc) (3.2.VIIId)

out on metal complexes of acetylacetone established the existence of certain critical symmetry elements in the chelate ring. In the crystal structure of tris-(acetylacetonate)-iron(III) (Roof, 1956) couples of C–Fe, C–C and C–O distances were found to be of equal length to within experimental error (C–C of 1.39 Å and C–O of 1.28 Å, both intermediate between single and double bonds). Strictly similar conclusions were drawn from the crystal structure of bis(*m*-bromobenzoyl)methane (Williams *et al.*, 1962) where the very short O···H···O bond [O···O distance of 2.46(1) Å] and the entire enolone fragment were found to be perfectly symmetric with respect to a crystallographic mirror perpendicular to the molecule without vibrational parameters focusing any sign of static disorder. A first interpretation of this symmetric behavior, originally suggested by Calvin and Wilson (1945) in terms of a supposed 'benzenoid resonance' (**3.2.VIIId**), was subsequently superseded by Holm and Cotton (1958) in terms of resonance between the **3.2.VIIIa** ↔ **3.2.VIIIb** canonical forms.

This last model is still widely prevailing today and is that extensively developed in the following sections trying to find a solution for the many inconsistencies that have plagued it all through its rather long history. To do that, the following questions must be answered: (*i*) which are the relationships between tautomerism and resonance; (*ii*) is it correct to assume resonant mixing between two canonical forms having the proton in two different positions; and (*iii*) by which mechanism the resonance within the π-conjugated fragment does so dramatically affect the contact O···O distance? It will be shown later (Section 3.2.2.5) that most of these problems may find a solution in the framework of more advanced VB treatments based on state-correlations (or avoided-crossing) diagrams (Gilli *et al.*, 2004).

3.2.2 *Evidence for RAHB from CSC studies of β-diketone enols*

3.2.2.1 *A survey of CSC results*

The first CSC study of a π-cooperative system was carried out on the crystal structures of β-diketone enols (Gilli *et al.*, 1989). These compounds were chosen for their constant interest to chemistry, the keto–enol equilibrium in β-diketones (**3.2.V**), the structures of the keto and enol forms, and the nature of the strong intra- or intermolecular O–H···O bond formed by the enol tautomer (**3.2.VI,VII**) having been already studied by a variety of methods, such as NMR, Raman and IR spectroscopies, X-ray and neutron diffraction, and theoretical calculations, as exhaustively reviewed by Emsley (1984). The study was performed on the best twenty structures of H-bonded 'simple β-enolones', that is enolones not carrying other hetheroatoms (except the O–R group of ketoesters) and not fused with aromatic rings, and a preliminary analysis ascertained that H-bond formation was considerably increasing the π-delocalization of the enolone fragment by greatly increasing the contribution of the polar form $^{-}$O–C=C–C=O^{+}–H to the ground state of the molecule. The nature and the entity of this effect is shown in Table 3.2 where *standard* distances are those tabulated for pure sp^2 single and double bonds (Allen *et al.*, 1987) while *unperturbed* ones are determined as an average of nine β-diketone enol methyl ethers that cannot be 'perturbed' by H-bond formation. While the unperturbed geometry can be expressed as a 86:14 mixture of the two resonance forms, inter- and intramolecular H-bond formation is seen to increase the contribution of the polar form up to 29 or 48%, respectively.

TABLE 3.2. Selected bond distances of the ···O=C–C=C–O–H··· fragment and corresponding Pauling bond numbers (in italics); *%polar* = per cent contribution of the polar form to the ground state of the fragment. Distances in Å and esds in parentheses. Data from Gilli *et al.*, 1989

	d(C–O) $= d_1$	d(C=C) $= d_2$	d(C–C) $= d_3$	d(C=O) $= d_4$	*% polar*
Standard	1.37	1.33	1.48	1.20	0
	1.00	*2.00*	*1.00*	*2.00*	
Unperturbed	1.353(2)	1.344(3)	1.454(5)	1.225(3)	14
	1.11	*1.88*	*1.17*	*1.82*	
Limiting perturbation by:					
intermolecular H-bond	1.316(2)	1.372(2)	1.431(2)	1.238(2)	29
	1.28	*1.67*	*1.29*	*1.73*	
intramolecular H-bond	1.281(4)	1.398(4)	1.410(4)	1.279(4)	48
	1.46	*1.49*	*1.42*	*1.48*	

A more evident connection between H-bond and π-delocalization is achieved by using symmetry coordinates, the most obvious being $q_1 = d_1 - d_4$ and $q_2 = d_3 - d_2$, whose scatter plot is reported in Fig. 3.8(a). The plot has central symmetry since data are plotted twice according to the equivalence of the ketoenolic (KE) and enolketonic (EK) forms and its center corresponds to the totally π-delocalised (TπD) form. The full squares marked KE and EK at the extremes represent the standard geometries of the fragment, while the two stars have similar meaning with reference to the geometries unperturbed by the H-bond (Table 3.2). The dashed line connecting the two squares has the equation $q_2 = 0.882\ q_1$ and it is seen to fit reasonably well the experimental points. As q_1 and q_2 are linearly dependent, a single coordinate $Q = q_1 + q_2 = d_1 - d_4 + d_3 - d_2$ can be adopted that has the double meaning of *coordinate of antisymmetric vibration* or *coordinate of π-bond delocalization* of the entire enolone fragment and assumes the values $Q = 0.00$ Å for the fully π-delocalized TπD structure and $Q = 0.320$ or -0.320 Å for the completely π-localized EK or KE forms. Since delocalization mixes KE and EK geometries, this mixing can be described by a *coupling parameter*, λ, the state of the fragment being expressed as λ(KE) + $(1 - \lambda)$(EK) with $\lambda = 0.0$, 0.5, and 1.0 for the EK, TπD, and KE forms, respectively. It is easy to show that λ is related to Q by the equation $\lambda = \lambda_Q = (1 - Q/Q_o)/2$ with $Q_o = d_{o,1} - d_{o,4} + d_{o,3} - d_{o,2} = 0.320$ Å. Finally, the degree of π-delocalization of the resonant chains can be expressed as *per cent* delocalization Del% = $100\ (1 - |2\ \lambda - 1|)$, which assumes values of zero for the EK and KE and 100 for the TπD forms. A summary of delocalization indices is given in the legends of Schemes **3.2.IX, X**. A more general method for the calculation of λ makes use of the Pauling bond numbers, n_i, instead of the bond distances, d_i, according to the relation $\lambda = \lambda_n = 1/N\ \Sigma_{i=1}^{N} \lambda_i$ where

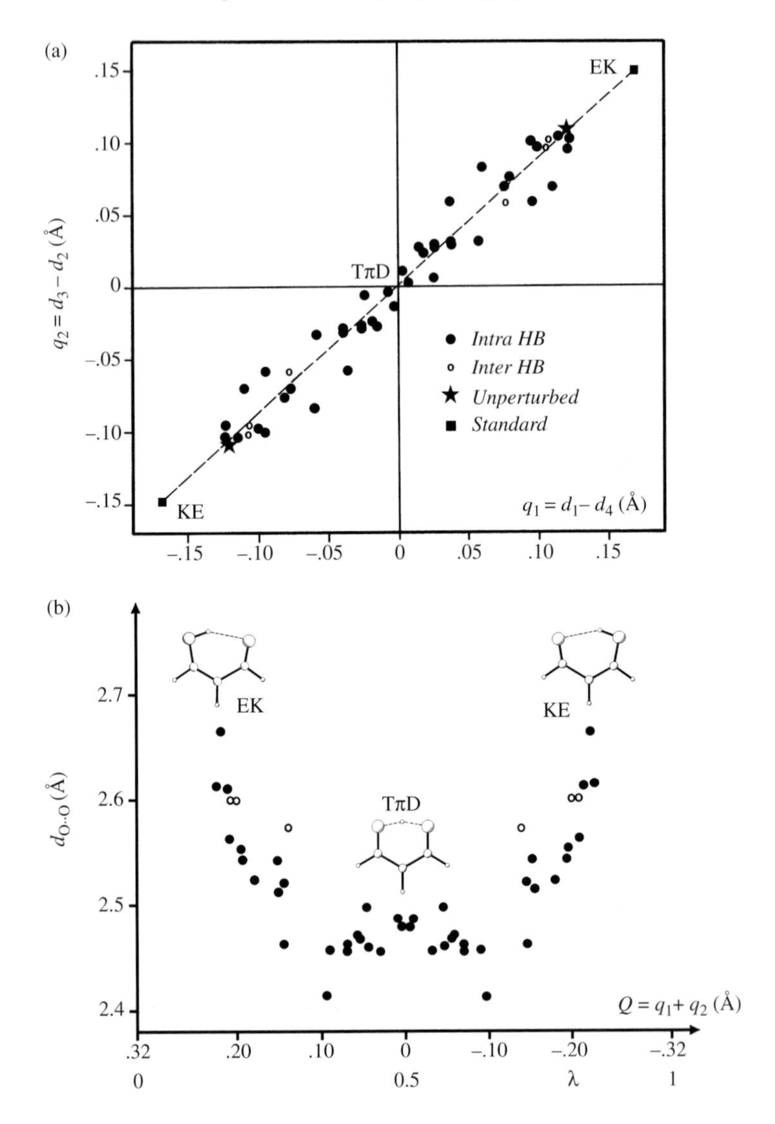

FIG. 3.8. (a) Scatter plot of $q_2 = d_3 - d_2$ versus $q_1 = d_1 - d_4$ (Å); full squares and stars refer to *standard* and *unperturbed* geometries, full and open circles to geometries perturbed by intra- or intermolecular H-bonds, respectively. (b) Scatter plot of $d(\text{O}\cdots\text{O})$ (Å) versus $Q = q_1 + q_2$ (Å) and $\lambda = (1 - Q/Q_o)/2$; full and open circles indicate intra- or intermolecular H-bonds, respectively. In both figures data points are doubled in view of the fragment symmetry. (Reproduced by permission from Gilli *et al.*, 1989)

$\lambda_i = \Delta n_i = n_i - 1$ (i odd) $= 2 - n_i$ (i even). This method gives the same results as $\lambda = \lambda_Q$ for homonuclear H-bonds and has the advantage that it can be also used for heteronuclear ones. Other delocalization parameters have been proposed (Grabowski, 2001).

The dependence of the O···O contact distance on the delocalization indices Q and λ is shown in the scatter plot of Fig. 3.8(b). Because of the duplication of points, the plot is symmetric with respect to the vertical line at $Q = 0$ and $\lambda = 0.5$, around which all very short H-bonds ($2.42 \leq d(\text{O}\cdots\text{O}) \leq 2.50$) are seen to gather. This shortening is remarkable when considering that the O···O distance is typically 2.76 Å in ice and, on average, 2.80 Å in $R_3C{-}OH\cdots O{=}CR_2$ contacts, clearly suggesting the presence of a *positive synergism between H-bond strengthening and increased delocalization of the π-conjugated system* that, accordingly, was called *resonance-assisted H-bond* (*RAHB*) (Gilli *et al.*, 1989).

The main drawback of this 1989 paper was the disparity between the small number of structures analyzed and the relevance of the conclusions drawn. For this reason, the same CSC analysis was repeated seven years later (Gilli *et al.*, 1996a,b) taking advantage of the increased number of enolone structures prompted by RAHB studies. A new CSD search for good-quality structures of β-diketones, β-ketoesters and β-ketoamides where the β-enolone fragment was not fused with

EK (3.2.IXa) — TπD (3.2.IXb) — KE (3.2.IXc)

(3.2.Xa) — (3.2.Xb) — (3.2.Xc)

$Q = q_1 + q_2 = d_1 - d_4 + d_3 - d_2$: 0.320 — 0.0 — −0.320 (Å)

$\lambda_Q = (1 - Q/Q_0)/2$, $Q_0 = 0.320$ Å: 0.0 — 0.5 — 1.0

$\lambda_n = 1/N \sum_i^N \lambda_i$, $\lambda_i = n_i - 1$ (i odd) or $\lambda_i = 2 - n_i$ (i even): 0.0 — 0.5 — 1.0

$\text{Del}\% = 100(1 - |2\lambda_n - 1|)$: 0 — 100 — 0

aromatic rings led to the identification of 54 structures and Fig. 3.9 shows the corresponding four-dimensional correlation plot between H-bond strength [d(O···O), d(O–H), and d(H···O)] and π-delocalization parameters (q_1 and q_2). The d(O···O) versus d(O–H) and d(H···O) plot on the left illustrates the well-known fact that shortening of the O···O distance is parallelled by a shortening of d(H···O) and a lengthening of d(O–H), these two distances becoming nearly equal for d(O···O) ≅ 2.40 Å. The three-dimensional [d(O···O), q_1, q_2] scatter plot is displayed on the right-side box. As usual, points representing each structure are doubled because of the symmetry of the fragment. The stars indicate the EK and KE forms unperturbed by the H-bond, corresponding to a barely delocalized π-conjugated enolone with only a 14% of ionic form (Table 3.2). The shortening of d(O···O) is associated with the decrease of both q_1 and q_2 until the central point at (0, 0) marked TπD is reached. All points essentially lie on the diagonal plane of the plot, showing that q_1 and q_2 are linearly dependent and that the d(O···O) versus (q_1, q_2) correlation can be replaced by the d(O···O) versus $Q = q_1 + q_2$ one. The latter is displayed in greater detail in Fig. 3.10 where, for comparison, the upper right rectangle shows the range of O···O distances for ordinary *non-resonant* H-bonds according to Kroon *et al.*, 1975.

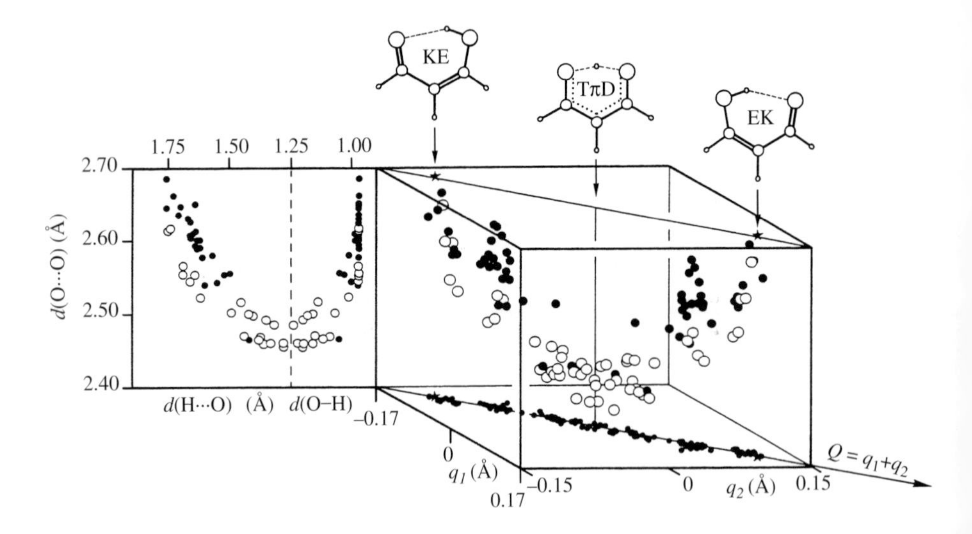

FIG. 3.9. Four-dimensional scatter plot and its projections for 54 enolone fragments forming intramolecular (open circles) or intermolecular (full circles) H-bonds. The variables q_1, q_2, d(O···O), d(O–H), and d(H···O) are defined in the text. The stars indicate the geometries of the enolone fragment not perturbed by the H-bond; KE = ketoenol, EK = enolketone, TπD = totally π-delocalized forms. (Reproduced by permission from Gilli *et al.*, 1996b)

These new findings support the definition of RAHB given above as a *synergism of H-bond strengthening and increased delocalization of the interleaving π-conjugated system.* Since H-bond strength and π-delocalization are actually measured by experimental parameters, RAHB can also be defined as an *intercorrelation among physical descriptors occurring when the H-bond donor and acceptor atoms are connected by a short resonant fragment.* This is to stress that RAHB is much more an *experimental finding* than a new H-bond theory.

When seen in the framework of a wider chemical-bond theory, however, RAHB can also be interpreted in terms of a specific binding model and, so far, three such interpretations have been proposed that are discussed in the following as: (*i*) the ionic model (Gilli *et al.*, 1989); (*ii*) the resonant model (Gilli *et al.*, 1994a); and (*iii*) the electron effective mass model (Munn and Eckhardt, 2001).

3.2.2.2 *RAHB interpretation: The ionic model*

The structural correlations shown in Figs. 3.8–10 are the basis for the qualitative ionic model sketched in Fig. 3.11. Let us firstly imagine breaking the resonance within the enolone fragment by substituting a single C–C in place of the double C=C bond. The H-bond formed will be the simple balance between the energies of the H-bond itself, E_{HB}, and that of the interatomic van der Waals interactions, E_{vdW}, including both attractive and repulsive terms. Re-establishing the double C=C bond, the resonance will now produce a certain delocalization of the π-conjugated system and the establishment of opposite charges on the terminal oxygens, a process that will stop when the sum $E_{RES} + E_{BP}$ reaches a minimum, where E_{RES} is the resonance energy and E_{BP} is the bond polarization energy needed to dissociate the partial charges. These charges have the correct sign for strengthening the hydrogen bond with consequent shortening of $d(\text{O}\cdots\text{O})$ and lengthening of $d(\text{O–H})$ and, since the movement of the proton to the right is equivalent to moving a negative charge to the left, the global effect will be the annihilation of the partial charges initially generated by resonance, which will allow a further delocalization of the π-system and a further strengthening of the H-bond itself. Iteration of such an imaginary process will inevitably lead to the complete delocalization of the π-conjugated system associated with a very short O···O distance of some 2.40 Å, unless stopped at an intermediate level by the increasing repulsion forces. In other words, the delocalization will proceed until the minimum of the function $E_{HB} + E_{RES} + E_{BP} + E_{vdW}$ is attained.

This simple model suggests that RAHB can lead to very strong and centered O···H···O bonds by a *mechanism of generation and annihilation of charges* or, in other words, that it can be conceived as a *negative-feedback mechanism* that maintains zero partial charges on the two opposite oxygens by quenching the increase of charge due to resonance through a parallel decrease caused by the proton shift.

In conclusion, it can be assumed that the equilibrium geometry of the H-bonded enolone fragment can be achieved by minimizing the function

$$E = E_{HB} + E_{RES} + E_{BP} + E_{vdW}, \tag{3.14}$$

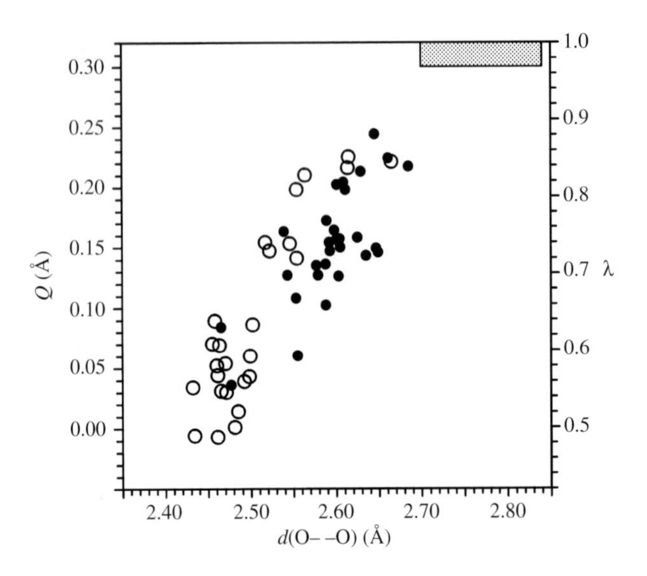

FIG. 3.10. Scatter plot of π-delocalization parameters (Q and λ) versus $d(\mathrm{O}\cdots\mathrm{O})$ for the 54 enolone fragments of Fig. 3.9 forming intramolecular (open circles) or intermolecular (full circles) H-bonds. The upper right rectangle indicates the range of ordinary *non-resonant* O–H···O bonds according to Kroon *et al.*, 1975. (Reproduced by permission from Gilli *et al.*, 1996a)

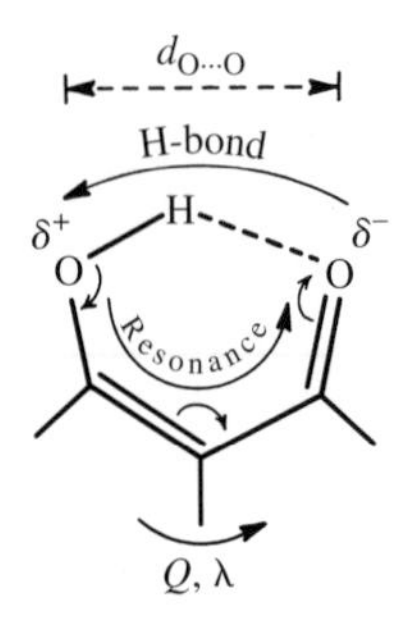

FIG. 3.11. A graphical representation of the RAHB model. (Reproduced by permission from Gilli *et al.*, 1989)

representing an approximate partitioning of the energy that has the great advantage that semiempirical treatments for the quantitative evaluation of the four terms on the right are available from the literature.

A complete evaluation of eqn 3.14 as a function of Q or λ (taken as measures of RAHB delocalization) and $n(\mathrm{O{-}H})$ or $d(\mathrm{O{-}H})$ (taken as indicators of

H-bond strength) has been reported (Gilli *et al.*, 1989). E_{HB} is the total H-bond energy of the O–H···O bond, including both attraction and repulsion terms, that can be obtained, together with d(O–H) = r values, as a function of d(O···O) = R by using the LS-HB method (Section 3.1.3.3). The values of E_{RES} were calculated by the expression $E_{RES} = f(Q)$ proposed by Krygowski *et al.* (1983), which was known to give resonance energies in good agreement with those obtained by more sophisticated methods. E_{BP}, the energy required to create the opposite fractional charges $\pm q$ on the two terminal oxygens, was evaluated by the expression $E_{BP}(q) = \Sigma_i^{\text{atoms}}(\alpha_i q_i + \beta_i q_i)$ where α_i and β_i are the coefficients of the atomic-ionization-energy versus electron-affinity curves tabulated for the main elements (Hinze and Jaffé, 1962, 1963; Hinze *et al.*, 1963). Charges q were calculated as the sum of two terms, that is $q = q_{RES}(Q) + q_{PT}(r)$, where $q_{RES}(Q)$ is the partial charge due to resonance, directly given by the weight of the polar resonance form used in the calculation of E_{RES}, and $q_{PT}(r)$ is the partial charge generated on the terminal oxygens by the transfer of the proton. For the evaluation of E_{vdW}, atom–atom potentials were taken from Giglio (1969) and the interactions between O–H and O= were neglected as they already were included in the E_{HB} term. Bending energies were omitted because they are barely relevant.

The final semiempirical potential-energy surface (PES) calculated for acetylacetone as a function of the π-delocalization parameters λ and Q and of the Pauling n(O–H) bond number is reported in Fig. 3.12. The map is centrosymmetric according to the fragment symmetry. By comparison with ice values (r = 1.01, R = 2.75 Å, $E_{HB} \cong$ 16.5 kJ mol^{-1}), the corresponding values for acetylacetone are shifted because of resonance (Q = 0.115 Å; λ = 0.68) to r = 1.08, R = 2.50 Å, and $E_{HB} \cong$ 53.4 kJ mol^{-1}. These values correspond reasonably well to those determined in a rather accurate g.e.d. molecular structure of acetylacetone (Iijima *et al.*, 1987) that has given Q = 0.122 Å, λ = 0.69, and $r = d$(O–H) and $R = d$(O···O) of 1.05 and 2.512 Å, respectively. This structure is indicated by two lozenges in the map and is seen to correspond reasonably well to the minima on the map itself. The most accurate geometries of the set of 54 enolones considered are shown as open (intramolecular) and full circles (intermolecular H-bonds), the shaded rectangles being clusters of 6 open and 17 full circles. All of them are correctly located inside the almost diagonal energy valley of the map, though, of course, shifted from the position of acetylacetone, the molecule for which the map was computed. As far as the origin of these shifts is concerned, two factors may be discerned. The first is the molecular symmetry, because very strong H-bonds are always observed in symmetrically substituted enolones. The second is steric in nature, strong bonds being systematically associated with bulky R_1 and R_3 substituents. This last effect can be simulated by computing the E_{vdW} term for $R_1 = R_3$ = t-butyl, which shifts (r, R) of the minimum to (1.11, 2.48 Å) and increases E_{HB} up to 82.5 kJ mol^{-1}.

3.2.2.3 *RAHB interpretation: The resonant model*

A. *The resonant RAHB model* The problem of resonance in the O–H···O RAHB of β-diketone enols has been recently reconsidered in considerable detail

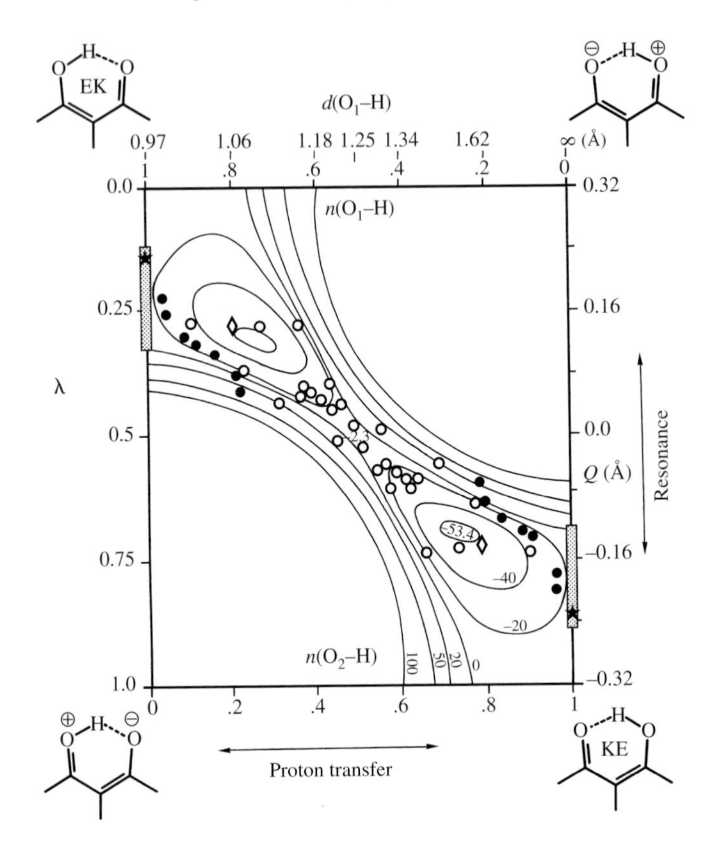

FIG. 3.12. Semiempirical potential-energy surface for acetylacetone (in kJ mol^{-1}) as a function of Q or λ (the π-delocalization parameters) and n(O–H) or d(O–H) (the H-bond strength parameters). The most accurate experimental geometries of the set of 54 enolones considered are added as open (intramolecular) and full circles (intermolecular H-bonds); the shaded rectangles are clusters of 6 open and 17 full circles; stars mark the unperturbed enolone geometry and lozenges the g.e.d. structure of acetylacetone (Iijima *et al.*, 1987). (Reproduced by permission from Gilli *et al.*, 1996a)

(Gilli *et al.*, 2004). Because of the symmetry of the enolone fragment, the two resonant covalent forms of the Coulson scheme (Chart 2.2) are doubled, giving so rise to the four VB canonical forms (EK, KE, $EK^{\pm}$, and $KE^{\pm}$) displayed in Scheme **3.2.XI** and to the corresponding four VB wavefunctions, Ψ(EK), Ψ(KE), $\Psi(EK^{\pm})$, and $\Psi(KE^{\pm})$. These can be combined according to the equations

$$\Psi(\text{O–H}\cdots\text{O}) = a\Psi(\text{EK}) + b\Psi(\text{KE}) + b^{\pm}\Psi(\text{EK}^{\pm}) \qquad (3.15a)$$

$$\Psi(\mathrm{O}\cdots\mathrm{H{-}O}) = b\Psi(\mathrm{KE}) + a\Psi(\mathrm{EK}) + a^{\pm}\Psi(\mathrm{KE}^{\pm}) \qquad (3.15b)$$

out of which only the first needs to be considered because enolone symmetry limits the canonical forms to EK, KE, and $EK^{\pm}$, an assumption strictly correct whenever treating planar molecules having symmetric substituents.

The application of this model has advantages and disadvantages, however. The positive aspect is that the coefficients of the three canonical forms (a, b, and $b^{\pm}$) can be directly derived from the crystal geometries (Gilli *et al.*, 2002, 2004) because EK and $EK^{\pm}$ are related by resonance (measured by the π-delocalization parameters Q or $\lambda = \lambda_Q$) and KE and $EK^{\pm}$ by proton transfer (PT), along the reaction coordinate $RC = d(\mathrm{O{-}H}) - d(\mathrm{H\cdots O})$. The difficulties arise because only high-quality structural data can provide proton locations accurate enough to evaluate RC and to distinguish between symmetric SW and DW potentials, a point which is mandatory because π-delocalization indices alone cannot perform such a SW/DW distinction. This is because a $Q = 0$ (or $\lambda = \lambda_Q = 0.5$) may be associated with two quite different situations: (*i*) a very strong, fully π-delocalized and proton-centered SW H-bond; or (*ii*) a weaker DW bond whose resonant chain *looks* fully delocalized because it is the unresolved superposition of the two tautomeric ···O=C–C=C–O–H··· and ···H–O–C=C–C=O··· groups, both occurring in the disordered crystal. Application of eqn 3.15a requires therefore that a truly reliable set of data has been previously selected.

B. *An accurate set of crystal data* The problem of selecting such a reliable set has been recently considered (Gilli *et al.*, 2004). A few structures of high quality are presently available, namely malondialdehyde (MDA; microwaves; Baughcum *et al.*, 1981), acetylacetone (ACAC; X-rays at 110 K; Boese *et al.*, 1998), dibenzoylmethane (DBM; neutrons at room temperature; Jones, 1976), and nitromalondiamide (NMD; neutons at 15 K; Madsen *et al.*, 1999). A further set of 34 structures of good accuracy has been retrieved from the CSD and, all together, define the 'standard set of 41 accurate β-diketone enols' to be used in

$\Psi(KE^{\pm})$ $KE^{\pm}$ (3.2.XId) — *Resonance* — (3.2.XIa) KE $\Psi(KE)$

Proton Transfer *Proton Transfer*

$\Psi(EK)$ EK (3.2.XIb) — *Resonance* — (3.2.XIc) $EK^{\pm}$ $\Psi(EK^{\pm})$

MDA (DW-LB) (3.2.XIIa)
DBMs (SW-NB) (3.2XIIIa)
BAC (DW-LB) (3.2.XIIIb)
ACAC (DW-LB) (3.2.XIIb)
R_2-ACACs (SW-NB) (3.2.XIId)
DCMs (DW-LB) (3.2.XIIIc)
Sym-ACACs (DW-LB) (3.2.XIIc)
Asym-ACACs (aSW-HB) $R_1 \neq R_3$ (3.2.XIIe)
R_2-DCMs (SW-NB) (3.2.XIIId)

the following analysis and whose chemical formulae are summarized in Schemes **3.2.XIIa–XIIId** (BAC = benzoylacetone; DCM = dicinnamoyl-methane).

Moreover, to get a deeper insight into the problem of determining the correct shape of PT profiles, three other compounds, whose crystal structures had been previously determined at room temperature (Bertolasi *et al.* 1991, 1996) suggesting they might span the complete range of PT pathways of interest, have been redetermined by variable-temperature X-ray crystallography (VTXRC) as the method of choice for the measurment of proton populations (or site-occupation factors, SOFs) in the solid state and then to assess the qualitative features of the PT pathway (see Section 6.3.2 for further details on the method). The compounds studied were two dibenzoylmethanes and one cyclohexanedione derivative that are respectively indicated by the letters **A**, **C**, and **B** in Fig. 3.13 and Table 3.3, which summarize the VTXRC results obtained.

Compound **A** is seen to form a very short RAHB [$d(\text{O}\cdots\text{O}) = 2.434$ Å at 100 K] associated with an almost completely π-delocalized enolone fragment ($\lambda = 0.43$) and a difference-Fourier map showing a single centered proton, indicative of a SW profile. Conversely, the RAHB mechanism is partially blocked in the other two molecules, in **B** because the H-bond shortening is sterically hindered and in **C** because of the N=O···C donor–acceptor interaction that spoils the symmetry of the enaminone fragment. For these reasons, they form longer H-bonds [$d(\text{O}\cdots\text{O}) = 2.573$ and 2.618 Å in **B** and 2.558 Å in **C**] with DW profiles

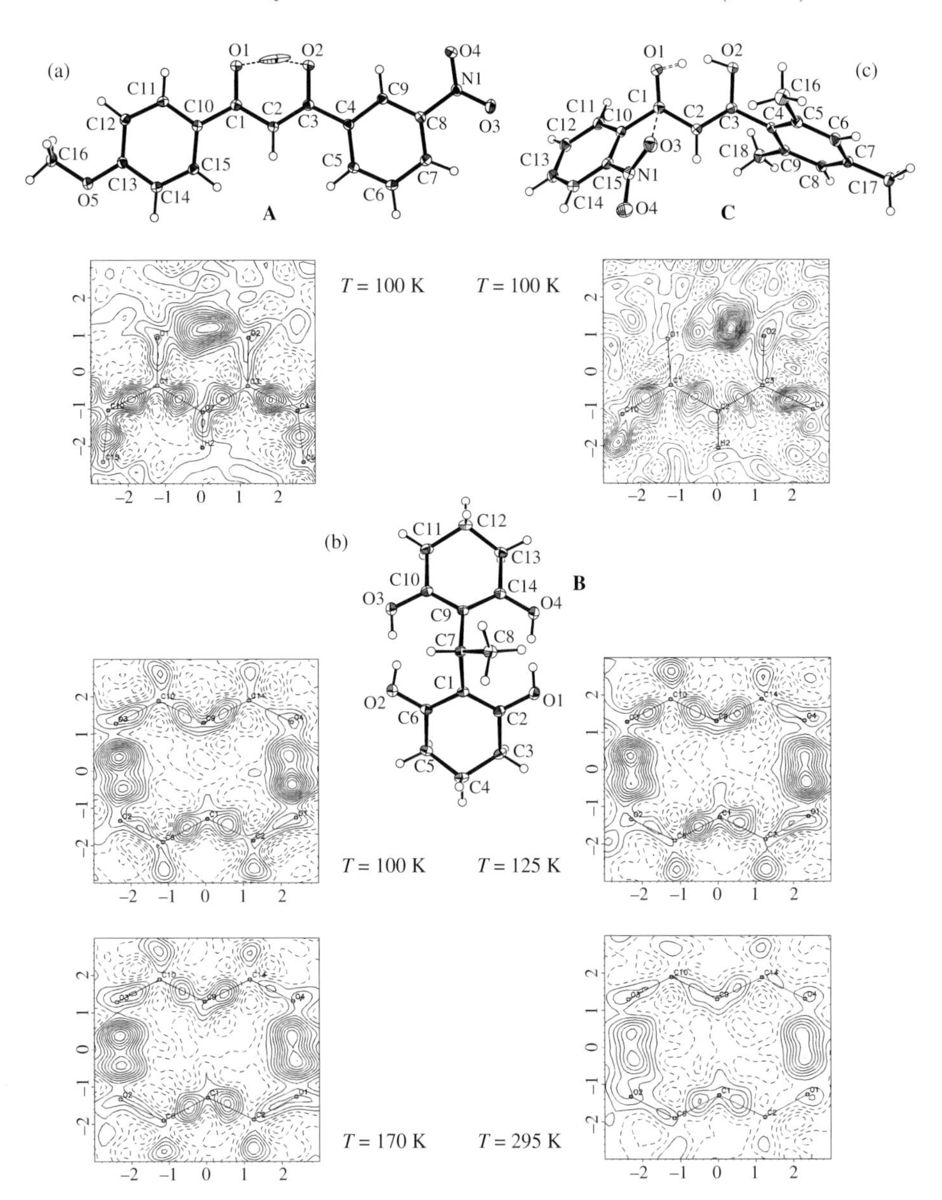

FIG. 3.13. ORTEP views (Burnett and Johnson, 1996) of compounds **A**–**C** as determined at 100 K and their corresponding difference-Fourier maps in the mean plane of the H-bonded chelate ring (a and c) or of the four oxygen atoms (b). The maps were computed after least-squares refinement carried out excluding the H-bonded proton. Positive (continuous) and negative (dashed) contours drawn at 0.04 e/Å^3. (Reproduced by permission from Gilli *et al.*, 2004)

that have, however, quite different features in the two compounds. In **B** the DW potential is nearly symmetric with an O–H···O/O···H–O *per cent* population ratio of 41:59 at 100 K becoming 48:52 at room temperature, while in **C** it is strongly asymmetric with a population ratio of 20:80 at 100 K, which turns out to be substantially independent of temperature.

C. *Determining the shape of the PT profile* The set of data collected has been sufficient to formulate a number of preliminary rules that may help to interpret the O–H···O bond in β-diketone enols in its generality:

(*i*) A DW potential inevitably causes the proton to be disordered in the crystal state. In RAHBs, this disorder is not localized on the proton position only but transmitted to the entire resonant fragment that becomes the superposition of the two tautomeric ···O=C–C=C–O–H··· and ···H–O–C=C–C=O··· moieties weighted by the corresponding proton populations. So far, neither X-ray nor neutron diffraction experiments have succeeded in resolving the heavy-atom disorder, but only that of the proton.

(*ii*) So far, any attempt to diagnose the disorder by analyzing the C and O vibrational parameters (Hirshfeld rigid-bond test: Hirshfeld, 1976; Rosenfield *et al.*, 1978) has been barely conclusive. For this reason, another method, called the λ-*test*, has been proposed (Gilli *et al.*, 2004) which is based on the comparison of the experimental π-delocalization parameter, λ_{EXP}, with the calculated one, λ_{CALC}, derived from the equation $\lambda_{CALC} = 3.47(3) - 1.25(10)\ d(\text{O}\cdots\text{O})$ obtained by linear regression of 38 accurate structures of β-diketone enols (Gilli *et al.*, 1993). The application of this method to compounds **A**–**C** is shown in Table 3.3. For **B** the large difference between experimental and calculated values ($\lambda_{EXP} = 0.44$ and $\lambda_{CALC} = 0.20$–0.25) clearly shows that the presumed large π-delocalization is an artifact due to disorder within the crystal, while in **A** these values are identical and the test confirms the presence of a true symmetric SW potential. Nothing,

TABLE 3.3. H-bond geometrical parameters (Å and degrees), proton populations, $p(\%)$, and experimental and calculated π-delocalization indices, λ_{EXP} and λ_{CALC} (e.s.d.s in parentheses) for compounds **A**–**C** of Fig. 3.13. Symbols *a* and *b* indicate the two H-bonded moieties of compound **B** and the pairs *a*, *a*′ and *b*, *b*′ the couples of tautomeric H-bonds. PT profiles: SW-NB = single-well/no-barrier; DW-LB = double-well/low-barrier; DW-MB = double-well/medium-barrier; ns and a = nearly symmetric and asymmetric (nomenclature according to **2.3.Ia–d**)

Compound	O···O	O–H	H···O	O–H–O	$p(\%)$	λ_{EXP}	λ_{CALC}	PT profile	Crystal disorder
A (100 K)	2.434(1)	1.15(3)	1.32(3)	159(2)	100	0.43	0.43	nsSW-NB	Ordered
B (100 K) *a*:	2.573(1)	0.84(3)	1.75(3)	170(3)	59(3)	0.44	0.25	nsDW-LB	Dynamically
a′:		0.85(4)	1.75(4)	165(4)	41(3)				disordered
b:	2.618(1)	0.85(3)	1.79(3)	164(3)	59(3)	0.44	0.20		
b′:		0.84(4)	1.81(4)	160(4)	41(3)				
C (100 K) *a*:	2.558(1)	0.89(2)	1.75(2)	150(2)	80(3)	0.25	0.27	aDW-MB	Statically
a′:		0.88(6)	1.72(7)	160(6)	20(3)				disordered

however, can be said for **C** because the too-low population of one of the two tautomers cannot alter heavy-atom distances in a significant way.

(*iii*) The disorder can be *static* or *dynamic* according to the height of the PT barrier that, in turn, depends on the symmetry of the PT profile, being higher or lower when the energies of the two minima are more different or alike. The nature of the disorder is detectable by VTXRC because proton populations depend on, or are independent of, temperature, according to whether the disorder is dynamic or static.

(*iv*) Crystals where proton populations, p, change smoothly with temperature are most probably dynamically disordered with a high exchange rate. In this case the enthalpy difference between the two minima can be determined by van't Hoff methods. If $K = p/(1 - p)$ is the ratio of the two O–H···O and O···H–O tautomers as derived from least-squares refinement, the standard enthalpy, $\Delta H°$, and entropy, $\Delta S°$, of the tautomeric equilibrium are obtained as the slope and intercept of the van't Hoff plot $\ln K = \Delta S°/R - \Delta H°/R\ (1/T)$. When applied to compound **B** this method gives values $\Delta H° = -88(6)$ cal mol^{-1} and $\Delta S° = -1.2(4)$ cal K^{-1} mol^{-1}.

D. *Interpreting the results* The simple rules above permit assignment of the correct PT profile to all structures of the 'standard set of 41 accurate β-diketone enols' illustrated in **3.2.XIIa–XIIId** where the profile nomenclature follows Schemes **2.3.Ia–d**: aSW-HB = asymmetric single-well/high-barrier, DW-LB = double-well/low-barrier (LBHBs = low-barrier H-bonds), and SW-NB = single-well/no-barrier (the strongest H-bonds, often simply called SWHBs = single-well H-bonds). The knowledge of the profile allows evaluation of the correct λ value, while reasonable values of $RC = d(\text{O–H}) - d(\text{H}\cdots\text{O})$ can be obtained by normalizing O–H distances to 0.94 Å whenever shorter of this limit.

Figures 3.14(a) and 3.14(b) summarize the complete data analysis in the form of the two plots RC and $d(\text{O}\cdots\text{O})$ versus λ. Symbols are open squares, open triangles and open circles for molecules forming SW, DW and aSW H-bonds, respectively. Superimposed small full points mark the structures of higher accuracy, while diagonal crosses indicate the incorrect positions assumed by DW H-bonds when misinterpreted as SW ones because their large λ_{EXP} values were erroneously attributed to almost complete π-delocalization instead of averaging between the two disordered and antidromic O=C–C=C–OH and HO–C=C–C=O tautomeric groups. Both plots clearly show that the mispositioning of points due to mistaking DW for SW H-bonds is very large and such to impair the correlation $d(\text{O}\cdots\text{O})$ versus λ typical of RAHB.

Finally, also the mixing coefficients a, b, and $b^{\pm}$ of eqn 3.15a can be obtained from the RC versus $\lambda = \lambda_{\text{Q}}$ scatter plot of Fig. 3.14(a) by a simple technique already described (Gilli *et al.*, 2002; note 21). Average results obtained for the full dataset are as follows:

$$\text{SW-NB: } 2.39 \leq d(\text{O}\cdots\text{O}) \leq 2.47\ \text{Å};\ 0.49 \geq \lambda_{\text{EXP}} \geq 0.35;$$
$$\text{average } a : b : b^{\pm} = 56[5] : 40[5] : 4[4];$$

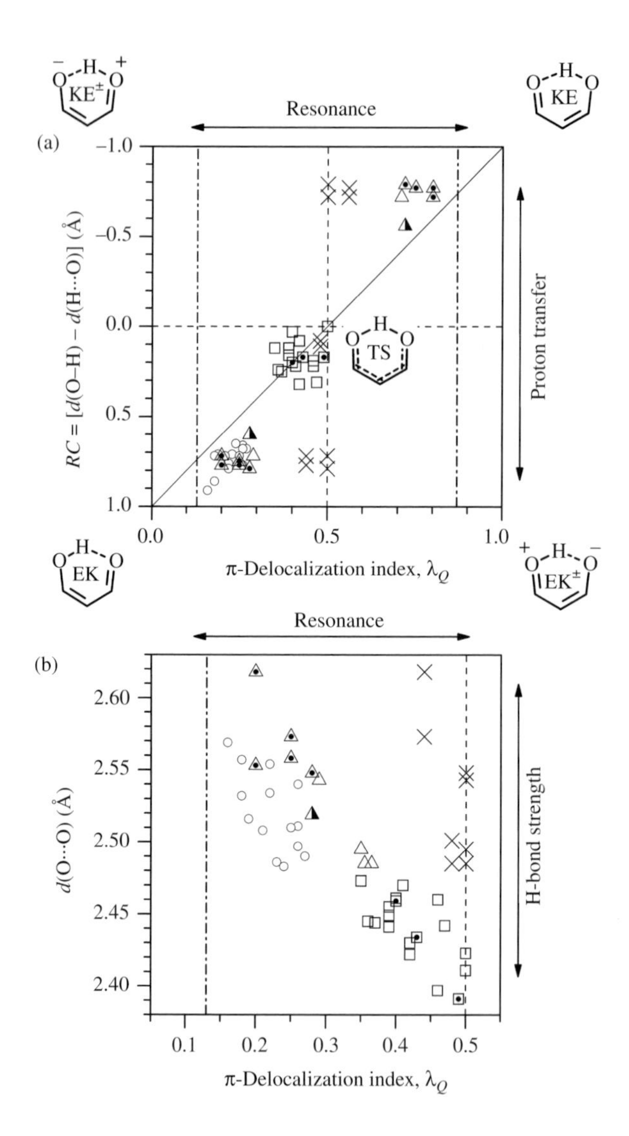

FIG. 3.14. (a) PT reaction coordinate $RC = [d(\text{O}-\text{H}) - d(\text{H}\cdots\text{O})]$ (Å) and (b) O···O contact distance (Å) plotted against the π-delocalization index $\lambda = \lambda_Q$ for a number of crystal structures of β-diketone enols forming intramolecular O–H···O RAHBs. Open squares (□), open triangles (△) and open circles (○) indicate SW-NB, DW-LB and aSW-HB H-bonds, respectively. Most accurate structures are marked by a superimposed smaller full point (•). Diagonal crosses (×) along the vertical line at $\lambda = 0.5$ mark the incorrect positions assumed by DW H-bonds when the tautomeric disorder is misinterpreted. Non-crystallographic DFT-optimized geometries (Schiøtt *et al.*, 1998) are shown by triangular half-filled symbols. The two vertical dash-dot lines at $\lambda = 0.13$ and 0.87 indicate the normal enolone π-delocalisation in its non-H-bonded state. (Reproduced by permission from Gilli *et al.*, 2004)

$$\text{DW-LB: } 2.48 \leq d(\text{O}\cdots\text{O}) \leq 2.55\ \text{Å};\ 0.36 \geq \lambda_{\text{CALC}} \geq 0.20 \text{ and}$$
$$0.64 \leq \lambda_{\text{CALC}} \leq 0.80;$$
$$\text{average } a : b : b^{\pm} = 75[4] : 13[3] : 12[4];$$
$$\text{aSW-HB: } 2.48 \leq d(\text{O}\cdots\text{O}) \leq 2.57\ \text{Å};\ 0.27 \geq \lambda_{\text{EXP}} \geq 0.16;$$
$$\text{average } a : b : b^{\pm} = 77[4] : 14[3] : 9[3]$$

Data show that the three types of PT pathways display strictly contiguous metric and bonding properties but that very short O···O distances and large $a : b$ covalent mixing are only associated with SW-NB bonds, while DW-LB and aSW-HB bonds have rather similar O···O distances and differ only in the degree of π-delocalization, which is slightly greater for DW-LB tautomeric pairs.

To complete this analysis, H-bond energies, E_{HB}, have been evaluated by DFT methods at the B3LYP/6–31+G(d,p) level by comparing the total energies of the two H-bonded (or closed) and non-H-bonded (or open) forms. Two β-diketones, dibenzoylmethane (DBM) and malondialdehyde (MDA), and one β-ketoester (HAA = 3-hydroxy-acrylic acid) were taken as typical examples of molecules forming SW-NB, DW-LB and aSW-HB H-bonds. Calculated E_{HB}s amount to 19.84, 13.56 and 13.59 kcal mol^{-1} (to be decreased by some 0.5 kcal mol^{-1} after zero-point correction), respectively, with associated O···O distances of 2.475, 2.569 and 2.592 Å. These results confirm that very strong, as well very short, H-bonds can only occur in connection with nearly symmetric SW-NB PT profiles produced by almost complete EK $\leftrightarrow$ KE resonance mixing, but also that resonance can give a significant contribution to the H-bond energy (from some 5 of water to some 13.5 kcal mol^{-1} of HAA and MDA) irrespective of the degree of mixing (see also Dannenberg and Rios, 1994).

E. *Conclusions* We hope that the present analysis may put an end to the long controversy lasting from the times of Hunter's (1944, 1945, 1947, 1954) theory of the mesohydric tautomerism. *Tautomeric and resonance forms do not need to be mutually exclusive* but can coexist, though with different meanings. Resonance occurs between two imaginary forms, EK and KE, having interchanged *pure single or double bond* distances: let's call them A $\leftrightarrow$ B. The two tautomeric forms are resonance hybrids A(B) and B(A) (i.e. mixtures of a small part of B in A and of a small part of A in B). The entity of this mixing determines the shape of the PT pathway, increasing mixing causing the approaching of the two minima in the DW potential until coalescing in the single minimum of the SW-NB potential. The factors that can produce such a coalescence can be empirically determined by comparing the chemical structures of the compounds of the full set (Schemes **3.2.XIIa–XIIId**), that is:

(*i*) All strong H-bonds with SW-NB potential (**3.2.XIId, XIIIa, d**) are associated with symmetric molecules ($R_1 = R_3$) while dissymmetric ones ($R_1 \neq R_3$; **3.2.XIIe**) are seen to form weaker bonds with an aSW-HB profile. This seems reasonable because the KE $\leftrightarrow$ EK resonance mixing reaches a maximum when the two forms are isoenergetic, which certainly occurs when they are equivalent for symmetry;

(*ii*) The transition from DW-LB to SW-NB profiles is definitely fostered by bulky R_2 substituents that contribute to close the H-bonded ring by steric compression. This becomes evident when comparing ACAC with R_2-ACACs (**3.2.XIIb, d**) and DCMs with R_2-DCMs (**3.2.XIIIc, d**);

(*iii*) The only compounds that seem able to form SW-NB bonds without the assistance of the R_2 substituent are DBM derivatives (**3.2.XIIIa**). This could be due to the slight repulsion between the phenyls and the central H-bonded ring or, more reasonably, to the increased π-bond delocalizability associated with the two aromatic substituents. This could also account for the fact that BAC (**3.2.XIIIb**) gives a DW-LB bond at variance with all other dissymmetric enolones (**3.2.XIIe**) that form only aSW-HB ones.

3.2.2.4 *Appendix 3A: RAHB as a cybernetic effector*

The factors affecting the shape of the PT profile and the strength of the O–H···O bond in β-diketone enols are summarized in Fig. 3.15 in the form of a *cybernetic effector diagram.* The H-bond strength may be enhanced (positive sign) or reduced (negative sign) by steric forces, respectively producing compression or stretching of the O···O contact distance and, above all, is increased by the degree of symmetry of the resonant fragment prompted by both homonuclearity and molecular symmetry. The most typical RAHB feature is the presence of the positive (or deviation-amplifying) feedback connecting the effect (the H-bond strength) with one of the causes (the resonant-fragment symmetry) through the π-delocalization of the resonant fragment. This connection triggers the RAHB synergistic mechanism by which the H-bond strengthening induces an enhanced π-delocalization that, in turn, causes further strengthening until repulsion forces stop the process at its equilibrium position. The graph is completed by the dashed arrow marked 'aromaticity' that should account for the contribution of aromatic substituents to the π-bond delocalizability of the H-bonded resonant fragment, as suggested above for BAC and DBM derivatives.

3.2.2.5 *Appendix 3B: RAHB as a state-correlation diagram*

The simultaneous occurrence of tautomeric O–H···O $\rightleftharpoons$ O···H–O and resonant O–H···O $\leftrightarrow$ O···H–O forms is not the only problem that has afflicted the application of the resonance theory to H-bond studies. Another difficulty raised until very recently (e.g. Srinivasan *et al.*, 2004; Note 3) is whether it is really correct to mix by resonance two forms that differ for the positioning of the O–H σ-bond. This criticism is not groundless and needs to be considered in some way.

Figure 3.16 is an attempt to do that through a *state-correlation* (or *avoided-crossing*) *diagram* (Shaik *et al.*, 1992 and references therein) illustrating the ideal PT pathway of a *symmetric* β-diketone enol forming *intrinsic* (i.e. symmetric) intramolecular H-bond. The two ground-state (GS) forms EK and KE have different spin pairing and therefore cannot mix without the intervention of the two charge-transfer excited states $EK^{\pm}$ and $KE^{\pm}$ (see also Scheme **3.2.XIa–d**

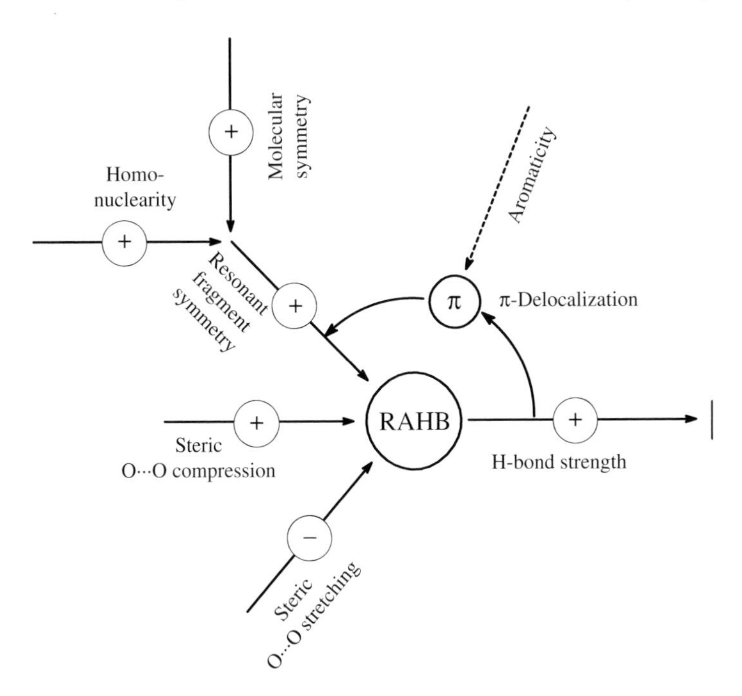

FIG. 3.15. The O–H···O RAHB logic described as a cybernetic effector with positive (or deviation-amplifying) feedback connecting backwards the H-bond strength (the effect) with the resonant-fragment symmetry (one of the input variables) through the π-delocalisation of the O=C–C=C–OH enolone fragment, π. (Reproduced by permission from Gilli *et al.*, 2004)

for comparison). The PT occurs at the crossing point where the two states, having identical spin pairing and energy, can mix, lowering the TS energy, while increasing that of the corresponding excited state, by the quantity $\Delta E_{\rm RES}$ that can be regarded as the quantum-mechanical *TS resonance energy*. This is small (dashed curve HB) for DW-HB, larger for DW-LB (dashed curve LB) and much larger for SW-NB H-bonds, for which the TS is lowered in such a way as to become the thermodynamically stable form (dashed curve NB). This treatment suggests that the word 'resonance' often used to define RAHB as due to an 'interplay between H-bond strengthening and resonance' (Gilli *et al.*, 1989; Bertolasi *et al.*, 1991) can be perhaps intended as the 'TS resonance energy' in the avoided-crossing diagram of Fig. 3.16. It makes it also clear that a SW-NB H-bond is a bond that possesses, in its GS, the structure that is normally displayed by the TS in weaker bonds, and that it can become short and essentially covalent just because it can participate, in a way, of the nature of the TS that is always the shortest, the most symmetrical, and most covalent bond achievable

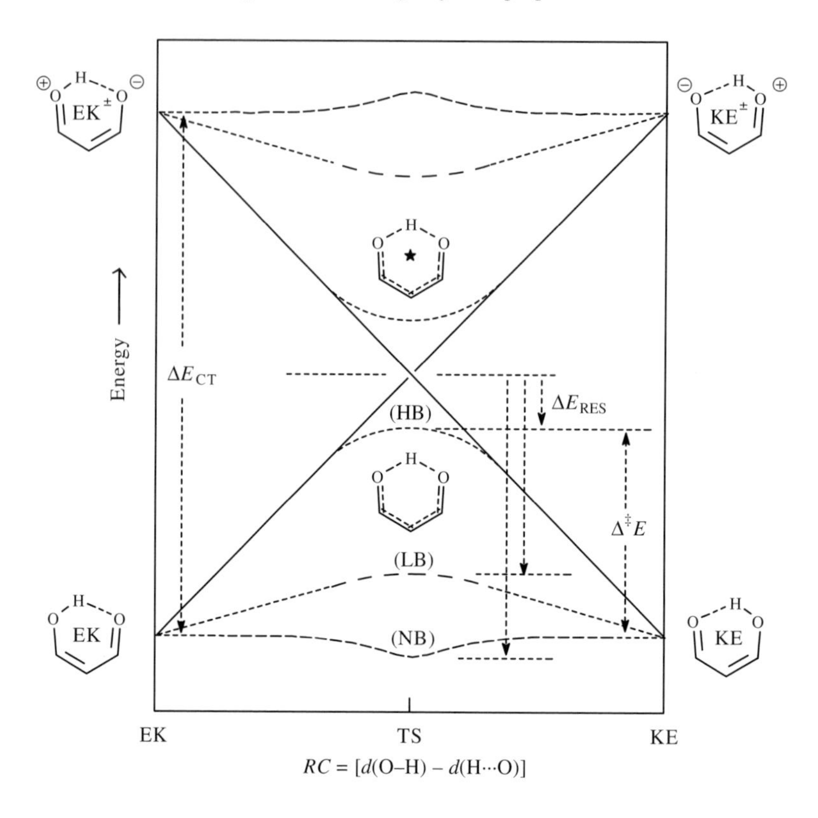

FIG. 3.16. State-correlation (or avoided-crossing) diagram (Shaik *et al.*, 1992) for the PT reaction in intramolecularly H-bonded and symmetrically substituted β-diketone enols. Because of the different spin pairing EK and KE forms cannot correlate without intervention of the charge-transfer excited states $EK^{\pm}$ and $KE^{\pm}$ that are higher in energy by ΔE_{CT}. PT occurs at the crossing point where the two states of identical spin pairing and energy can mix by resonance, lowering the TS energy by ΔE_{RES} and then decreasing the PT activation energy $\Delta^{\ddagger}E$. ΔE_{RES} is small for DW-HB, larger for DW-LB, and much larger for SW-NB H-bonds for which the TS is transformed into the thermodynamically stable configuration (curves HB, LB and NB, respectively). (Reproduced by permission from Gilli *et al.*, 2004)

in any H-bonded system (Gilli *et al.*, 2002). This point will be taken up again in Chapter 6.

An alternative explanation of how KE and EK can mix to give the fully π-delocalized enolone (TπD; **3.2.IXb**) has been given by Haddon (1980) by using a method that is the exact opposite of a state-correlation diagram. He, in fact, applied the simple Hückel-MO theory (Streitwieser, 1961) to the analysis of

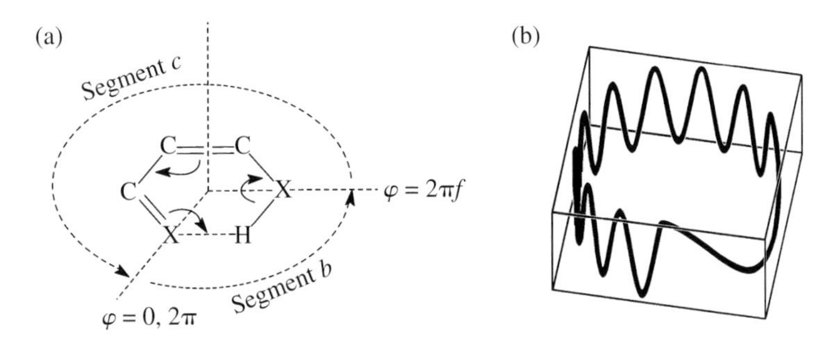

FIG. 3.17. (a) Structural pattern for resonance-assisted hydrogen bonding; (b) Real part of the wavefunction for $f = 1/3$ and $\rho = 5$. (Reproduced by permission from Munn and Eckhardt, 2001)

the factors that may perturb, or not, the fully π-delocalized geometry characteristic of proton-centered sSW-NB H-bonds by a second-order Jahn–Teller effect (Pearson, 1969) prompted by HOMO–LUMO configuration interaction. A large HOMO-LUMO gap was found to be the essential condition to avoid such a perturbation and then to form strong and symmetric H-bonds, in complete agreement with the conclusions drawn from the avoided-crossing diagram of Fig. 3.16.

3.2.2.6 *Appendix 3C: RAHB electron effective-mass model*

The two RAHB models discussed in Sections 3.2.2.2 and 3.2.2.3 essentially rely on the VB theory of the chemical bond. On the other hand, valuable insight into these conjugated systems can also be obtained from other quantum-mechanical models such as the particle in a box or in a ring. The particle in a ring model has been often used in understanding NMR shielding and deshielding effects in proximity of a benzene ring in terms of ring current (Pople, 1956) and in classifying the excited states of aromatic hydrocarbons (Platt, 1949). This same model has been applied by Munn and Eckhardt (2001) to the intramolecular RAHB in β-diketone enols (Fig. 3.17) that can also be thought of as an electronic current running inside a ring. The treatment was then extended to dimers and larger rings up to infinite chains with cyclic boundary conditions.

The basic model consists of a ring of radius R divided into two segments, segment b representing the X–H···X bond and segment c the π-conjugated system that connects the two X atoms. An electron moves in segment c with an effective mass m_c and in segment b with an effective mass m_b. The concept of *effective mass,* normally indicated as m^*, is borrowed from solid-state physics (Hall, 1974) where it is used to account for the dynamics of electrons travelling along a chain of atoms (a crystal). In analogy with the electron travelling in vacuum with an energy $\varepsilon = p^2/2m$ (p = momentum; m = mass of the electron), it is said that electrons travel freely through the crystal like particles with a mass m^*. This mass is determined by the inverse of the curvature of the band and may differ markedly from the normal inertial mass of the free particle. Here, it allows for the fact that the mobile particle is not a free electron but an electron interacting with other electrons and with nuclei. In particular, it enables us to use the electron transfer within the H-bond as a surrogate or alias for the opposite proton movement from one side to the other of the H-bond. We expect that $m_b > m_c$, because the electron transfer through the H-bond region requires contemporary movement of the much heavier proton. We also expect m_b to change markedly by replacing hydrogen by deuterium, providing us with a method for studying the isotopic effect.

If the electron position on the ring is described by the angular variable φ ($0 \leq \varphi \leq 2\pi$), the bond segment b corresponds to $0 \leq \varphi \leq 2\pi f$ and the conjugated segment c to $2\pi f \leq \varphi \leq 2\pi$ (Fig. 3.17(a)), where f is the fraction of the ring occupied by the H-bond. The whole ring comprises an even number of atoms n and the same number of bonds. Two of these bonds comprise segment b, and hence, $f = 2/n$. If we suppose that all bonds have the same length d, then the circumference is nd and the ring radius is $R = nd/2\pi$.

A particle of mass m moving in a ring of radius R with moment of inertia $I = mR^2$, classical angular momentum L, and energy $E = L^2/2I = L^2/2mR^2$ has a time-independent Schrödinger equation

$$-\frac{\hbar^2}{2mR^2}\frac{\partial^2\psi}{\partial\varphi^2} = E\psi, \tag{3.16}$$

with solution

$$\psi = \psi^\circ \exp(iL\varphi/\hbar). \tag{3.17}$$

In this treatment there are two different segments along the ring (b and c) where the electron has different effective masses (m_b and m_c), moments of inertia, and angular momenta. Accordingly, we can write for the energy of the system

$$E = L_b^2/2m_b R^2 = L_c^2/2m_c R^2, \tag{3.18}$$

where we shall set $L_b = \rho L_c$ so that $\rho^2 = m_b/m_c$. At the same time, the wavefunction will split into the system of equations

$$\psi_b = \psi_b^0 \exp(iL_b\varphi/\hbar) \tag{3.19a}$$

$$\psi_c = \psi_c^0 \exp(iL_c\varphi/\hbar). \tag{3.19b}$$

The solutions are subject to the boundary conditions that the wavefunctions must be equal at the ends where the segments join. These boundary conditions are $\psi_b(0) = \psi_c(2\pi)$ and $\psi_b(2\pi f) = \psi_c(2\pi f)$ and lead to the conclusion that the angular momentum L_c in segment c must be quantised in the form

$$L_c = \hbar k/(1-f+\rho f), \tag{3.20}$$

where k is a non-zero integer. Hence, the energy of an electron in state k is $E_k = L_c^2/2m_c R^2$ and that in its lowest level of energy ($k = 1$)

$$E_1 = \frac{\hbar^2}{2m_c R^2 (1-f+\rho f)^2}. \tag{3.21}$$

Given the quantized value of the angular momentum L_c in segment c, the corresponding values in segment b is obtained as $L_b = \rho L_c$ while the wavefunctions 3.19, expressed in terms of L_c, become

$$\begin{aligned} \Psi_c &= \Psi_c^0 \exp(iL_c \varphi/\hbar) \\ \Psi_b &= \Psi_c^0 \exp(2\pi i L_c/\hbar) \exp(iL_c \rho\varphi/\hbar). \end{aligned} \tag{3.22}$$

The real part of the total wavefunction for $k = 1$ is plotted in Fig. 3.17(b) for parameters $f = 1/3$ (corresponding to one hydrogen bond in a six-membered ring) and $\rho = 5$ (evaluated below). In the H-bonded segment b, which lies at the bottom right-hand corner of the figure, the wavefunction Ψ_b varies slowly while the variation of Ψ_c in the conjugated segment c is much more rapid owing to the additional factor ρ that multiplies φ in eqn 3.22 for Ψ_b .

The next stage of the analysis consists in evaluating the RAHB stabilization energy by comparing the energy of the electron within the resonant ring (E_1 in eqn 3.21) with the corresponding energy in an isolated X–H···X bond. In this second case the H-bond comprises just segment b, for which the label B can now be used, and its wavefunction is still given by eqn 3.19a but with boundary conditions $\psi_B(0) = \psi_B(2\pi f) = 0$, which means that the solutions are no longer travelling but standing waves obtained by combining degenerate solutions with equal and opposite L_B. The solution that satisfies the first boundary condition is a simple sine function

$$\Psi_B = \Psi_B^0 \sin(L_B \varphi/\hbar), \tag{3.23}$$

while the second boundary condition $\psi_B(2\pi f) = 0$ yields the quantization condition $L_B = \hbar K/f$, where K is a non-zero integer. The corresponding energy $E^0{}_K = L_B^2/2m_b R^2$ for the ground state with $K = 1$ becomes

$$E_1^0 = \frac{\hbar^2}{2m_b R^2 f^2} = \frac{\hbar^2}{2m_c R^2 f^2 \rho^2}, \tag{3.24}$$

where the superscript zero denotes the unconjugated H-bond.

Subtraction of E_1^0 (without conjugation) from E_1 (with conjugation) gives the energy lowering *per* electron due to RAHB formation, that is:

$$\Delta E_1 = \frac{\hbar^2}{2m_c R^2}\left[\frac{1}{\rho^2 f^2} - \frac{1}{(\rho f + 1 - f)^2}\right], \tag{3.25}$$

$$\Delta E_1 = \frac{\hbar^2}{2m_b R^2}\left[\frac{1}{f^2} - \frac{\rho^2}{(\rho f + 1 - f)^2}\right]. \tag{3.26}$$

This is zero when $f = 1$ because then there is no more conjugated segment. Increasing m (as happens on deuteration) and R are seen to reduce the energy lowering produced by the conjugated segment. Moreover, by expressing f and R in terms of the number of bonds n and their average bond length d, the energy lowering *per* electron becomes

$$\Delta E_1 = \frac{h^2}{8m_b d^2}\left[1 - \frac{4\rho^2}{(n - 2 + 2\rho)^2}\right]. \tag{3.27}$$

This is the *conclusive equation* on which to base any attempt of a quantitative appreciation of ρ from the RAHB stabilization energy ΔE_1. Substituting $m_b = \rho^2\, m_c = \rho^2\, m_e$ (m_e = inertial electron mass) and $h^2/8m_e d^2 = 480.7$ kcal mol^{-1} for an average $d = 1.344$ Å, we obtain the *simplified equation*

$$\Delta E_1 = \frac{480.7}{m_b}\left[1 - \frac{4m_b}{(n - 2 + 2\sqrt{m_b})^2}\right], \tag{3.28}$$

where m_b is now in m_e units and whose trend is illustrated in Fig. 3.18 for three different dimensions (n = 6, 8, and 10) of the RAHB ring. The increase of the H-bond stabilization ΔE_1 is associated with an ever-decreasing effective mass of the electron in the H-bond segment. This seems reasonable because stronger H-bonds are associated with ever flatter PT potentials in the X–H···X region when going from strongly asymmetric to symmetric SW through nearly symmetric DW potentials. The increasing flattening makes the motion of the electron easier, so decreasing its effective mass, which is substantially the expression of the lowering of the speed of the electron when crossing a strongly bent potential region. Then, both the increase of the H-bond energy and the decrease of the effective mass are manifestations of the X–H···X potential flattening induced by connecting the H-bond donor and acceptor atoms by the interleaving resonant chain. The effect of the length of this chain is shown in Fig. 3.18 by the three curves with n = 6, 8, and 10, which can be taken as representative of the intramolecular H-bond in β, δ, and ζ-diketone enols. The results can be interpreted by saying that increasing the length of the conjugated chain also increases the H-bond stabilization energy, which is in agreement with a number of experimental observations (Gilli *et al.*, 1996a,b). This treatment has been extended to sequences of conjugated monomers, each consisting of n atoms, which form N-membered rings by closing N H-bonds. In the limit $N \to \infty$ this ring will become an infinite chain of monomers within the usual cyclic boundary condition. Two examples are shown in Fig. 3.18 concerning dimers and infinite chains of carboxylic acids. They illustrate the general facts that H-bonds in dimers are stronger than in chains and that both are considerably weaker than the intramolecular bond formed by β-diketone enols, in substantial agreement with the experimental findings.

In the original paper (Munn and Eckhardt, 2001) the effective electron mass was estimated for a ring of six atoms ($n = 6$; $f = 1/3$) with average $d = 1.344$ Å starting from the reasonable assumption of an isolated H-bond with an energy $E_{\mathrm{HB}} = 4.1$ kcal mol^{-1} that becomes 16.5 kcal mol^{-1} by inserting it in the six-membered ring. The RAHB stabilization energy is therefore 12.4 kcal mol^{-1} to be equated to $2\Delta E_1$ for the two electrons in the bond, which, through eqn 3.27, leads to a value of $\rho = 5.85$ and to an effective mass of the electron in the H-bond segment $m_b = 34.2m_e$. The authors remark that this mass is much higher than the electron inertial mass but much lower than the inertial mass of the proton, $m_p = 1836m_e$, and that all this is consistent with the use of the electron motion as a surrogate for the proton motion to represent the flattening of the H-bond potential induced by RAHB.

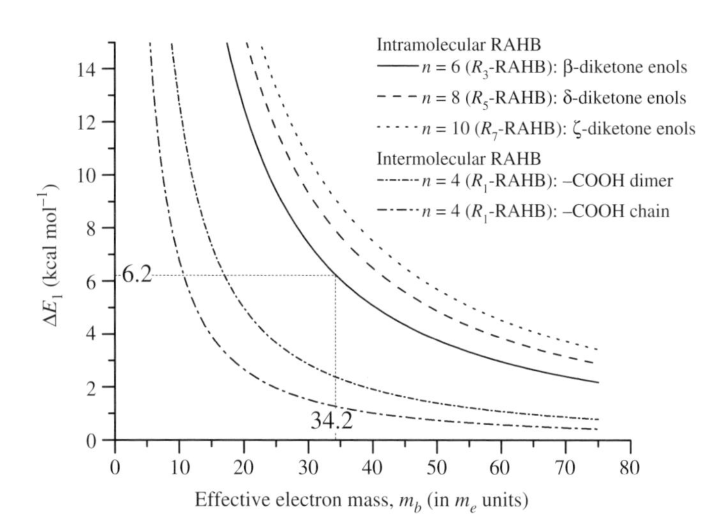

FIG. 3.18. Plot of the intramolecular RAHB stabilization energy, ΔE_1, as a function of the effective electron mass (in units of the inertial electron mass, m_e) as calculated from eqn 3.28. The three continuous, dashed, and dotted curves correspond to rings having $n = 6$, 8, and 10 atoms, as, respectively, found in H-bonded β, δ, and ζ-diketone enols, and the dash-dotted ones to dimers and chains of carboxylic acids.

The authors conclude by remarking that using effective mass in a context of H-bond is certainly unusual but is helpful conceptually. For example, it serves the purpose to give the proton an implicit role in RAHB by its indirect effect on the electron dynamics, the effective mass being an indicator of the relative flatness of the PT potential, which presumably also takes some account of proton tunnelling. The energy lowering induced by resonance (eqns 3.27, 28) applies not only to a simple intramolecular H-bond but also to H-bonded dimers and larger rings that in the limit become infinite chains. Though geometrical constraints on bonding will govern which structure may reasonably occur, present results indicate that an intramolecular ring or a dimer will tend to be favored, consistently with experimental findings. In this model, the chain is three times less favored than the intermolecular ring on a monomeric base, though its length can easily account for its stability and frequent occurrence in molecular crystals. The present treatment is limited to X–H···X homonuclear H-bonds and its extension to the X–H···Y heteronuclear case would be certainly valuable in providing new insight into the applicability of electron effective mass to H-bond studies.

3.2.3 *RAHB generalization and systematics*

3.2.3.1 *RAHB generalization*

A. *Generalities* In general, resonance-assisted H-bonds can be represented as

$$\cdots \mathrm{A}{=}R_n{-}\mathrm{D}{-}\mathrm{H}\cdots, \tag{3.29}$$

where D and A are the H-bond donor and acceptor atoms that are connected through R_n, a *resonant spacer* of n atoms (normally carbons or nitrogens) built up by a short chain of alternating single and double bonds. The integer number n can assume only *odd values*, for instance $n = 1$ in carboxylic acids (···O=C–OH···) or amides (···O=C–NH···), $n = 3$ in β-diketone enols (···O=C–C=C–OH···), β-enaminones (···O=C–C=C–NH···) or ketohydrazones (···O=C–C=N–NH···), and $n = 5$ and 7, respectively, in the few cases of H-bonded δ- and ζ-diketone enols so far known. These resonant fragments are interconnected head-to-tail to give the patterns shown in Fig. 3.19.

RAHBs can be either homonuclear (D = A) or heteronuclear (D $\neq$ A) and, at variance with all other H-bonds, must fulfill the further condition that the atoms at the two ends of the heteropolyene must be able to form double bonds. This obviously rules out all the halogens. As discussed in more detail below, the most frequently observed types of RAHB are –O–H···O=, –N–H···N=, –N–H···O=, and –O–H···N=, though a very limited number of resonant bonds involving sulphur and selenium have been reported (see Section 2.4.4.2).

Since, in the following, both homo- and heteronuclear RAHBs will be considered, the π-bond delocalization index, λ, will be exclusively calculated starting from the Pauling bond numbers, n_i, through the equation (see Section 3.2.2.1 and Schemes **3.2.IX, X**)

$$\lambda = \lambda_n = 1/N \Sigma_{i=1}^{N} \lambda_i \tag{3.30}$$

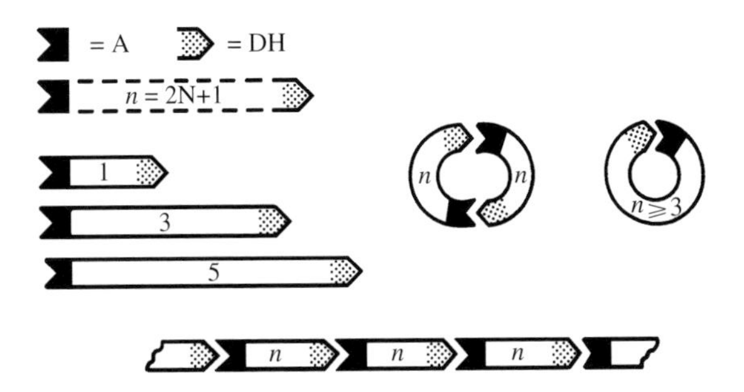

FIG. 3.19. The RAHB unit can be represented by the formula $\cdots$A=R_n–D–H$\cdots$, where A and D are the H-bond donor and acceptor atoms and R_n is the resonant spacer with n odd. The fragments can interconnect in chains or dimers or, for $n \geq 3$, in intramolecularly H-bonded rings. (Reproduced by permission from Gilli *et al.*, 1993)

where $\lambda_i = \Delta n_i = n_i - 1$ (i odd) $= 2 - n_i$ (i even). This index λ_n represents a better estimate of λ and has several advantages, in particular: (i) it has the same meaning and the same domain ($0 \leq \lambda_n \leq 1$) as λ_Q; (ii) for homonuclear H-bonds λ_Q and λ_n are numerically identical; (iii) it can be used with heterodienic systems of any type and, in particular, when some of the $n_{i,o}$ are not exactly 1 or 2 because the resonant fragment is fused with aromatic rings such as benzene or naphthalene.

B. *A general RAHB classification* The most significant examples of homonuclear RAHB are illustrated in Chart 3.1. The scenario is rather complex, being inclusive of three types of H-bond (O–H$\cdots$O, N–H$\cdots$N, and S–H$\cdots$S), four lengths of the resonant fragment (n = 1, 3, 5, and 7), and three main types of H-bond connectivity (intramolecular rings and intermolecular dimers or chains).

Data distribution is quite uneven. Dimers and chains of carboxylic acids (C3.1.a1,a3) are very well documented together with a small number of other oxy- and thioacids, such as the dimer of dimethylarsenic acid C3.1.a2, the chain of phosphoric acid units C3.1.a4, and the two dithiophosphinic acid derivatives C3.1.c1,c2 already discussed in Section 2.4.3.2. Also, rings (C3.1.b1–b3) and chains (C3.1.b4,b5) of β-diketone enols are widely represented, even because the first RAHB proposal (Gilli *et al.*, 1989) prompted a considerable number of new accurate crystal structures in the field. As for enaminoimines, a fairly good coverage of intramolecular H-bonded rings (C3.1.d1–d7) is available while intermolecular bonds cannot be considered for lack of sufficient data. The occurrence of R_5- and R_7-RAHBs in δ- and ζ-diketone enol derivatives (C3.1.e1,f1) does not rely on more than a dozen crystal structures.

CHART 3.1. HOMONUCLEAR π-BOND COOPERATIVE OR RESONANCE-ASSISTED HYDROGEN BONDS (RAHBs)

Reference X···X distances for ordinary non-resonant and non-charged C(sp^3)–X–H···X bonds

Bond	$D = d$(X···X) (vdW cutoff)	$D = d$(X···X) (intermolecular)	$D = d$(X···X) (intramolecular)
O–H···O	3.70	2.70–2.84	2.65–2.90
N–H···N	3.72	3.00–3.20	2.80–3.00
S–H···S	4.24	~4.00	–

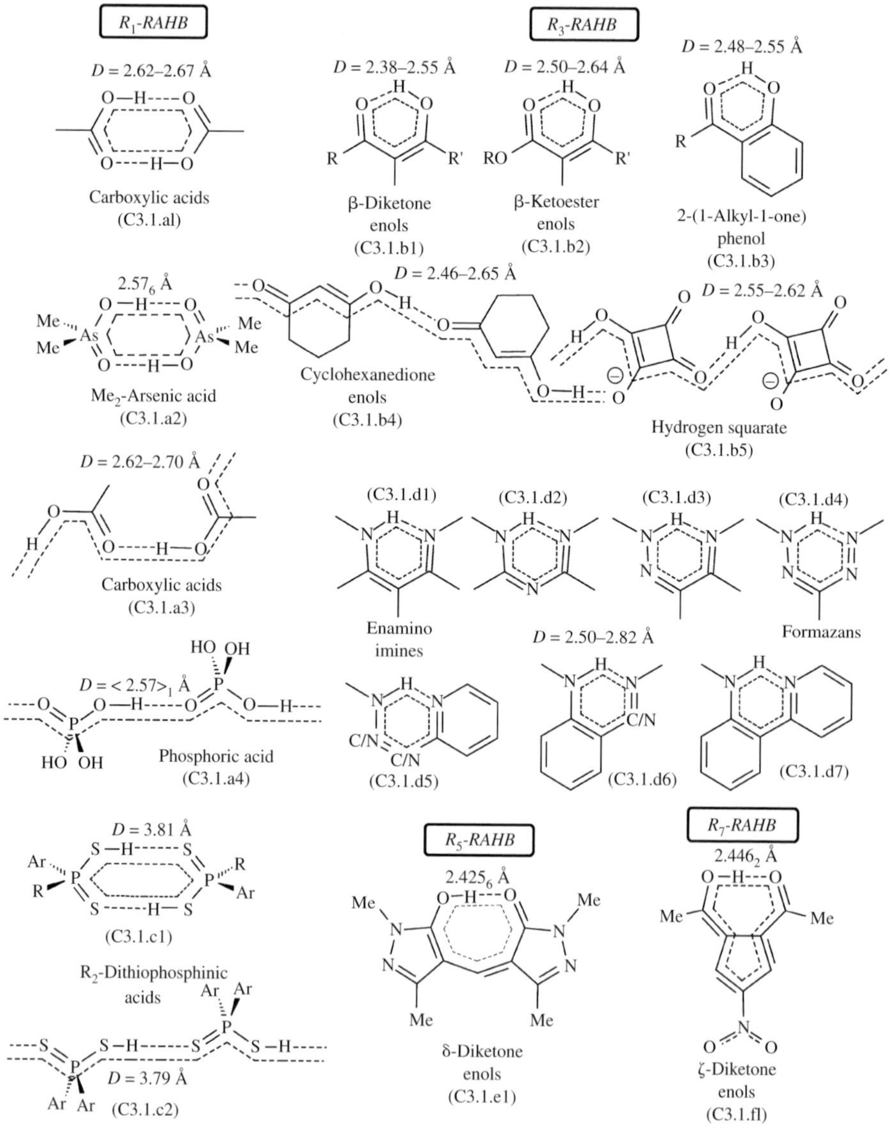

R_1-RAHB

D = 2.62–2.67 Å

Carboxylic acids (C3.1.al)

2.57$_6$ Å

Me_2-Arsenic acid (C3.1.a2)

D = 2.62–2.70 Å

Carboxylic acids (C3.1.a3)

D = <2.57>$_1$ Å

Phosphoric acid (C3.1.a4)

D = 3.81 Å

(C3.1.c1)

R_2-Dithiophosphinic acids

D = 3.79 Å

(C3.1.c2)

R_3-RAHB

D = 2.38–2.55 Å

β-Diketone enols (C3.1.b1)

D = 2.50–2.64 Å

β-Ketoester enols (C3.1.b2)

D = 2.48–2.55 Å

2-(1-Alkyl-1-one) phenol (C3.1.b3)

D = 2.46–2.65 Å

Cyclohexanedione enols (C3.1.b4)

D = 2.55–2.62 Å

Hydrogen squarate (C3.1.b5)

(C3.1.d1) (C3.1.d2) Enamino imines

(C3.1.d3) (C3.1.d4) Formazans

D = 2.50–2.82 Å

(C3.1.d5) (C3.1.d6) (C3.1.d7)

R_5-RAHB

2.425$_6$ Å

δ-Diketone enols (C3.1.e1)

R_7-RAHB

2.446$_2$ Å

ζ-Diketone enols (C3.1.fl)

TABLE 3.4. Contraction of crystallographic D···A contact distances in homonuclear RAHBs with respect to ordinary bonds (OHBs). χ_P = Pauling electronegativity; $d_{D\cdots A}$(vdW)= van der Waals D···A cutoff in the absence of H-bond; $d_{D\cdots A}$(OHB,min) and $d_{D\cdots A}$(RAHB,min) = minimum D···A distances in OHBs and RAHBs; SHR% = per cent $d_{D\cdots A}$ shrinking with respect to $d_{D\cdots A}$(vdW) (distances in Å; *see Table 3.5)

D–H···A	χ_P	$d_{D\cdots A}$ (vdW)	H-bond	$d_{D\cdots A}$ (OHB,min)	SHR% (OHB,min)	$d_{D\cdots A}$ (RAHB,min)	SHR% (RAHB,min)
O–H···O	3.44	3.70	inter	2.70	–27	2.46	–34
			intra	2.65	–30	2.38	–36
N–H···N*	3.04	3.76	intra	2.70	–28	2.45	–33
S–H···S	2.58	4.24	inter	4.00	–6	3.79	–11

The H-bond shortening induced by resonance is considered in Table 3.4 in terms of the minimum D···A distance achievable and corresponding per cent shrinkage (SHR%) with respect to the van der Waals cutoff. On average, ordinary (that is neither charged nor resonant) O–H···O and N–H···N bonds display a shrinking of 28%, which is increased to a 35% because of resonance. Comparison between intra- and intermolecular O···O distances is possible only for O–H···O bonds. The former are shorter than the latter (2.38 against 2.46 Å) though this difference practically vanishes when considering the sum $d' = d(\text{O–H}) + d(\text{H}\cdots\text{O})$ (2.48 against 2.46 Å) because of the quite different O–H–O mean angle in the two cases (some 148 against 170–180°).

Intramolecular RAHB shortening is greater in O–H···O than N–H···N bonds (36 against 33%), which could be imputed to the smaller electronegativity of the nitrogen as well as to the fact that, while oxygen can form strong H-bonds with a sSW PT profile (Section 3.2.2.3), nitrogen only forms longer bonds having DW profile (Section 3.2.3.3). The importance of decreasing electronegativity is more evident for sulphur which can only give weak OHBs with a shrinking of 6%, increased to some 11% because of resonance.

These considerations could well explain why homonuclear RAHBs are practically confined to O, N, and, rarely, S atoms. The halogens are to be ruled out because they are unable to form double bonds, while all other potentially involved atoms, except O and N, are less electronegative than sulphur (Table 2.1). A possible exception could be selenium, whose electronegativity of 2.55 is not far from that of sulphur (2.58), but this would need new experimental data.

3.2.3.2 *Intra- and intermolecular O–H···O RAHB*

A. *Effect of molecular symmetry on O–H···O RAHBs* It has been shown in Section 3.2.2.3 that strong and fully π-delocalized bonds are normally found in symmetrically substituted molecules. This aspect has been reinvestigated comparing the CSD structures of β-diketone, β-ketoester and β-ketoamide enols (**3.2.XIVa–c**; R, R′ = alkyl or aryl) forming intramolecular H-bonds, for which 56, 5 and 37 cases were respectively found (Gilli *et al.*, 1996a,b). The results are summarized in the histograms of Fig. 3.20 that show that the distributions

(3.2.XIVa) (3.2.XIVb) (3.2.XIVc)

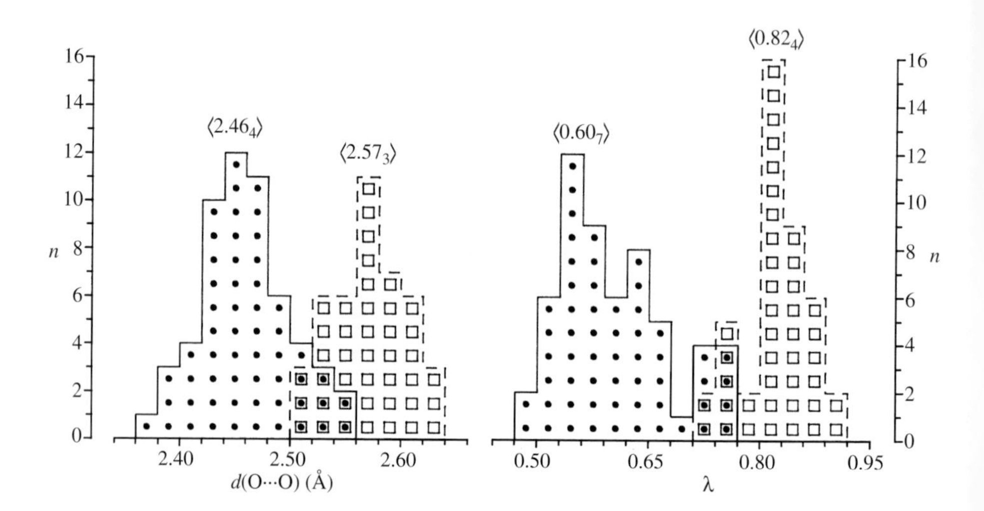

FIG. 3.20. Histograms of the O···O contact distances (a) and of the π-delocalization parameters λ (b) for a series of intramolecularly H-bonded β-diketone (full points), β-ketoester and β-ketoamide (open squares) enols. Average values with e.s.d.s are given on top of each distribution. Amides and esters were statistically indistinguishable and then grouped together. (Reproduced by permission from Gilli *et al.*, 1996b)

of both $d(\text{O}\cdots\text{O})$ and λ are clearly bimodal, indicating that RAHBs in esters and amides are significantly longer and less π-delocalized than in ketones. This seems a further indication that the RAHB strengthening is strictly associated with molecular symmetry, in agreement with the proposed model of a resonance mixing between the O–H···O and O···H–O forms that is the greater the more alike (in other words, isoenergetic) the two forms are.

B. *Effect of resonant-chain length on O–H···O RAHBs* As seen before, all O–H···O RAHBs can be written in the form ···O=R_n–O–H···, where R_n is the *resonant spacer* of *length n* (n odd), and what we want to know here is how the strength of such bonds is affected by the value of n. The problem has been addressed (Gilli *et al.*,

1996a,b) starting from CSD selected H-bond geometries transformed into energies by the LS-HB method already discussed in Section 3.1.3.3.

While the R_1 and R_3 terms are well represented by carboxylic acids (C3.1.a1,a3) and β-diketone enols (C3.1.b1–b5), R_5 (δ-diketone enols) and R_7 (ζ-diketone enols) terms are quite difficult to find out. The only examples of δ-diketones are three derivatives of the rubazoic acid (C3.1.e1) characterized by nearly linear H-bonds with very short O···O distances (2.42–2.45 Å) and an almost total delocalization of the O=C–C=X–C=C–OH heterotriene (λ = 0.52–0.53) (Drück and Littke, 1980). Direct evidence of ζ-diketones is unavailable but it can be shown that some vicinal diacetyl-cyclopentadienes (C3.1.f1) behave like them. These molecules look like γ-diketones but inspection of bond distances shows that they are instead conjugated ζ-diketone enols where the heterotetraene is bridged by a single C–C bond not taking part in conjugation. Five structures are known (Ferguson *et al.*, 1975), all showing very short ($2.43 \leq d(\mathrm{O\cdots O}) \leq 2.51$ Å) and practically linear O–H···O bonds associated with an almost total π-delocalization ($0.51 \leq \lambda \leq 0.53$).

Figure 3.21(a) summarizes the distribution of the $d(\mathrm{O\cdots O})$ distances for O–H···O RAHBs of different chain lengths. The geometries of the shortest H-bonds in each class have then been transformed into H-bond energies, E_{HB}, by using the LS-HB method (Section 3.1.3.3). Final results are displayed in Fig. 3.21(b) where E_{HB} values are seen to increase in the order: non-resonant < carboxylic acids < β-diketone enols (intra and inter) < δ- and ζ-diketones. This increase is not constant, being small from non-resonant bonds to carboxylic acids ($n = 1$), maximum from carboxylic acids to β-diketone enols ($n = 3$) and only slightly smaller when going from these to δ-diketone enols ($n = 5$), while no further increment is observed for ζ-diketone enols ($n = 7$). Data suggest that the factor that enhances H-bond energies is to be sought in the increasing π-polarizability of the conjugated chain due to the larger number of interleaving sp^2 atoms. Moreover, the intrinsic weakness and poor π-delocalizability of carboxylic acids suggest that a similar effect can be expected in all short resonant systems and, in particular, in the biologically relevant amide group.

C. *Morphology of intermolecular O–H···O R_3-RAHBs* Since resonant H-bonds are typically more energetic than ordinary ones, they naturally tend to take control of intermolecular recognition patterns. Three classes of neutral O–H···O bonds are most frequently found in molecular crystals, non-resonant OHBs, R_1-RAHBs of carboxylic acids and R_3-RAHBs of β-diketone enols, whose relative energies can be evaluated from their geometries by the LS-HB method. It is easy to see that, if the shortest O···O distances are respectively 2.70, 2.62, and 2.46 Å, the associated maximum H-bond energies are of the order of 4.5, 6.7, and 16.2 kcal mol^{-1}. Given this scale of energies, it is not surprising that most of the intermolecular patterns that later were considered typical of RAHB have been known ever since the first crystal-packing studies. This is true for the packing of carboxylic acids whose analysis dates back to the seminal paper by

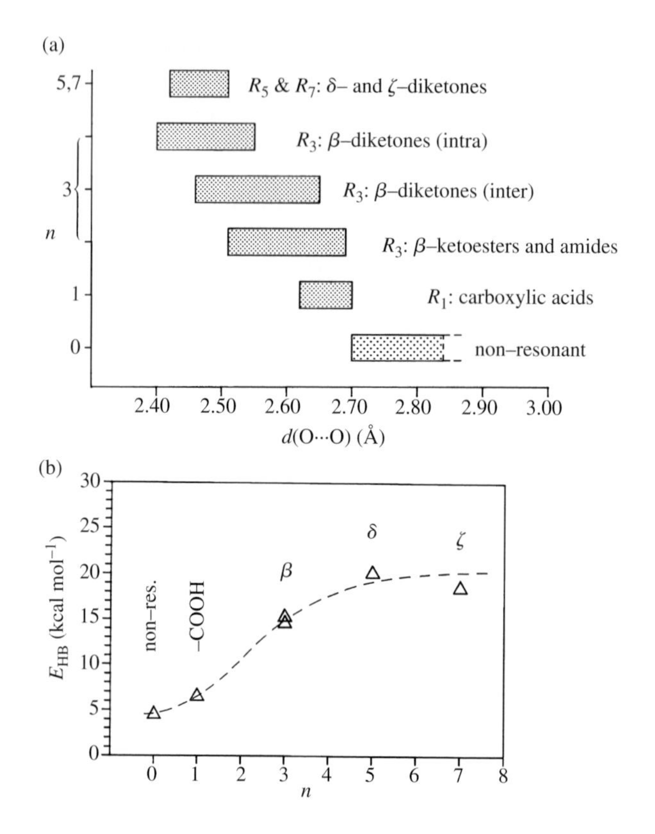

FIG. 3.21. (a) Bar chart of the O···O distances in inter- and intramolecular O–H···O RAHBs arranged according to the length of the resonant spacer R_n (non-resonant bonds added for comparison). (b) H-bond energies, evaluated by the LS-HB method, of the strongest bond observed for each class of O–H···O R_n-RAHBs as a function of the length n of the resonant spacer R_n. (Reproduced by permission from Gilli *et al.*, 1996b)

Berkovitch-Yellin and Leiserowitz (1982) as well as for chains of β-diketone enols (*resonant β-chains*), firstly discussed from a general point of view by Etter *et al.* (1986). Later, these latter compounds were systematically investigated in two further publications that are briefly discussed below (Gilli *et al.*, 1993; Bertolasi *et al.*, 1996).

Molecules involved in the formation of β-chains belong to two main classes, namely enolone (**3.2.XVa**: 2-en-3-ol-1-one) and endiolone (**3.2.XVb**: 2-en-2, 3-diol-1-one) derivatives. The chains are normally classified according to the crystallographic symmetry operations relating subsequent molecules in the chain. The most frequent symmetries are g for glide, 2_1 for twofold screw axis,

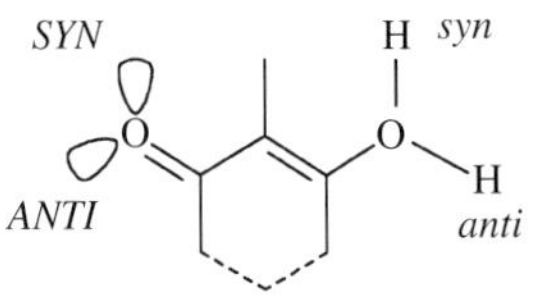

Enolones (2-en-3-ol-1-ones)
(3.2.XVa)

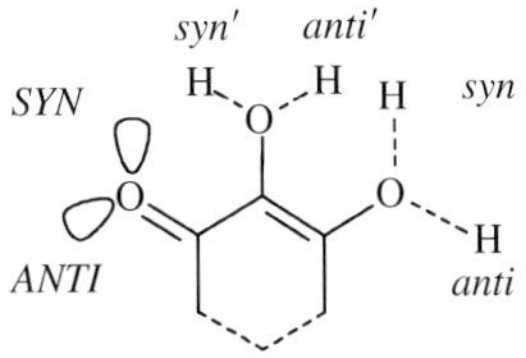

Endiolones (2-en-2,3-diol-1-ones)
(3.2.XVb)

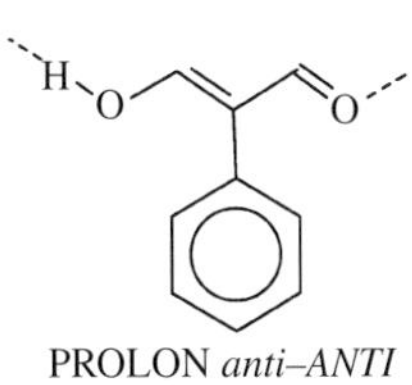

PROLON *anti–ANTI*
(3.2.XVIa)

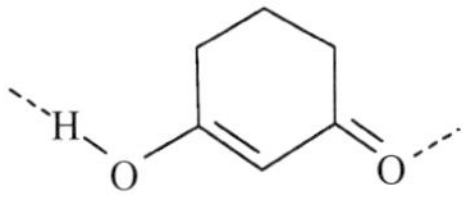

FACRIK *anti–ANTI*
(3.2.XVIb)

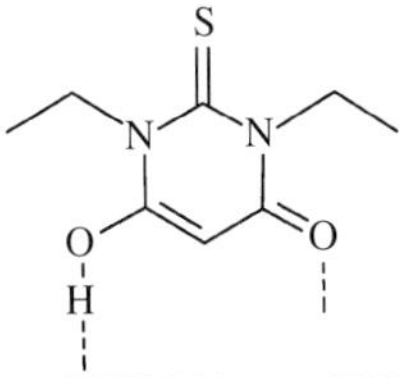

DETSBR01 *syn–SYN*
(3.2.XVIc)

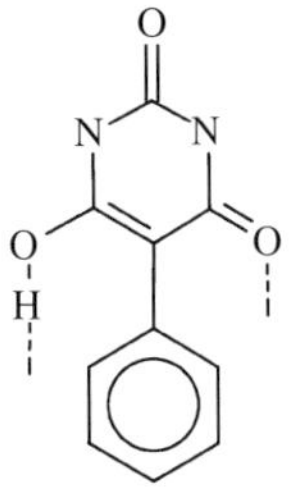

FERBUZ *syn–SYN*
(3.2.XVId)

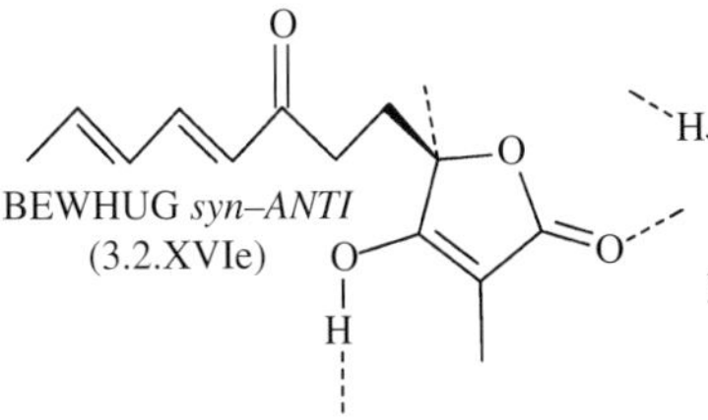

BEWHUG *syn–ANTI*
(3.2.XVIe)

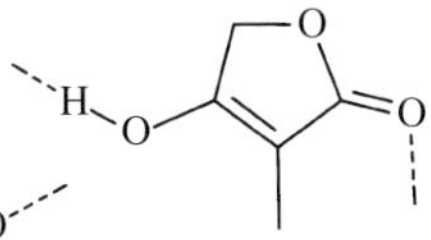

MTETAC02 *anti–SYN*
(3.2.XVIf)

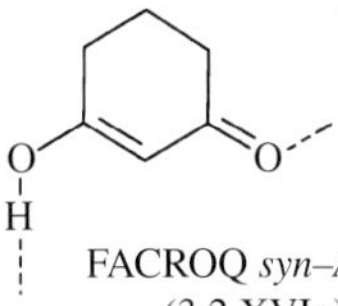

FACROQ *syn–ANTI*
(3.2.XVIg)

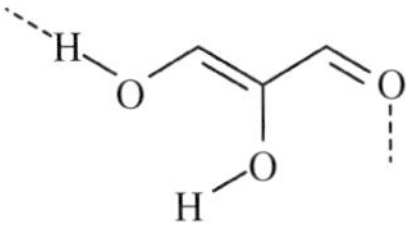

TRIRED *anti–SYN*
(3.2.XVIh)

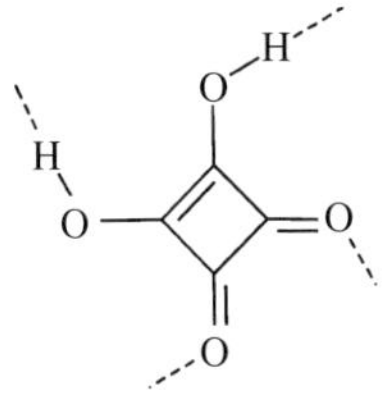

KECYBU06 *syn–ANTI*
(3.2.XVIi)

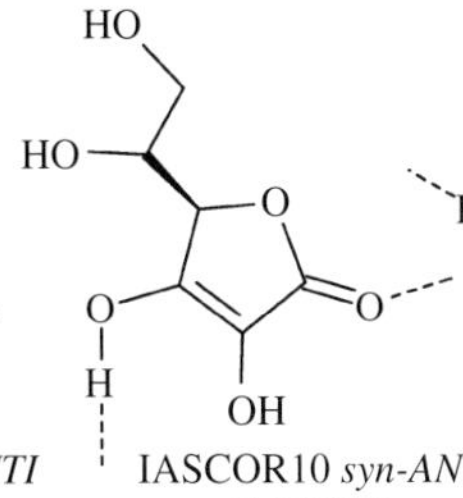

IASCOR10 *syn-ANTI*
(3.2.XVIj)

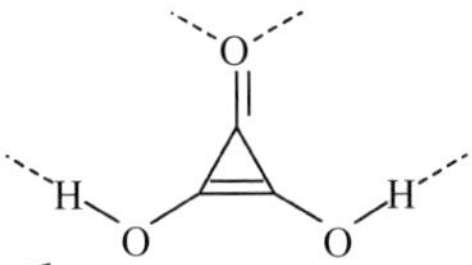

FUDPOJ *anti–SYN*
(3.2.XVIk)

t for simple translation along the chain, and t(A–B) for translational propagation of a double and internally H-bonded asymmetric unit. Approximate non-crystallographic symmetries may also occur, such as *pseudo* 2_1. In all chains the whole enolone fragment, together with its O–H bonds and carbonyl lone pairs, tends to be planar by conforming to the trigonal geometry of the sp^2 hybrid. This is the basis of the stereochemical *syn/anti-SYN/ANTI* nomenclature illustrated in **3.2.XV** and originally proposed by Etter *et al.* (1986) for describing the configurations around the C–O and C=O bonds. Finally, the conformation of the β-chain is determined by rotation around the intermolecular C–O–H···O=C linkage that can be measured by the C–O---O–C torsion angle according to the

Klyne and Prelog (1960) convention [*sp* = synperiplanar, *sc* = synclinal, *ac* = anticlinal, *ap* = antiperiplanar for torsion angles in the ranges $\pm 30°$, $\pm$ (30–90)°, $\pm$ (90–150)° and (150–210)°, respectively].

As a rule, only molecules whose resonant fragment is fused in a ring, such as 1,3-cyclohexane- or 1,3-cyclopentane-diones, give origin to chains because the intramolecular bond formation is preferred whenever possible, with the only exceptions of PROLON (**3.2.XVIa**) and TRIRED (**3.2.XVIh**). The reason for this preference is not enthalpic, because it can be shown that strong intra- and intermolecular O–H···O bonds are nearly isoenergetic when corrected for the O–H–O angle, but most probably entropic and due to the large decrease of entropy produced, in the intermolecular association, by the decreasing number of particles. This phenomenon, often called the *chelate effect*, has been extensively studied for the substitution of ammonia by ethylenediamine in metal complexes, finding $T\Delta S$ contributions around 2.4 kcal mol^{-1} at 300 K for each chelate ring formed (Huheey *et al.*, 1993). Hence, an equivalent energy contribution can be supposed to stabilise intramolecular against intermolecular H-bonds in enolones.

Some typical molecules involved in resonant β-chain formation are collected in Schemes **3.2.XVIa–k**, while the corresponding association patterns are shown in Fig. 3.22. In enolones (**3.2.XVIa–g**; Fig. 3.22(a–g)) the different angular combinations can give rise to *flat chains* of *t*, 2_1 and *g* symmetry, *spiral helices* of 2_1 symmetry, or *large rings*, of which FACROQ (Fig. 3.22(g)) is the only example, consisting of six cyclohexanediones that surround a solvated benzene molecule (not shown in the figure). Solvent removal induces the molecule to form the more common β-chain of FACRIK (Fig. 3.22(b)).

The packing of endiolones (**3.2.XVIh–k**; Fig. 3.22(h–k)) is made more complex by the additional α-hydroxyl group. In TRIRED the resonant β-chains (marked by shading) are intersected by a second non-resonant chain (indicated by rows of small open squares) which may be called an α-chain. The O···O distance is shorter in the former (2.635(2) Å) than in the latter (2.744(2) Å) according to expectation. The packing of deltic acid (FUDPOJ) and squaric acid (KECYBU06) give a still more complex net of intertwined β-chains and α-rings with rather short O···O distances of 2.555(1) and 2.553(1) Å, respectively. It can be shown that the H-bond net is always controlled by RAHB through two β-chains crossing perpendicularly in squaric acid and intertwined in a Greek fret in deltic acid (Gilli *et al.*, 1993).

From a more practical point of view, the most important property of β-chains is that they give rise to monodimensional two-state (or bistable) systems of infinite length because of the ···O=C–C=C–OH···O=C–C=C–OH··· $\rightleftharpoons$ ···HO–C=C–C=O···HO–C=C–C=O··· ketoenol $\rightleftharpoons$ enolketo tautomerism, which can switch from one configuration to the other with inversion of the large dipole moment of the chain. Accordingly, they can display (anti)ferroelectric–paraelectric behavior and undergo order–disorder transitions at some critical temperature (Curie temperature). For these reasons, they are considered potential non-linear dielectrics and, as such, have aroused a large interest in the field of non-linear optics (NLO), particularly as second-harmonic generation (SHG) materials (Tellgren, 1975; Chemla and Zyss, 1987; Prasad and Williams, 1991; Katrusiak, 2003).

The low-temperature structure of squaric acid (KECYBU06) consists of planar ferroelectric layers, antiferroelectrically stacked along b in the monoclinic space group $P2_1/m$ (Semmingsen *et al.*, 1977) and undergoes an antiferroelectric–paraelectric transition (Semmingsen and Feder, 1974) at the Curie temperature of $T_c = 375$ K, which is raised to 527 K by deuteration of the enolic proton. Though this transition has been variously classified as first- or second-order, its order–disorder nature is unquestionable because the high-temperature $I4/m$ neutron structure (Hollander *et al.*, 1977) clearly shows the proton disordered over two equivalent positions with site occupancy of 0.5 and averaged C–C, C–O, and d(O···O) distances, which is consistent with an O–H···O bond whose proton experiences a symmetrical DW potential.

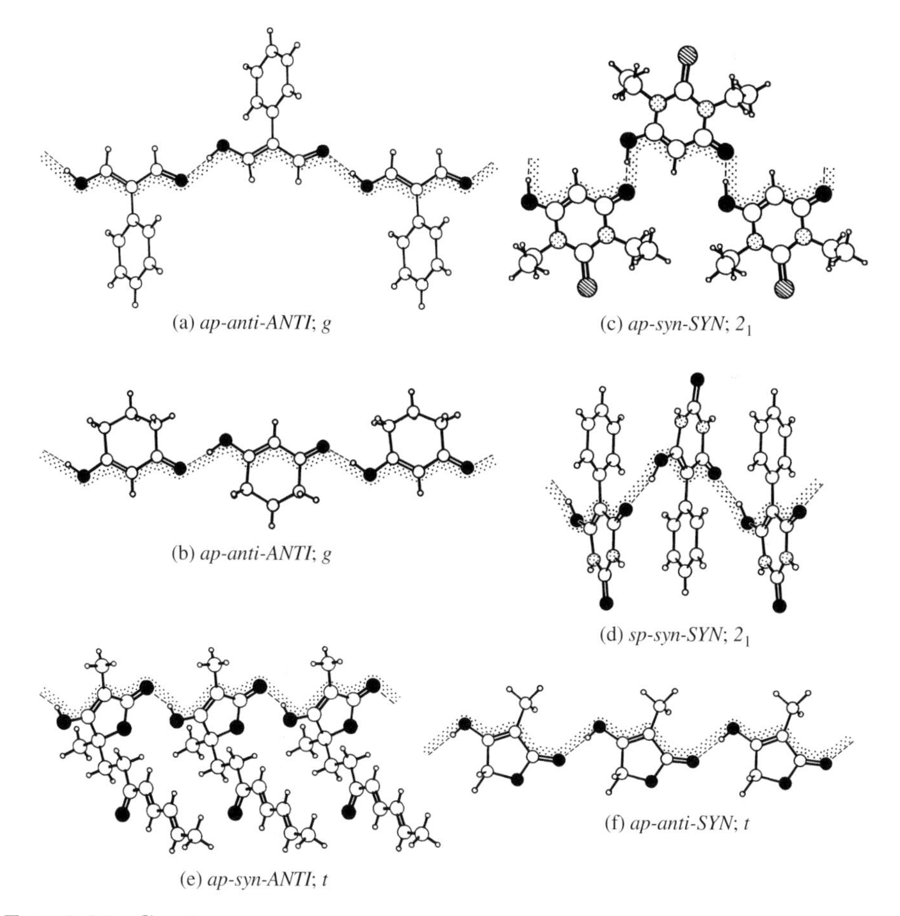

FIG. 3.22. Continues

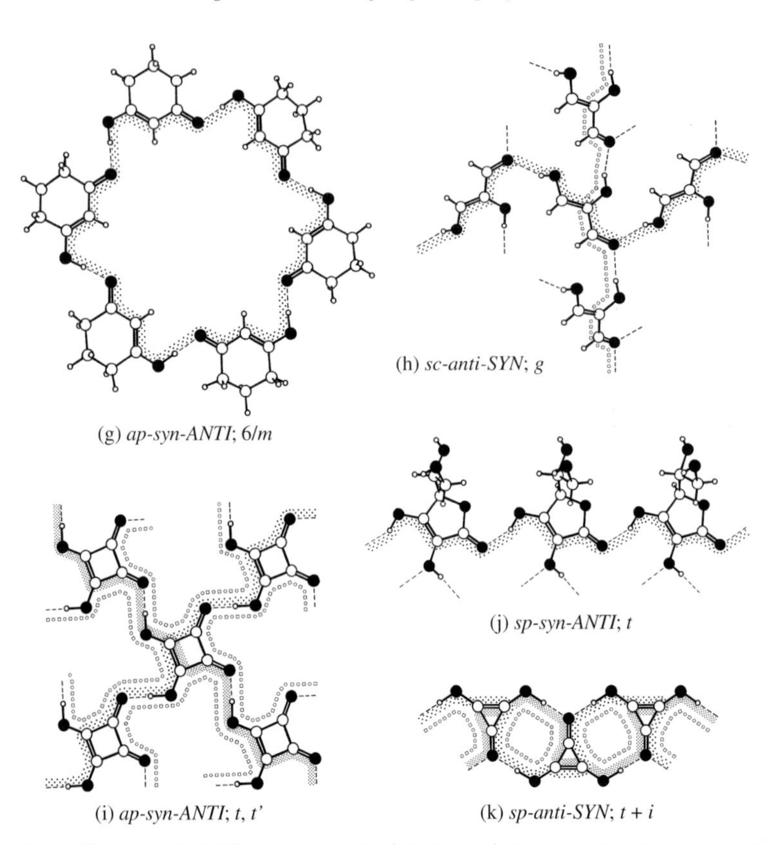

FIG. 3.22. Chains of β-diketone enols (β-chains) formed in the crystal structures of compounds **3.2.XVI**. The β-chains are marked by shadings and α-chains or rings by rows of small open squares. References: (a) PROLON, Semmingsen, 1977; (b) FACRIK, Etter *et al.*, 1986; (c) DETSBR01, Bideau *et al.*, 1977; (d) FERBUZ, de Meester *et al.*, 1986; (e) BEWHUG, Trifonov *et al.*, 1982; (f) MTETAC02, Krogh Andersen and Krogh Andersen, 1975; (g) FACROQ, Etter *et al.*, 1986; (h) TRIRED, Semmingsen, 1974; (i) KECYBU06, Semmingsen *et al.*, 1977; (j) IASCOR10, Azarnia *et al.*, 1972; (k) FUDPOJ, Semmingsen and Groth, 1987. (Reproduced by permission from Gilli *et al.*, 1993)

This interpretation is supported by the facts that: (*i*) an increase of pressure causes a decrease of T_c by shortening $d(\mathrm{O}\cdots\mathrm{O})$ and lowering the PT barrier; and (*ii*) deuteration causes an increase of T_c by lowering the vibrational deuteron zero-point energy.

In principle, it could be expected that all β-chain-based structures display a similar order–disorder behavior with ferroelectric ordering when the space group is polar and the chain is oriented along the polar axis and an antiferroelectric one when the crystal is not polar, such as in the centrosymmetric case. It is not easily understood why such order–disorder transitions are not observed in simple β-chains in spite of the several attempts done (Katrusiak, 1990, 1991). A possible explanation comes from the statistical-mechanics consideration that monodimensional lattices cannot undergo phase transitions, which turn out to become possible only for at least two-dimensional lattices (Landau and Lifschitz, 1967). This would be in agreement with the fact that order–disorder transitions have been actually observed in crystals with an at least two-dimensional net of H-bonds, such as squaric acid (Semmingsen and Feder, 1974), the C form of naphthazarin (Herbstein *et al.*, 1985) and, in particular, the well-known ferroelectrics of the KDP (potassium dihydrogen phosphate) type, which will be further discussed in Section 8.4.2. Conversely, crystals containing isolated β-chains are to be considered monodimensional systems because their intrachain interactions are much stronger than the interchain ones. The design of ferroelectrics of this type would therefore require to plan crystals having, at a same time, the β-chains iso-oriented along a polar axis and a sufficient coupling constant among them.

D. *PT profiles in intermolecular O–H···O R_1- and R_3-RAHBs* PT profiles in intermolecular RAHBs have been investigated by X-ray or neutron diffraction methods at variable temperature for both carboxylic acid dimers and β-diketone enol chains. A typical DW behavior was found in carboxylic acid dimers, and exemplified by the neutron structures of benzoic (Wilson *et al.*, 1996) and 2,4,6-trimethylbenzoic acids (Wilson and Goeta, 2004). In the latter, the O···O distance at 100 K is rather long (2.640 Å), proton population ratios are 94:6, 84:16, and 65:35 at 100, 170 and 240 K, respectively, and application of the van't Hoff equation gives a ΔH° difference between the two minima of 600 cal mol^{-1}. In the former, the O···O distance is remarkably shorter (2.606 Å) with population ratios of 87:13, 79:21, 67:33, and 62:38 at 20, 50, 100, and 175 K, respectively, which lead to a lower ΔH° difference of 120 cal mol^{-1}. Both structures are indicative of dynamical disorder within an asymmetric DW-LB potential whose two minima become more alike, while the H-bond becomes shorter.

Conversely, structures of β-diketone enols (**3.2.XVIa–d**) have been often described as forming asymmetric single-well H-bonds, typical of weak bonds, in spite of their remarkably short O···O distances (2.46–2.64 Å; cf. Fig. 3.21(a)). To try to elucidate the question, the structure of DETSBR (**3.2.XVIc** and Fig. 3.22(c)), the compound forming β-chains with the shortest O···O distance (2.460(1) Å at 100 K), has been recently redetermined by VTXRC methods (Bertolasi, 2007). The difference-Fourier map at 100 K, plotted in the mean plane of the $O_1 \cdots O_{2'}$ and $O_2 \cdots O_{1'}$ H-bonds (Fig. 3.23), is indicative of a dissymmetric DW potential with proton populations in the ratio 66:34, as determined by full-matrix refinement. At variance with carboxylic acid dimers, however, this ratio is unaffected by changing the temperature to 150, 200, and 295 K, suggesting that the disorder is of static nature, that is that the two minima are separated by a considerably high PT barrier. These results are in clear disagreement with what was previously observed in intramolecular O–H···O RAHBs, where the transition from static to dynamic disorder was definitely prompted by the shortening of the H-bond (Section 3.2.2.3). Following the previous discussion on order–disorder transitions in ferroelectric crystals, this behavior could perhaps be explained by saying that monodimensional β-chains cannot undergo dynamic PT transitions, irrespective of their O···O distance, because this process would require the *contemporary shift* of all H-bonds of the chain together with the interchange of all single and double bonds of the resonant fragments.

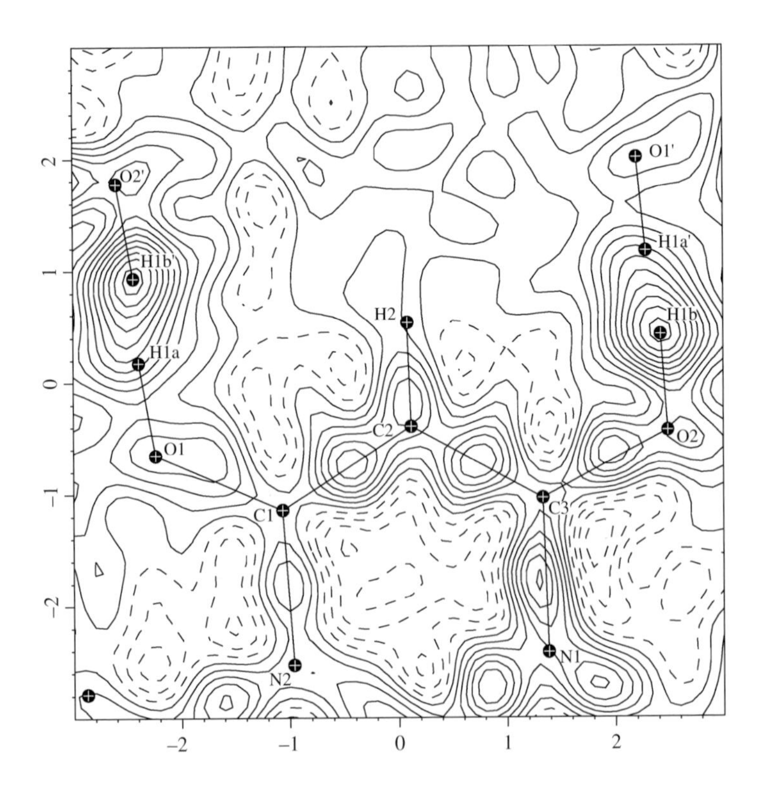

FIG. 3.23. Difference-Fourier map for 1,3-diethyl-2-thiobarbuturic acid (DETSBR) at 100 K plotted in the mean plane of the $O_1 \cdots O_{2'}$ and $O_2 \cdots O_{1'}$ H-bonds. Partial protons $H_{1b'}$ and H_{1a} (as well as H_{1b} and $H_{1a'}$) have a population ratio of 66:34 as determined by least-squares refinement. The map was computed after refinement without the H-bonded proton. Positive (continuous) and negative (dashes) contours at 0.04 e/Å^3. (Courtesy of Prof. V. Bertolasi, 2007).

3.2.3.3 *Intramolecular N–H···N RAHB*

Besides O–H···O, intramolecular N–H···N is the only type of resonant bond for which a large number of crystal structures are easily available. However, no systematic investigation has ever been performed on these bonds, so that such an investigation is carried out here by using the last version of the CSD database (November 2007). The analysis is limited to the selected classes of molecules listed in Fig. 3.24 that can form either resonant or non-resonant H-bonds by closing a six-membered ring, for a total of 1445 H-bonds in 763 crystal structures. Particular attention has been paid to select structures of good quality without serious perturbations of a steric nature. According to Fig. 3.24, compounds are divided into three main classes. The first includes

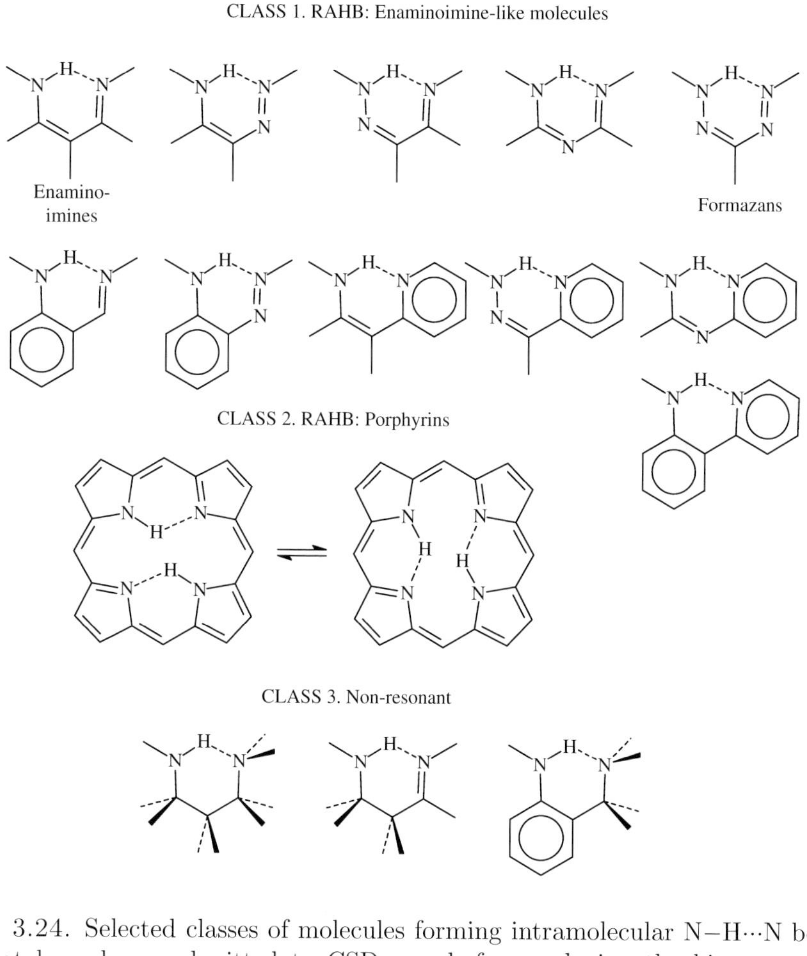

CLASS 1. RAHB: Enaminoimine-like molecules

Enamino-imines

Formazans

CLASS 2. RAHB: Porphyrins

CLASS 3. Non-resonant

Fig. 3.24. Selected classes of molecules forming intramolecular N–H···N bond that have been submitted to CSD search for producing the histograms of Fig. 3.25.

β-enaminoimine-like molecules that, being structurally similar to β-diketone enols, are natural candidates for forming N–H···N RAHBs. The second one concerns porphyrins, molecules that are still compatible with RAHB formation, though probably weakened by the steric constraints within the tetrameric ring and the internal H-bond tautomerism. The third class, finally, includes the molecules whose N–H···N bonds cannot be assisted by resonance because the resonant fragment is interrupted by the presence of one or more sp^3 atoms. These three classes contain, respectively, 393, 722, and 330 H-bonds derived from 308, 208, and 247 crystal structures.

Histograms of Figs. 3.25(a–e) show the distributions of N···N distances, d(N···N), for the different cases. Histogram (a), which is comprehensive of all the 1445 H-bonds (763 structures) of the global sample, is reduced to (c) and then to (e) by stepped subtraction of porphyrins, (b), and non-resonant molecules, (d). The global distribution is then the sum of three subdistributions whose statistical parameters are shown in Table 3.5 and out of which (e) can be taken as the *true N–H···N RAHB distribution* having both minimum, $d_{\text{N}\cdots\text{N}}(\text{min}) = 2.45$ Å, and mean, $d_{\text{N}\cdots\text{N}}(\text{mean}) = 2.67(6)$ Å, values some 10% shorter than those of the non-resonant distribution (d) (2.70 and 2.96(13) Å, respectively).

To conclude, the statistical analysis of N···N distances confirms without any reasonable doubt that RAHB is really effective in shortening the intramolecular N–H···N bond when not hindered by steric circumstances (as, for instance, in porphyrins). This shortening amounts to 10 per cent of the N···N distance and is almost identical to that previously observed for the O–H···O bond in β-diketone enols (Table 3.4). A preliminary analysis shows, moreover, that shorter bonds are increasingly delocalized up to λ values of 0.4–0.5, in agreement with what previously observed in O–H···O bonds. The only basic difference between N–H···N and O–H···O RAHBs so far observed is that the former seems unable to form SW-NB but only DW-LB bonds (Perrin and Ohta, 2001), a fact needing, however, further experimental verification.

3.2.3.4 *Heteronuclear X–H···Y RAHB*

A. Generalities The most significant heteronuclear R_n-RAHBs derivable from crystallographic databases are summarized in Chart 3.2. Most of them are N–H···O/O–H···N bonds, the few examples of O–H···S/Se and N–H···S/Se bonds having been already treated in Section 2.4.4.2. The most common R_n lengths are $n = 1$ or $n = 3$, prototypically represented by *amides* (···O=C–N(R)H···) and *β-enaminones* (···O=C–C=C–N(R)H···), though with quite uneven occurrences because amides are common functional groups, while enaminones are practically synthesized on purpose, their chemical relevance having been recognized only recently.

As a general rule, *N–H···O RAHBs are considerably weaker than O–H···O ones* because oxygen is endowed with a much lower proton affinity (PA) than nitrogen and this fact hinders the resonance **3.2.XVIIa** ↔ **3.2.XVIIb** because of the much higher stability (lower energy) of the N–H form. Accordingly, while

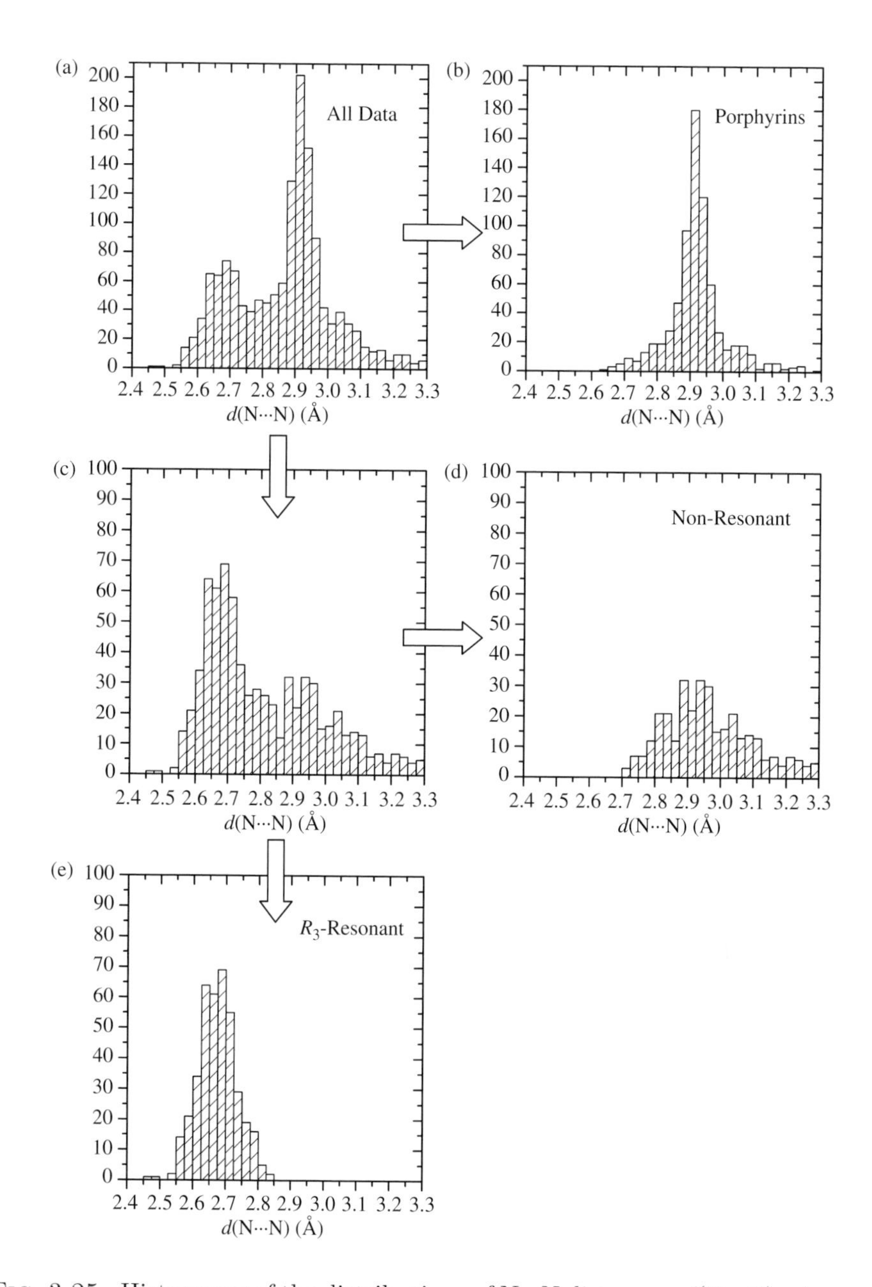

FIG. 3.25. Histograms of the distributions of N···N distances, d(N···N), observed in intramolecular N–H···N bonds formed by molecules of Fig. 3.24: (a) d(N···N) distribution for the full dataset, including 1445 H-bonds found in 763 CSD structures; (b) distribution for porphyrins (722 H-bonds); (c) remaining distribution after porphyrins removal; (d) distribution for non-resonant molecules (330 H-bonds); and (e) remaining distribution after non-resonant removal showing the final d(N···N) distribution for RAHB-forming molecules (393 H-bonds).

TABLE 3.5. Contraction of the crystallographic N···N contact distances in the two classes of N–H···N RAHBs (enaminoimines and porphyrins) with respect to the corresponding non-resonant values. $d_{N\cdots N}$(min) and $d_{N\cdots N}$(mean) = minimum and mean N···N distances; SHR% = per cent $d_{N\cdots N}$ shrinking with respect to the non-resonant value; n_{HB} and n_{SRT} = number of H-bonds and crystal structures in the sample (distances in Å; standard deviations in parentheses)

Class	$d_{N\cdots N}$ (min)	SHR% (min)	$d_{N\cdots N}$ (mean)	SHR% (mean)	n_{HB}	n_{SRT}
1. Enaminoimines	2.45	−9.3	2.67(6)	−9.8	393	308
2. Porphyrins	2.63	−2.6	2.92(9)	−1.5	722	208
3. Non-resonant	2.70	0	2.96(13)	0	330	247

homonuclear bonds in β-enolones are the stronger the more *symmetric* they are (Section 3.2.3.2), heteronuclear bonds in β-enaminones are strongly stabilized by any type of *asymmetric* R_1–R_3 *substitution* able to level out the PA difference between the donor and acceptor atoms. Hence, the role of chemical substitution becomes predominant in determining the features of N–H···O/O–H···N RAHBs, the modulation of the electronic properties of the substituents being able to modify the shape of the PT profile and to switch the bond from pure N–H···O to pure O–H···N through tautomeric N–H···O ⇌ O–H···N bonds in a more or less symmetric DW potential.

The other important factor is the length of the resonant chain R_n. The H-bond strengthening is a minimum for $n = 1$, so that the geometrical changes produced by resonance in amide dimers (C3.2.a1) and chains (C3.2.a2) are near the limit of detectability and therefore difficult to ascertain by sheer crystallographic methods in spite of the many studies prompted by the great importance of the N–H···O bonds in determining the secondary structure of proteins and the thymine–adenine (C3.2.a3) and cytosine–guanine (C3.2.a4) coupling in DNA. While a convincing proof of the real importance of RAHB in the coupling of nucleic bases has never been achieved, effects in protein chains have been recently substantiated by quantum-mechanical methods, as briefly discussed in one of the next paragraphs.

The analysis of the N–H···O/O–H···N system requires definition of quantitative indicators of H-bond strength and π-delocalization of the resonant fragment. In analogy with O–H···O bonds (Section 3.2.3.1), suitable π-delocalization indices are $\lambda = \lambda_n$ ($0 \leq \lambda_n \leq 1$) calculated from the Pauling bond orders, n, and Del% = 100 (1− |2 λ_n− 1|) (0 ≤ Del% ≤ 100), while λ_Q cannot be used for lack of fragment symmetry. C=O distances are sometimes used as a simplified delocalization index ranging from 1.200 to 1.275 Å while n(C–O) changes from 2 to 1.5. H-bond strengths can be monitored by geometrical quantities ($D = d$(N···O) or $D' = d$(N–H) + d(H···O) as well as by quantum-mechanical H-bond energies, E_{HB},which require, however, to be shortly commented upon.

CHART 3.2. HETERONUCLEAR π-BOND COOPERATIVE OR RESONANCE-ASSISTED HYDROGEN BONDS (RAHBs)

Reference X···Y distances for ordinary non-resonant and non-charged $C(sp^3)$–X–H···Y bonds			
Bond	$D = d(X\cdots Y)$ (vdW cutoff)	$D = d(X\cdots Y)$ (intermolecular)	$D = d(X\cdots Y)$ (intramolecular)
N–H···O/O–H···N	3.73	2.87–3.30	2.71–2.75

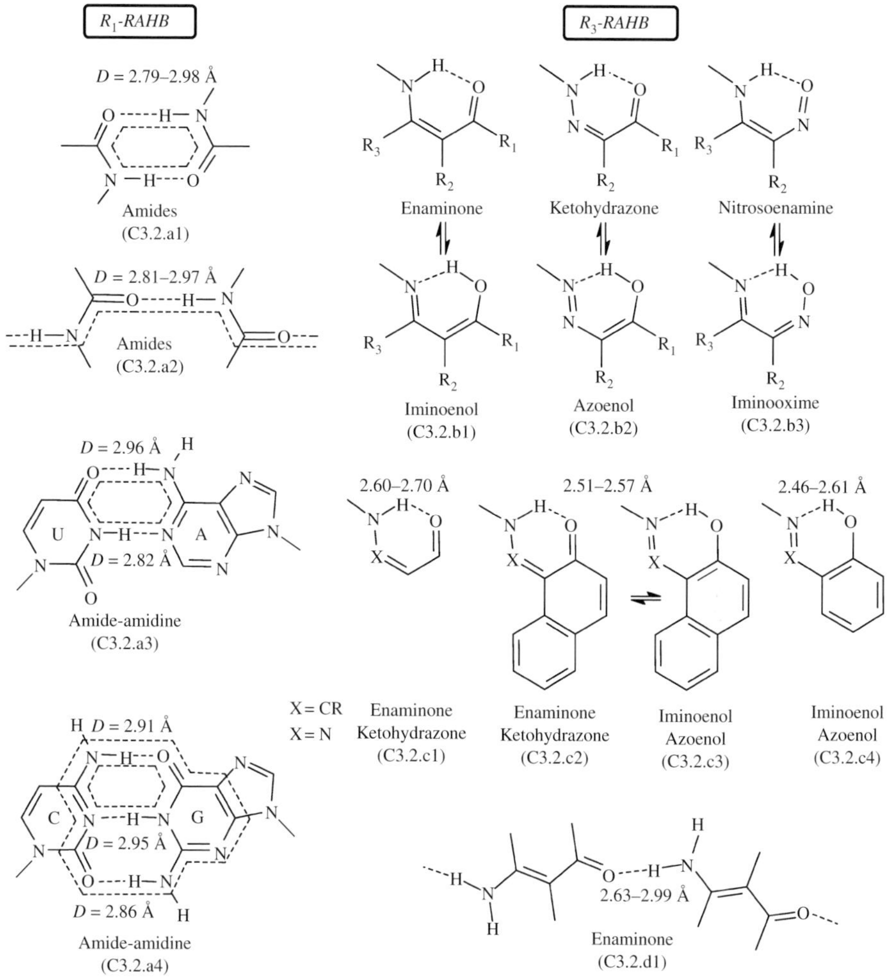

R₁-RAHB

D = 2.79–2.98 Å
Amides (C3.2.a1)

D = 2.81–2.97 Å
Amides (C3.2.a2)

D = 2.96 Å, D = 2.82 Å
Amide-amidine (C3.2.a3)

D = 2.91 Å, D = 2.95 Å, D = 2.86 Å
Amide-amidine (C3.2.a4)

R₃-RAHB

Enaminone ⇅ Iminoenol (C3.2.b1)

Ketohydrazone ⇅ Azoenol (C3.2.b2)

Nitrosoenamine ⇅ Iminooxime (C3.2.b3)

2.60–2.70 Å
X = CR Enaminone
X = N Ketohydrazone
(C3.2.c1)

2.51–2.57 Å
Enaminone Ketohydrazone (C3.2.c2) ⇌ Iminoenol Azoenol (C3.2.c3)

2.46–2.61 Å
Iminoenol Azoenol (C3.2.c4)

2.63–2.99 Å
Enaminone (C3.2.d1)

It is known that RAHB geometries cannot be well reproduced at the Hartree–Fock level (Frisch *et al.*, 1985) but that good agreements between calculated and experimental values are generally attained by *ab initio* MP2 (or higher) or DFT B3LYP (or similar) methods by using 6–31+G(d,p) basis set (or larger). The evaluation of intramolecular RAHB energies poses particular problems.

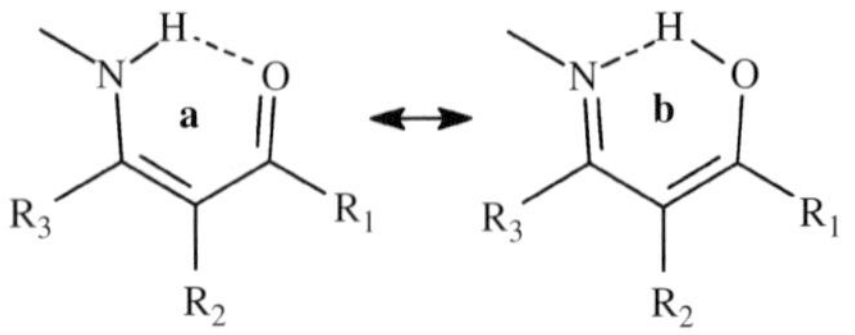

Enaminone (3.2.XVIIa) ↔ Iminoenol (3.2.XVIIb)

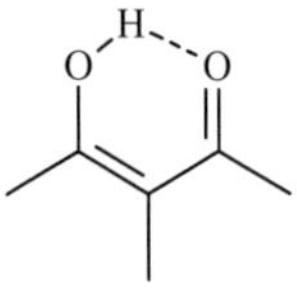

Closed *cC* (3.2.XVIIIa)

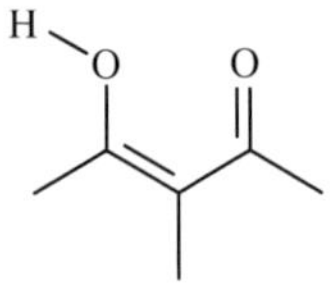

Open *cC* (3.2.XVIIIb)

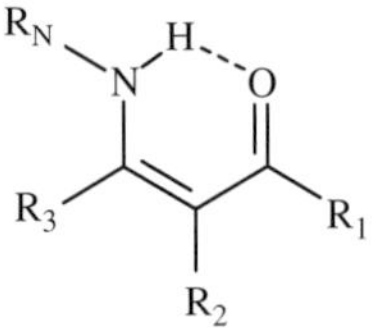

Closed *cC* (3.2.XIXa)

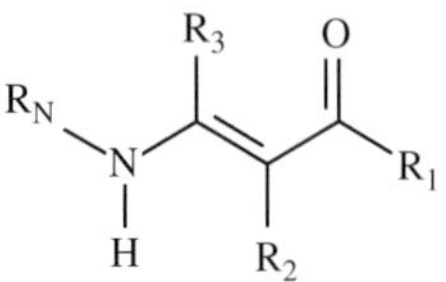

Open *cT* (3.2.XIXb)

Since these energies cannot be measured, no comparison with experiments is possible and we can only guess that, as long as geometries are well reproduced by calculations, the energies will also be, particularly if account is taken of vibrational corrections, at least at the zero-point level. A more serious problem comes from the definition of the H-bond energy itself (Gilli *et al.*, 2000; Buemi, 2006). In β-enolones, binding energies are normally evaluated as $E_{HB}(cC) = E(\text{open } cC) - E(\text{closed } cC)$, that is as the difference between the open (non-H-bonded) *cC* and closed (H-bonded) *cC* configurations, where the open form is obtained by simple 180° proton rotation around the C–OH bond (**3.2.XVIII**). This quantity is believed to be a good strength indicator as it is the energy difference between two molecules with and without H-bond that *maintain comparable interatomic repulsions*. This same quantity cannot be computed for N–H···O bonds because of steric hindrance of the NHR group, so that the only energy difference that can be evaluated (**3.2.XIX**) is $E_{HB}(cT) = E(\text{open } cT) - E(\text{closed } cC)$, a quantity hardly interpretable as a pure H-bond energy because the opening of the H-bonded form also modifies the pattern of the other interatomic interactions. Hence, intramolecular O–H···O and N–H···O bonds cannot be compared without knowing the value of the $\Delta E_{HB} = E_{HB}(cC) - E_{HB}(cT)$ difference, which, though unattainable for the N–H···O bond, can be approximately evaluated from that of the O–H···N tautomer coexisting in an even formal DW potential. Preliminary DFT calculations on 1-alkylimino-2-naphthols (C3.2.c3; Gilli *et al.*, 2000) suggest that such a ΔE_{HB} can be as large as 4.30 kcal mol^{-1}, a value clearly too important to be neglected. These few considerations suggest that a better way for obtaining a correct estimate of $E_{HB}(cC)_{N-H\cdots O}$ consists in computing $E_{HB}(cC)_{O-H\cdots N}$ (which is always possible) and then in evaluating the H-bond energy of the other tautomer as $E_{HB}(cC)_{N-H\cdots O} = E_{HB}(cC)_{O-H\cdots N} - \Delta E_r$, where ΔE_r is the thermodynamic energy difference between the two tautomers (see Section 6.3 for a more complete analysis). This strategy will give intramolecular N−H···O energies that are strictly comparable with the O−H···O ones, though at the expense of doubling the computing time.

B. *Substituent effects in intramolecular N−H···O/O−H···N R_3-RAHBs* Systematic screening of crystal databases shows that these bonds are mostly associated with the enaminone ⇌ iminoenol (C3.2.b1), ketohydrazone ⇌ azoenol (C3.2.b2), and, less frequently, nitrosoenamine ⇌ iminooxime (C3.2.b3) tautomeric systems. Which tautomer actually occurs and how strong the H-bond may be does not depend, however, on the tautomeric system itself but is rather a consequence of the substituents present, the most important type of substitution being the fusion (or less) of the resonant fragment with an aromatic ring. The fusion gives rise to three definite classes [non-fused or *simple* (C3.2.c1), naphthalene-fused (C3.2.c2,c3), and phenylene-fused (C3.2.c4) enaminone-like compounds] that widely differ in their N···O distances and for their ability of forming N−H···O, tautomeric N−H···O ⇌ O−H···N, or O−H···N bonds, respectively. This peculiar behavior needs to be accounted for.

It has been already remarked that non-fused derivatives (C3.2.c1) tend to form rather weak N−H···O bonds with strongly asymmetric PT profiles (aSW) because the large oxygen-nitrogen PA difference makes the N−H···O resonance form much more stable than the O−H···N one. This latter, however, becomes the stabler one after fusion with a phenylene moiety (C3.2.c4) because N−H···O formation would now imply the loss of the large resonance energy of the aromatic ring, which is greater than the PA difference. The resulting PT profile is still an aSW, but symmetrically reversed. Fusion with a naphthalene, having an intermediate value of resonance energy *per* ring, leads to the more interesting situation of two roughly isoenergetic N−H···O and O−H···N forms (C3.2.c2,c3) that, collapsing by resonance mixing, could in principle generate either sSW or sDW PT profiles. Only experience can tell us which profile actually occurs and data so far collected indicate that N−H···O/O−H···N RAHBs, at variance with the O−H···O ones, are very seldom so strong as to produce sSW profiles instead of the more or less symmetric DW potentials whose associated tautomeric equilibrium can be finely tuned by chemical substitution.

This simple scheme may be made more complex by other factors, such as: (*i*) steric and electronic effects of other substituents; (*ii*) the practical difficulty of detecting tautomeric DW N−H···O ⇌ O−H···N bonds by X-ray diffraction techniques and, in particular, of measuring the relative p(N−H) and p(O−H) partial proton populations; and (*iii*) accounting for the strict intermixing of tautomerism and resonance effects occurring in these compounds. Fortunately, X-ray crystallography of tautomeric species has been greatly improved by the recent introduction of CCD counters and low-temperature methods, though the problem of hundreds of older room-temperature structures where DW bonds are often misinterpreted as SW ones still remains. Also, the problems arising from the strict interplay between tautomerism and resonance mixing can be considered to have found a definitive solution, as illustrated in Fig. 3.26 for, e.g. the enaminone $\leftrightarrow$ iminoenol resonance mixing of Scheme **3.2.XVII**. Let $E(\mathbf{a})$ and $E(\mathbf{b})$ be the energies of the two N–H and O–H resonant forms. When $E(\mathbf{a}) \ll E(\mathbf{b})$ (case A) or $E(\mathbf{a}) \gg E(\mathbf{b})$ (case A′) the two resonant forms cannot mix and the pure and weak N–H···O, on the left, or O–H···N, on the right, bonds will be observed. When, however, this energy difference becomes smaller, i.e. $E(\mathbf{a}) < E(\mathbf{b})$ (case B) or $E(\mathbf{a}) > E(\mathbf{b})$ (case B′), the $\mathbf{a} \leftrightarrow \mathbf{b}$ mixing, and the $\mathbf{b} \leftrightarrow \mathbf{a}$ one on the opposite side, will grow larger, causing a shrinking of the N···O distance because the geometries of the two opposite N–H···O and O–H···N tautomers have become more similar. Since, nevertheless, one of the two forms is still considerably lower in energy, only the more stable one will be experimentally observed and will appear as an N–H···O (case B) or O–H···N (case B') bond that has been shortened (strengthened) for the effect of those substituents that are able to stabilize the opposite resonant form. Finally, when $E(\mathbf{a}) \cong E(\mathbf{b})$, the resonance mixing between **a** and **b** becomes a maximum, the N···O distance a minimum and what will be observed is either a tautomeric N–H···O $\rightleftharpoons$ O–H···N equilibrium finely tunable by substituents (case C) or a very strong and symmetrically centered N···H···O bond (case D).

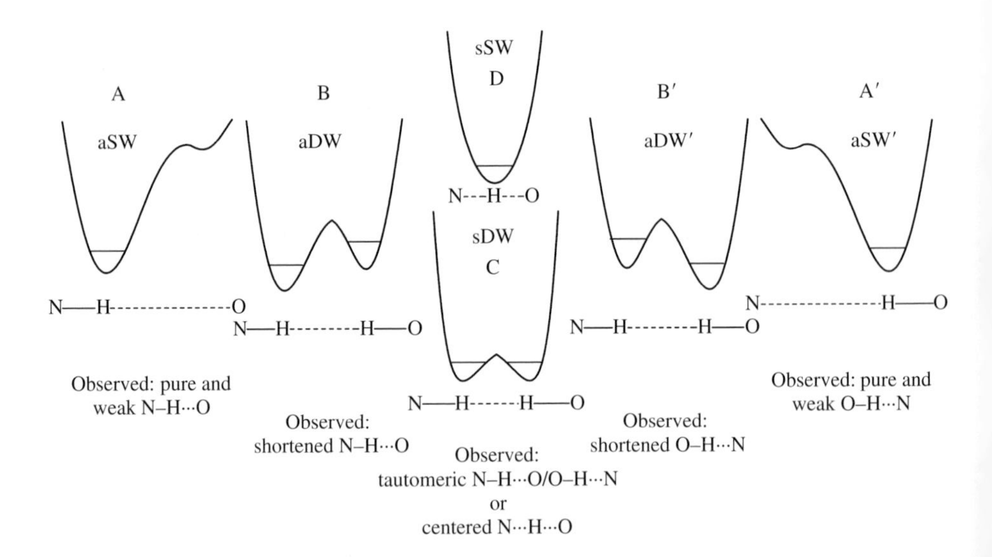

FIG. 3.26. The different X-ray diffraction outcomes (lower line) that are expected to result from potentially tautomeric N–H···O/O–H···N bonds having different shapes of their PT profiles (upper-line sketches).

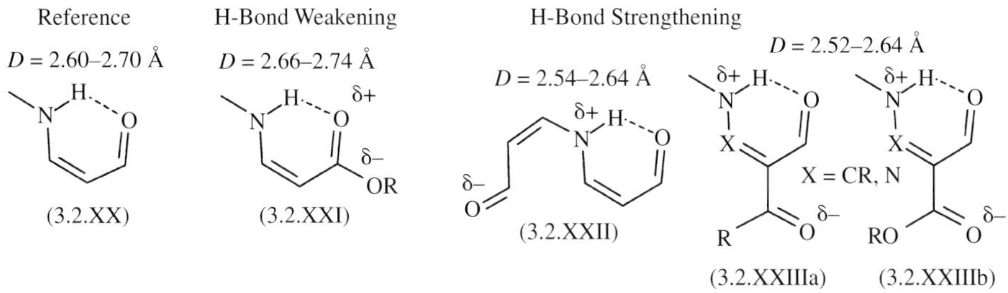

Reference $D = 2.60–2.70$ Å (3.2.XX)

H-Bond Weakening $D = 2.66–2.74$ Å (3.2.XXI)

H-Bond Strengthening $D = 2.54–2.64$ Å (3.2.XXII); $D = 2.52–2.64$ Å, X = CR, N (3.2.XXIIIa) (3.2.XXIIIb)

The effects of substituents on N–H···O strengths have been the object of systematic studies (Bertolasi *et al.*, 1993, 1994a,b, 1999; Gilli *et al.*, 2000; Gilli and Gilli, 2000). The comparison of N···O distances in substituted derivatives (**3.2.XXI–XXIII**) with those of simple enaminones (**3.2.XX**) indicates that the H-bond is strengthened (or weakened) according to whether the substitution increases (or decreases) the acidity of the N–H group and then favors (or disfavors) the resonance mixing between the N–H···O and N···H–O forms. The effects are only moderately large and are in agreement with the $E_{HB}(cT)$ values (Gilli *et al.*, 2000) calculated by DFT methods for the same compounds that are seen to increase monotonically, from left to right, from 4 to 9 kcal mol^{-1}. No substituent is electron attracting enough, however, to induce the aSW → DW transition from the asymmetric N–H···O to the tautomeric N–H···O ⇌ O–H···N bond.

The structural properties of phenylene-fused derivatives have been recently reviewed (Bertolasi *et al.*, 2009). The most complete series suited to study the effects of substituents is that of *N*-phenyl-salicylideneanilines **3.2.XXIVa–f**. The reference term is compound **3.2.XXIVa** that is found to form a rather strong intramolecular O–H···N bond with d(N···O) = 2.615 Å. Substitution effects at the *N*-phenyl turn out to be driven by the electron-attracting or -donating properties of the substituents as measured by their Hammett constants, σ_m or σ_p. The 13 structures available show a continuous shortening of the N···O distance with the decrease of σ in the range $0.78 \geq \sigma \geq -0.60$, supporting the idea that the H-bond is strengthened by electron-donating groups that stabilize the N–H form. The extreme terms of the series are **3.2.XXIVb** [$-NO_2$, $\sigma_p = 0.78$, d(N···O) = 2.651 Å] and **3.2.XXIVc** [–OH, $\sigma_p = -0.37$, d(N···O) = 2.546 Å and $-NMe_2$, $\sigma_p = -0.60$, d(N···O) = 2.564 Å].

Similar strengthening effects can be produced by destabilizing the O–H form, that is by making more acidic the phenol hydroxyl. This result is achieved through strong electron attractors on the phenylene moiety as well as through H-bonded contacts of the type –O–H···O–H···N. Heavier perturbations can sometimes produce, besides simple O–H···N strengthening, larger effects including the transition to the O–H···N ⇌ N–H···O tautomeric form or to the pure aSW N–H···O bond. This is illustrated in **3.2.XXIVd**, where the *o*-NO_2 group induces a very short (2.519–2.533 Å) tautomeric bond by combined electron

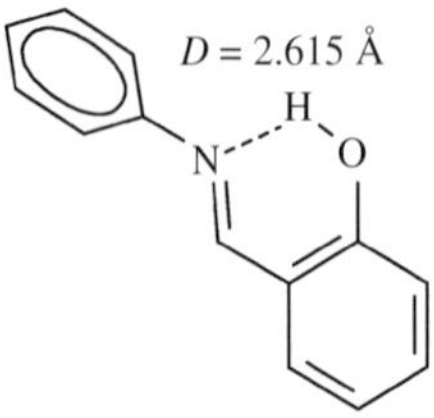

SALCAN01 (3.2.XXIVa) — D = 2.615 Å

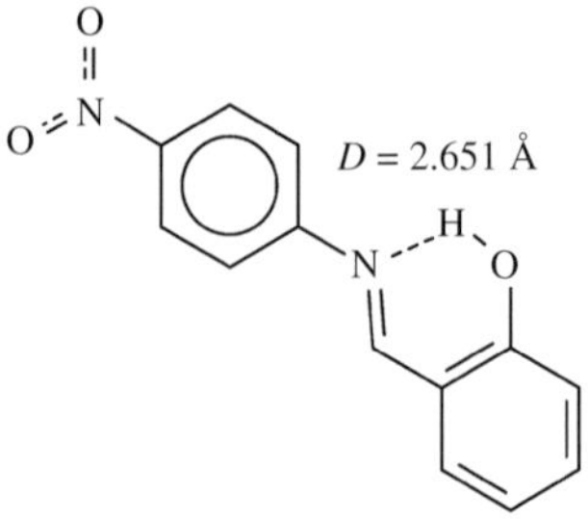

CAVQOF (3.2.XXIVb) — D = 2.651 Å

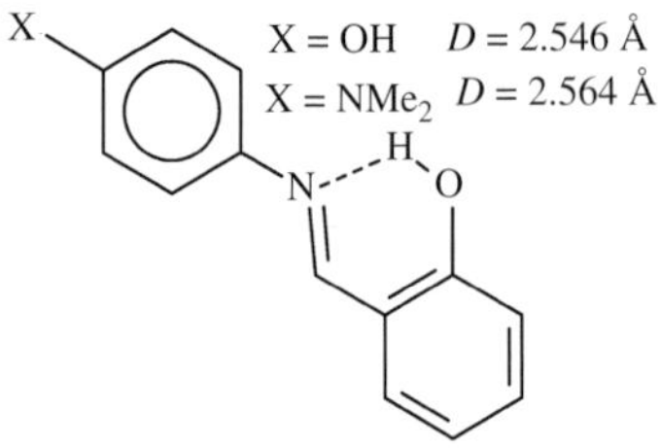

ETEYUX, CEBKEZ (3.2.XXIVc) — X = OH D = 2.546 Å; X = NMe_2 D = 2.564 Å

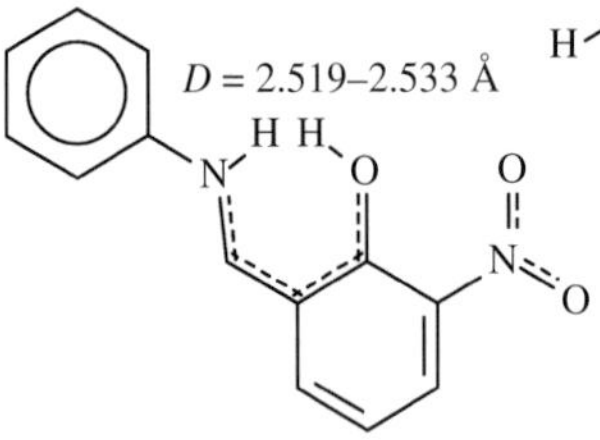

LIXRIT01-03 (3.2.XXIVd) — D = 2.519–2.533 Å

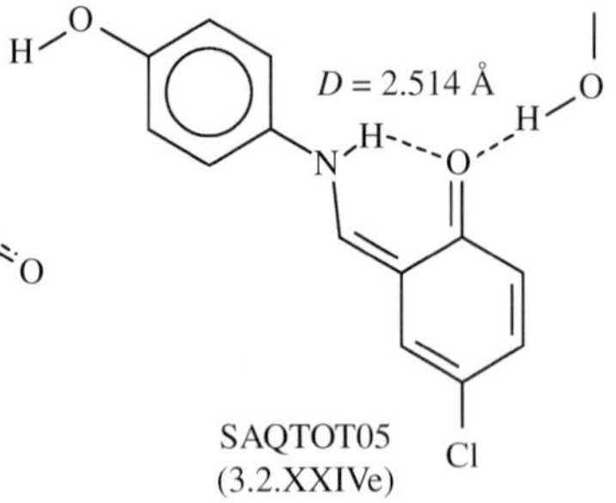

SAQTOT05 (3.2.XXIVe) — D = 2.514 Å

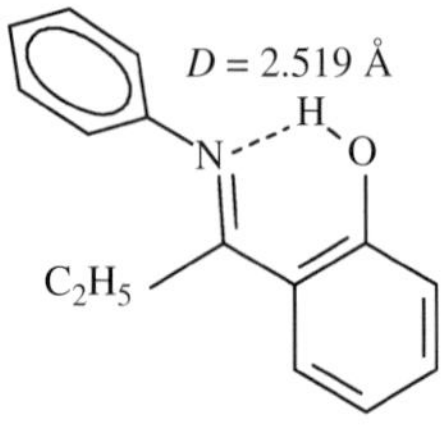

BAHQUX (3.2.XXIVf) — D = 2.519 Å

attraction and steric compression, and in **3.2.XXIVe**, where a pure and very short N–H···O bond (2.514 Å) is produced by the combined action of three different effects, the *p*-Cl and *p*-OH substitutions and the additional –O–H···O–H···N intermolecular contact.

Finally, in analogy with what happens in β-diketone enols, N–H···O/O–H···N bonds can be markedly shortened by substituents exerting a sterical compression on the bond itself, a characteristic example being **3.2.XXIVf** where the ethyl group squeezes the bond down to 2.519 Å in the absence of a significant electronic effect. Compression effects have been particularly investigated by the Koll and Filarowski group in Wrocław (*e.g.*, Filarowski *et al.*, 1998, 1999) who, by a proper choice of the substituents, have produced H-bond contractions down to 2.46 Å. It seems significant that these shortened bonds preserve the O–H···N form, suggesting that only electronic factors are actually able to produce the aSW → DW transition.

Also, naphthalene-fused derivatives have been recently reviewed (Bertolasi *et al.*, 2009). Since they mostly form DW H-bonds with proton populations modulated by substituents, only high-quality and low-temperature structures can be considered and the most complete series fulfilling these requirements concerns the aryl-substituted 1-(arylazo)-2-naphthols sketched in **3.2.XXVa,b** (Gilli *et al.*, 2002, 2005a, 2006). These compounds (also considered in Section 6.3) can easily form intramolecular H-bonds that are switchable from pure N–H···O

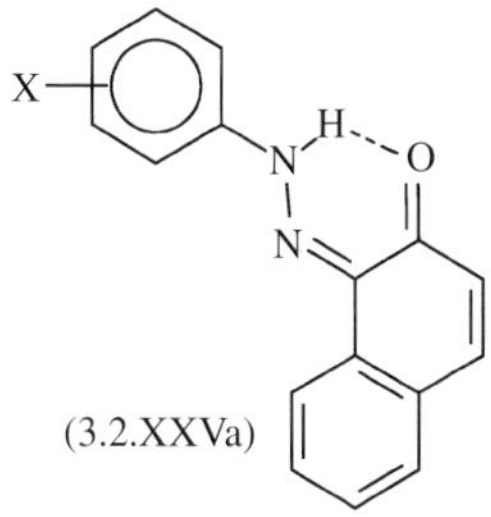

–X	p(N–H)/p(O-H)	d(N···O)	$\sigma^o{}_R$
–H	100/0	2.553 Å	0.00
m-OMe	100/0	2.548 Å	–

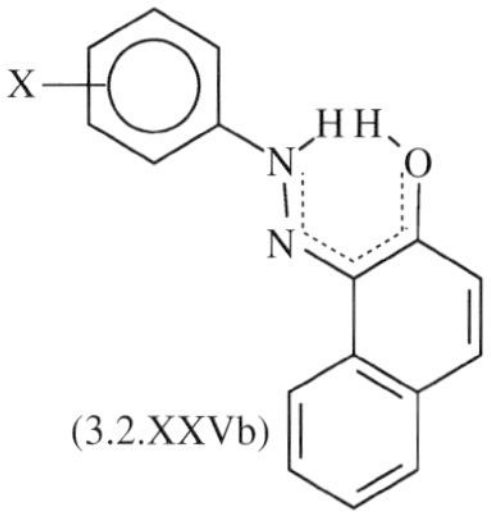

–X	p(N–H)/p(O–H)	d(N···O)	$\sigma^o{}_R$
o-F	77/23	2.546 Å	–
p-Cl	69/31	2.516 Å	–0.29
p-F	64/36	2.535 Å	–0.40
p-NMe$_2$	21/79	2.534 Å	–0.53

to almost pure O–H···N through tautomeric N–H···O $\rightleftharpoons$ O–H···N bonds by tuning the electronic properties of the aryl substituent, X. The modelling quantity turned out to be the mesomeric constant $\sigma°_R$ instead of the usual σ_p and σ_m Hammett constants, a result that might seem logical in view of the resonance-assisted nature of the bond formed. The unsubstituted simple ketohydrazone (X = H) makes a strong N–H···O bond [d(N···O) = 2.553 Å] with a p(N–H)/p(O–H) population ratio of 100/0 and quite similar properties are associated with the m-OMe derivative. However, substituents having decreasing $\sigma°_R$, that is having an increasing π-electron-donating proficiency, can induce the aSW → DW transition by continuously increasing the p(O–H) population, the limit being achieved for the p-NMe$_2$ derivative where the population ratio is 21/79. These data seem to suggest that the p-O$^-$ substituent ($\sigma°_R = -0.60$) could induce pure O–H···N bond formation but, unfortunately, the crystal structure has not been determined.

Figure 3.27 illustrates, as a typical example of these compounds, the molecular structure, the difference–Fourier map in the H-bond region, and the van't Hoff plot of the proton populations derived from least-squares refinement for 1-(p-chlorophenylazo)-2-naphthol, a molecule that is found to form a DW H-bond with p(N–H)/p(O–H) ratio of 69:31 at 100 K that decreases with continuity with the increasing temperature down to 58:42 at 295 K. The calculated $\Delta S°$ of 0.19(17) cal mol^{-1} K^{-1} is not significantly different from zero, as expected for an intramolecular process. The small enthalpy difference between the two minima, $\Delta H°$ = –146(25) cal mol^{-1}, together with the substantial linearity of the van't Hoff plot, seems a clear indication of a fast-exchange equilibrium process of dynamic nature occurring in the solid state. To notice that RAHBs that are disordered in the H-bond proton position are also disordered in the resonant fragment, which then becomes the population-weighted average of two ···HN–N=C–C=O··· and ···N=N–C=C–OH··· tautomeric groups simulating an almost complete

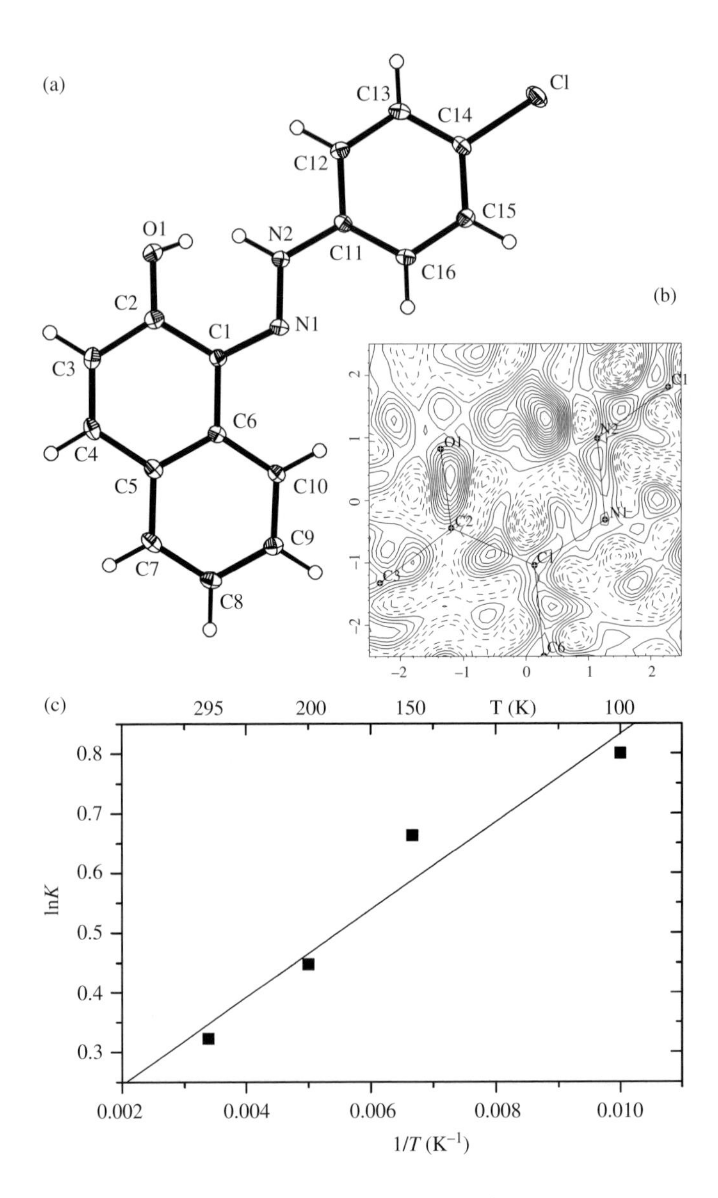

FIG. 3.27. (a) ORTEP (Burnett and Johnson, 1996) view of the molecular structure of 1-(*p*-chlorophenylazo)-2-naphthol as determined at 100 K with thermal ellipsoids at 40% probability; (b) difference-Fourier map in the mean plane of the H-bonded ring at 100 K. The map was computed after least-squares refinement carried out without the H-bonded proton. Positive (continuous) and negative (dashed) contours drawn at 0.06 e/Å^3 intervals. (c) van't Hoff plot $\ln K = \Delta S°/R - \Delta H°/R\ (1/T)$. $K = p/(1-p) = p(\text{N–H})/p(\text{O–H})$ is the ratio of the proton populations as derived from least-squares refinement. $\Delta H° = -0.146(25)$ kcal mol^{-1}, $\Delta S° = 0.19(17)$ cal mol^{-1} K^{-1} ($n = 4$, $r = 0.972$). (Reproduced by permission from Gilli *et al.*, 2005a)

π-delocalization ($\lambda = 0.52$ in the present case). This disorder, at variance with the simultaneous disorder of the proton, is practically unresolvable by diffraction methods and can only be appreciated by quantum-mechanical emulation of the geometries of the two tautomers (see below).

C. *A resonant model for intramolecular N–H···O/O–H···N R_3-RAHBs* In strict analogy with the resonant model developed in Section 3.2.2.3 for O–H···O RAHBs, the behavior of substituted 1-(arylazo)-2-naphthols can be rationalized in terms of simplified VB theory by representing each of the two N–H···O and O–H···N tautomers as a mixture of the four VB resonance forms drawn at the corners of Fig. 3.29(d), two of which are neutral (**a** and **b**) and two ionic (**a**$^\pm$ and **b**$^\pm$), according to the following combinations of VB wavefunctions

$$\Psi(\text{N–H}\cdots\text{O}) = a\Psi(\mathbf{a}) + b\Psi(\mathbf{b}) + b^{\pm}\Psi(\mathbf{b}^{\pm}) \quad (3.31a)$$

$$\Psi(\text{O–H}\cdots\text{N}) = b\Psi(\mathbf{b}) + a\Psi(\mathbf{a}) + a^{\pm}\Psi(\mathbf{a}^{\pm}), \quad (3.31b)$$

a formulation having the great advantage that the coefficients of the canonical forms (a, b, $a^\pm$, and $b^\pm$) can be straightforwardly derived from the crystal geometries, as already discussed for the O–H···O system.

What is different in this example is that the input geometries are not taken from crystal structures but *emulated by DFT methods* at the B3LYP/6–31+G(d,p) level with full optimization of all geometries along the full PT pathway and computed for five test molecules designed in such a way to form H-bonds that change smoothly from N–H···O to O–H···N through tautomeric N–H···O/O–H···N bonds. The PT profiles obtained are shown in Fig. 3.28. In agreement with the empirical considerations above, the profile is strongly asymmetric in favor of the N–H···O form for the non-fused ketohydrazone **1**, becomes increasingly symmetric for the naphtalene-fused derivatives **2**, **3**, and **4**, and finally reverses in favor of the O–H···N form for the phenylene-fused azophenol **5**. Calculations confirm the experimental observation that this particular RAHB system does not form sSW bonds but only DW ones, whose symmetry depends on the choice of the substituents. In this series the most symmetric DW profile is associated with the 1-(p-fluorophenylazo)-2-naphthol **4** having a zero-point-corrected energy difference between the two minima, ΔE_{ZPC}, as low as 0.29 kcal mol^{-1} and a PT barrier of only 0.5–0.6 kcal mol^{-1} (Gilli *et al.*, 2002). These energetic aspects will be discussed in more detail in Section 6.3.

Here, the point of interest is that the geometries of compounds **1**–**5** calculated along the full PT profile can be used to draw a number of *emulated-structure correlations (ESCs)* that are the equivalent of the crystal–structure correlations (CSCs) obtained from crystal data but with the advantage of an unprecedented degree of resolution. Figs. 3.29(a, b) represent the d(N–H) versus d(O–H) and d(N···O) versus $RC = [d(\text{O–H}) - d(\text{N–H})]$ plots. Curves for the five test molecules are seen to have similar shapes and wide overlapping, differing mostly

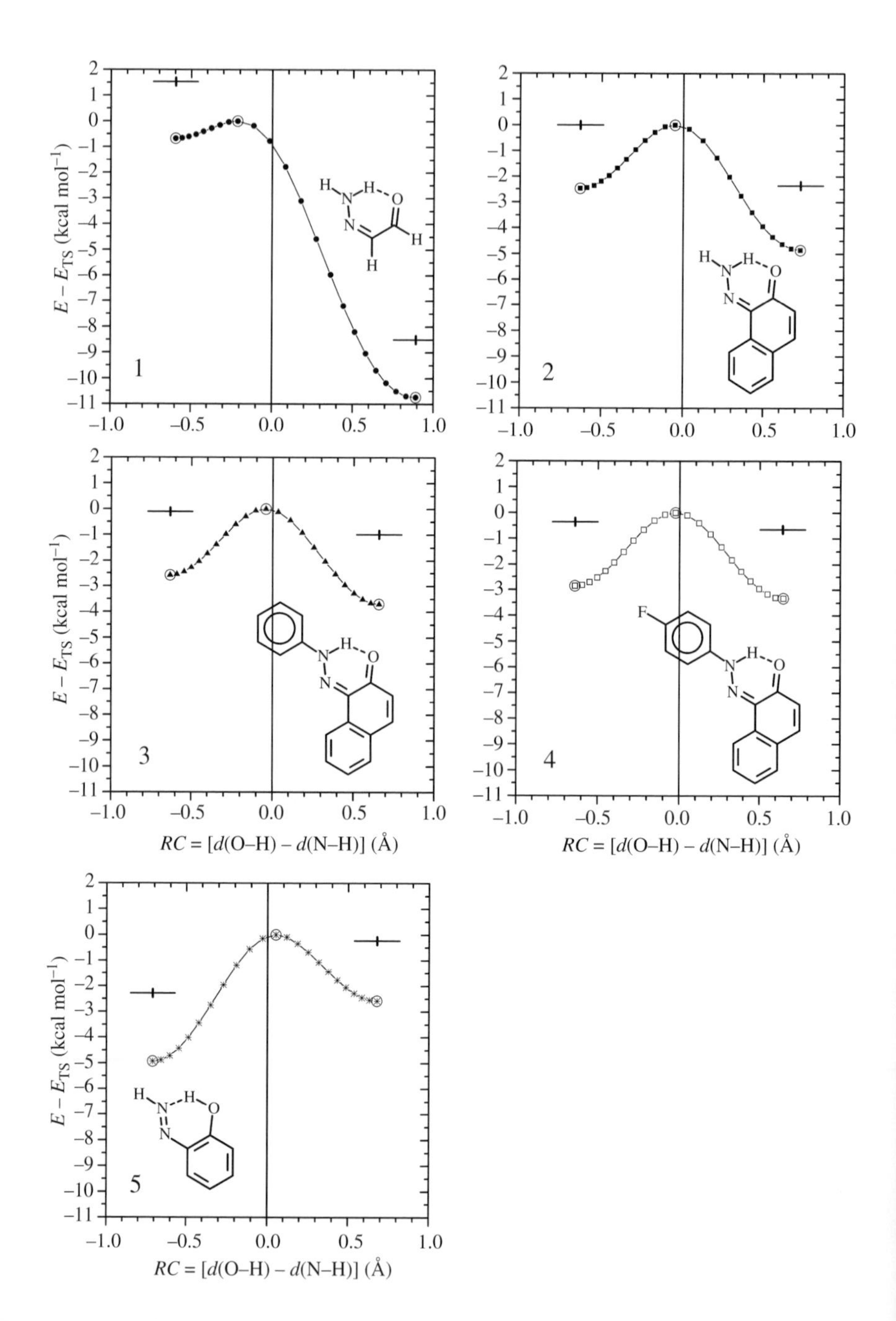

FIG. 3.28. Relative energies, $\Delta E = E - E_{TS}$, in kcal mol^{-1} referred to the TS taken as zero and plotted along the PT reaction coordinate $RC = [d(\text{O–H}) - d(\text{N–H})]$ (Å) as calculated by DFT methods for the test molecules **1–5** indicated. The short horizontal strokes over the minima mark their zero-point-corrected energies, ΔE_{ZPC}. DFT-optimized stationary points are marked by encircled symbols. (Reproduced by permission from Gilli *et al.*, 2002)

for the positions of the ground and transition states, and compare well with the dashed curves that represent the crystal correlations obtained by Steiner (1998a) by using the most accurate neutron data. The correlations $d(\text{N}\cdots\text{O})$ and *RC* versus λ of Figs 3.29(c, d) are the most important for discussing the VB treatment of RAHB. In particular, the plot of *RC* versus λ allows an easy determination of the coefficients a, b, $a^{\pm}$, and $b^{\pm}$ of eqn 3.31 because the couples $\mathbf{a}$–$\mathbf{b}^{\pm}$ and $\mathbf{b}$–$\mathbf{a}^{\pm}$ are related by simple resonance (measured by λ), while the couples $\mathbf{a}$–$\mathbf{a}^{\pm}$ and $\mathbf{b}$–$\mathbf{b}^{\pm}$ are related by pure proton transfer (measured by *RC*) (Gilli *et al.*, 2002; note 21). The mixing coefficients computed for all the stationary points of the plot are listed in Table 3.6.

Data calculated can help to clarify the basic problem of all N–H···O/O–H···N RAHBs, that is their prevalent covalent or ionic character. In recent papers, the N···O shortening occurring in these bonds has been qualitatively assigned to $\mathbf{a} \leftrightarrow \mathbf{b}$ neutral (covalent) mixing (Gilli and Gilli, 2000; Gilli *et al.*, 2000) or $\mathbf{a} \leftrightarrow \mathbf{b}^{\pm}$ (or $\mathbf{b} \leftrightarrow \mathbf{a}^{\pm}$) charge-transfer (ionic) mixing (Krygowski *et al.*, 1997; Filarowski *et al.*, 1998, 1999) of the corresponding VB forms. The data of Table 3.6 allow a more quantitative analysis of the problem from which the following conclusions can be drawn (for brevity, the abbreviations KH = ketohydrazone, AE = azoenol, and NHB = non-H-bonded form are used):

(*i*) The heteronuclear N–H···O/O–H···N bond is dissymmetric in all its manifestations (Figs 3.29(c, d)). The total λ range is significantly shifted towards the AE form and the TS does not occur at $\lambda = 0.5$ (as would be expected for homonuclear RAHBs) but is shifted in the same direction in the restricted range $\Delta\lambda = 0.637 \pm 0.005$ indicated in the figures by the vertical shadowed band;

(*ii*) This dissymmetry originates from the fact that the two tautomers are differently π-delocalized also in their NHB ground states, sketched on top of Fig. 3.29(c) and marked by the two vertical dash-dot lines at $\lambda = 0.205$ and 0.911, respectively;

(*iii*) The mixing coefficients (Table 3.6) confirm the basic lack of symmetry between KH and AE forms. Compound **1**, a non-fused simple KH that can be considered the epitome of a 'normal' N–H···O RAHB with strongly aSW profile (Fig. 3.28(1)), forms a stable N–H···O bond with a strong ionic component, its $a{:}b{:}b^{\pm}$ ratio amounting to 68:5:27. Conversely, the O–H···N bond in compound **5**, the most characteristic phenylene-fused AE, is mostly covalent with a $a{:}b{:}a^{\pm}$ ratio of 15:83:2. Therefore, any controversy about the covalent or ionic nature of the N–H···O/O–H···N RAHB is out of place because, among the N–H···O and O–H···N tautomers, the former is mostly ionic while the latter is mostly covalent in nature. These different characteristics are maintained for all five molecules irrespective of the energy difference between the two minima.

(*iv*) The intrinsic difference of electronic properties of the N–H···O and O–H···N forms could be the reason why *this particular resonant system does not seem able to form symmetric single-minimum H-bonds, but only double-well ones, irrespective of chemical substitution.*

(v) An interesting fact is that the TS geometries are surprisingly similar for all five compounds (Fig. 3.29) in spite of their quite different PT profiles (Fig. 3.28). All TSs practically display the same degree of π-delocalization ($0.630 \leq \lambda \leq 0.645$) and, excluding **1,** which does not form a true DW bond, the remaining four compounds have almost identical proton-centering ($-0.049 \leq RC \leq 0.051$ Å), N···O distances ($2.374 \leq d$(N···O) ≤ 2.386 Å) and the same mixing ratio of VB coefficients, on average a: b: $b^{\pm}$ = 36: 52:12. This constant ratio fits the basic Pauling (1939, 1940, 1960) and

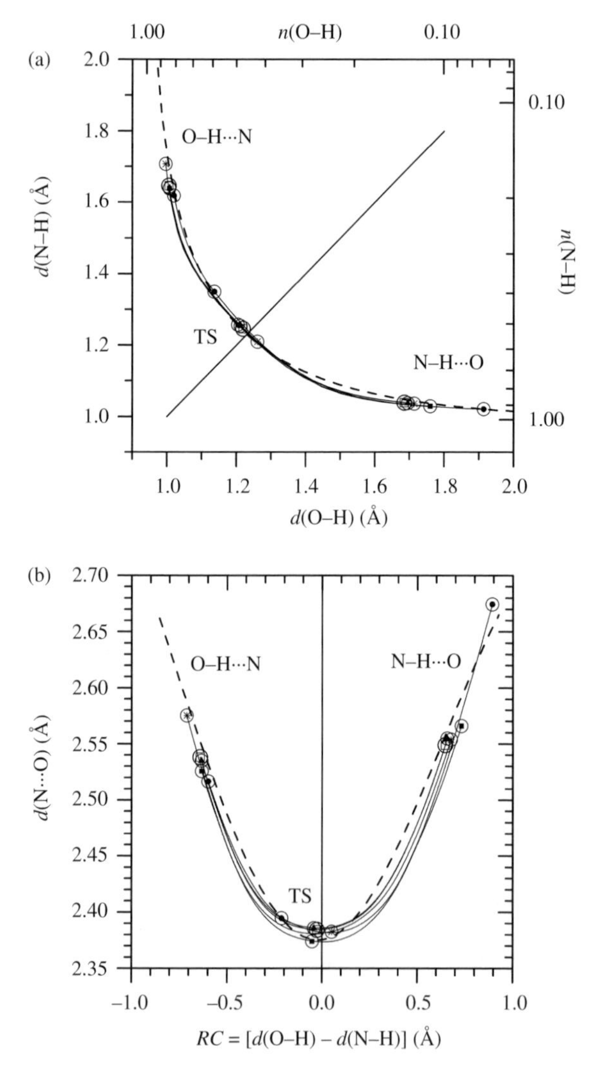

FIG. 3.29. Continues

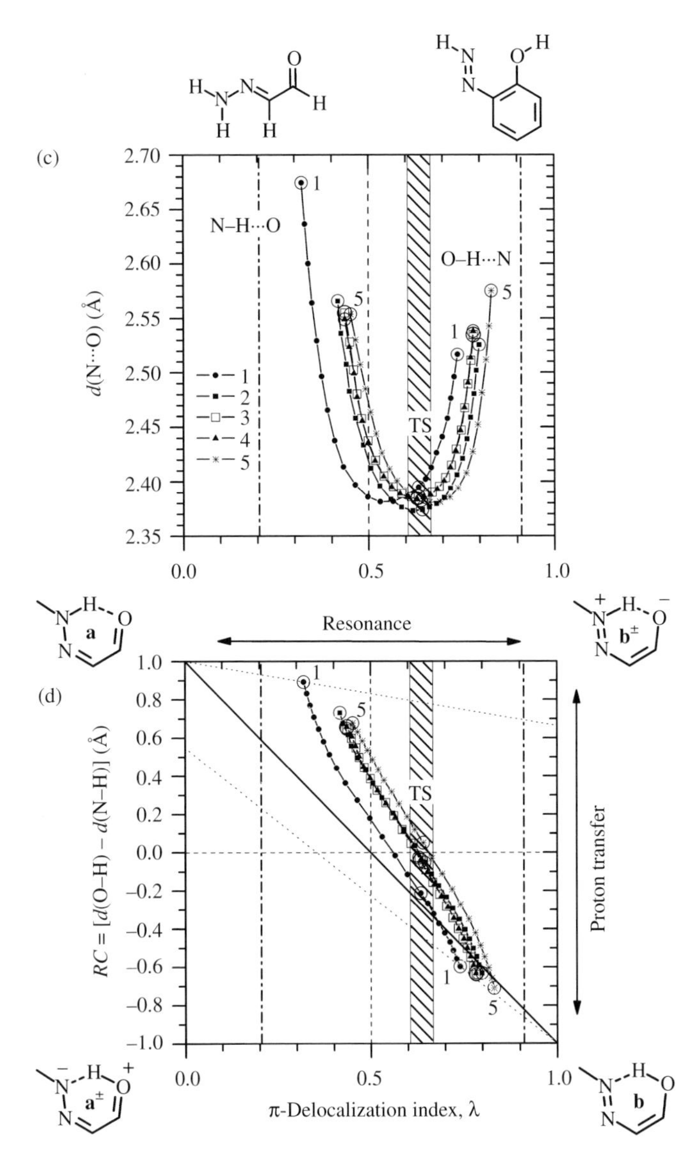

FIG. 3.29. (a) Scatter plot of the N–H against O–H bond distances (Å), and (b) of the N···O contact distances versus the reaction coordinate $RC = [d(\mathrm{O{-}H}) - d(\mathrm{N{-}H})]$ (Å). (c) Scatter plot of the N···O contact distances and (d) of the reaction coordinate $RC = [d(\mathrm{O{-}H}) - d(\mathrm{N{-}H})]$ (Å) against the π-delocalization index λ. DFT-optimized stationary points are marked by encircled symbols, and dashed curves in (a) and (b) map the structure correlations derived from neutron crystallography (Steiner, 1998a). (Reproduced by permission from Gilli *et al.*, 2002)

TABLE 3.6. Mixing coefficients (×100) in eqn 3.31 for the four canonical forms **a**, **b**, $\mathbf{a}^{\pm}$ and $\mathbf{b}^{\pm}$ displayed at the corners of Fig. 3.29(d). NHB-KH and NHB-AE are, respectively, the non-H-bonded ketohydrazone (KH) and azoenol (AE) forms shown on the top of Fig. 3.29(c).

Test molecule	a (×100)	b (×100)	$a^{\pm}$ (×100)	$b^{\pm}$ (×100)
NHB-KH	79.5	0.0		20.5
1 N–H···O	68.1	5.4		26.6
2 N–H···O	58.3	13.4		28.3
3 N–H···O	56.4	17.1		26.4
4 N–H···O	56.4	17.7		25.9
5 N–H···O	54.8	16.1		29.1
1 TS	36.5	60.6		2.9
2 TS	35.5	52.6		11.9
3 TS	36.6	52.1		11.2
4 TS	37.0	51.3		11.7
5 TS	35.9	47.4		16.7
1 O–H···N	20.1	74.0	5.9	
2 O–H···N	18.5	79.9	1.7	
3 O–H···N	18.4	78.3	3.3	
4 O–H···N	18.0	78.3	3.7	
5 O–H···N	14.5	83.2	2.4	
NHB-AE	0.0	91.1	8.9	

Sanderson (1967) idea of a heteronuclear bond where oxygen and nitrogen electronegativities must be levelled out by the $^{+}$N···H···O^{-} charge transfer before the two electron-neutral forms so produced can mix with a (36 + 12): 52 = 48: 52 ≅ 1:1 coefficient ratio.

(*vi*) It may be concluded that *the TS is always characterized by the shortest donor–acceptor distance, by the greatest proton symmetry, and by the largest possible degree of covalency* and that these properties are weakly affected by chemical substitution but appear to be intrinsic to the H-bonded system considered.

D. *Appendix. AIM analysis of intramolecular N–H···O/O–H···N R_3-RAHBs*
R_3-RAHB is the only H-bond where a complete Coulson VB analysis of covalent and ionic contributions (eqn 3.31) is actually practicable. This is due to the somewhat fortuitous circumstance that, in this particular type of bond, H-bond strength and π-delocalization are interlinked by the RAHB mechanism, so allowing direct appreciation of the mixing coefficients from the geometry of the resonant fragment. Other more general methods of H-bond analysis are available that are based either on the dissection of electron densities or on the partitioning of bonding energies. In the following, the AIM method (Bader, 1990) of topological

analysis is applied to the DFT electron densities computed for compounds **1**–**5** of Fig. 3.28 to show that its results are in substantial agreement with those derived from the VB analysis. As for AIM methods, it is sufficient to recall here that, in this approach, the H-bond does not substantially differ from other types of chemical bonds and is to be classified as mostly covalent (shared interaction) or electrostatic (closed-shell interaction) according to whether the Laplacian, $\nabla^2\rho(cp)$, at the H-bond (3,–1) critical point (cp) is negative or positive, while the corresponding electron density, $\rho(cp)$, assumes values more or less compatible with those of ordinary covalent bonds.

Though covalency is not expected to dominate heteronuclear N–H···O/O–H···N RAHBs in view of their prevalent DW character, the AIM method can still be used to assess the bonding variations occurring along the PT pathway by plotting the changes undergone by $\rho(cp)$ and $\nabla^2\rho(cp)$ along the reaction co-ordinate, RC (Fig. 3.30). Notice that only the values for the stationary points have been actually calculated, while the continuous curves have the pure graphical function of connecting the computed values at $cp1$ (the cp between N and H) and $cp2$ (the cp between O and H) for the three stationary points that are, from right to left, the N–H···O, TS, and O–H···N states. As expected, the cps associated with covalent O–H and N–H bonds have large $\rho(cp)$ (from 2.05 to 2.25 $e\text{Å}^{-3}$) and large and negative $\nabla^2\rho(cp)$ values (from –39.0 to –45.0 $e\text{Å}^{-5}$). Conversely, the cps associated with the N···H and O···H contacts are generally characterized by low $\rho(cp)$ values of 0.21–0.44 $e\text{Å}^{-3}$ and slightly positive $\nabla^2\rho(cp)$ values of 2.3–3.8 $e\text{Å}^{-5}$, results that could be considered indicative of weak interactions with predominant ionic (closed-shell) character.

These strong differences between 'bonds' and 'contacts' vanish when considering the properties of the N···H···O TS for which both electron densities and Laplacians are seen to cluster around the same point with average coordinates $\rho(cp) = 1.18(2)$ $e\text{Å}^{-3}$ and $\nabla^2\rho(cp) = -10(1)$ $e\text{Å}^{-5}$ (excluding compound **1**). Two points seem particularly important: (*i*) the value of $\rho(cp)$ of the TS is nearly one half of that of a normal N–H or O–H covalent bond, confirming that the H-bond formed is really a true three-center–four-electron covalent bond consisting of two bonds having bond numbers of one half; and (*ii*) the value of $\nabla^2\rho(cp)$ is negative, confirming that the TS has to be classified as a shared interaction, that is, again, as a mostly covalent bond.

E. *Systematics of intermolecular N–H···O RAHBs* Intermolecular N–H···O bonds are ubiquitous in biological systems, where they exert essential functions of control in many processes such as DNA base pairing, secondary structure of proteins, enzymatic catalysis, and ligand–receptor interactions (Jeffrey and Saenger, 1991; Jeffrey, 1997; Klebe, 1994), and their peculiar skills as builders of natural systems have prompted a large number of studies as potential controllers of molecular recognition processes in solids and in solution or in the building of new crystalline materials having specific properties (Rebek, 1990; Hamilton, 1990; Schneider, 1991; MacDonald and Whitesides, 1994). Moreover,

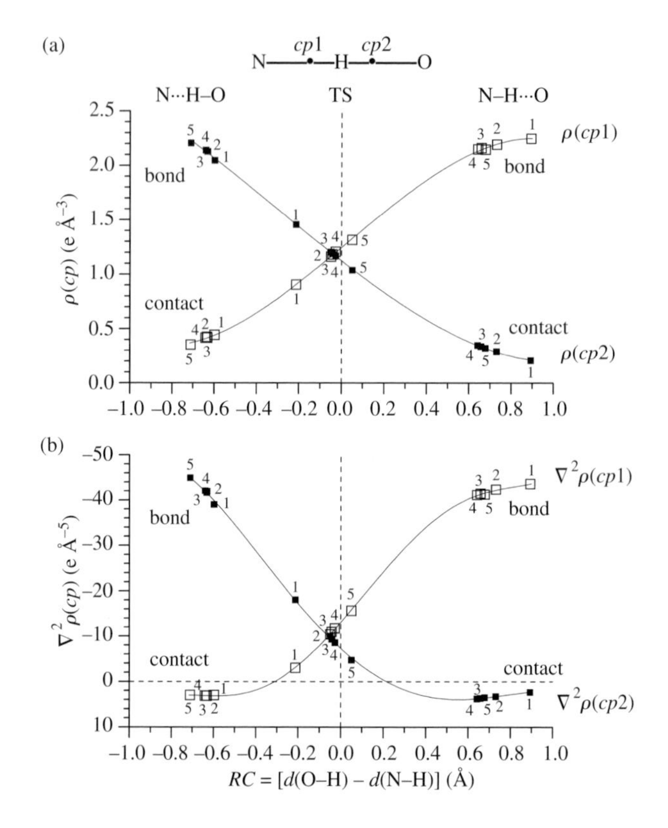

FIG. 3.30. H-bond critical-point parameters. (a) electron density $\rho(cp)$ (eÅ^{-3}) and (b) Laplacian $\nabla^2\rho(cp)$ (eÅ^{-5}) plotted against the reaction coordinate RC = [d(O–H) − d(N–H)] (Å) as calculated by AIM analysis (Bader, 1990) of DFT-optimized stationary points for the test molecules **1**–**5**. Interpolating continuous lines have purely graphical purposes. (Reproduced by permission from Gilli *et al.*, 2002)

neutral intermolecular N–H···O bonds are a common driving force in the aggregation of molecular crystals and, for this reason, have been the subject of several investigations (Etter, 1982, 1990; Taylor and Kennard, 1984; Berkovitch-Yellin *et al.*, 1983).

Including RAHB, the overall interval of N···O distances associated with neutral N–H···O bonds is quite large (typically 2.63–3.30 Å) and not easy to rationalize for a number of reasons that are intrinsic to the bond itself: (*i*) the H-bonds formed are weakened by the different proton affinities of the O and N atoms; (*ii*) amides, as well as enaminones, are π-delocalized by themselves before H-bond formation, so that C–N shortening and C=O lengthening effects are smaller and more difficult to detect; (*iii*) most of the strongest H-bonds are formed by unsaturated heterocycles with multiple π-conjugated pathways compatible with R_1, R_3 and R_5 as well as (R_1+R_5) or (2 × R_3) resonance-assisted bonds; (*iv*) last but not least, steric effects may be relevant because of the repulsions exerted by bulky *N*-substituents.

A first systematic analysis of intermolecular N–H···O RAHBs was based on the structures of 39 molecules, selected from the CSD for forming bonds with various lengths of the resonant spacer, R_n (Bertolasi *et al.*, 1995). This study led to the identification of the surprising variety of intermolecular binding patterns shown in Fig. 3.31 and Schemes **3.2.XXVI–XXVIII** and whose structural parameters are listed in Table 3.7. Amides are found to form R_1-chains (*a*), R_1-rings (*b*), and more complex patterns (*c* and *d*) with N···O distances in the range 2.81–3.02 Å that, though not particularly short, is on the short side of the interval of 2.90–3.30 Å determined on a sample of 42 non-resonant $(Csp^3)_2N-H\cdots O=C(Csp^3)_2$ bonds. It has been often reported (Jeffrey, 1984, 1985; Cheeseman *et al.*, 1988; Popelier *et al.*, 1989; Asakawa *et al.*, 1992) that N–H···O bond formation is associated with the setting up of partial charges, negative on the oxygen and positive on the NH group, and an enhanced π-delocalization of the O=C–N fragment with lengthening of the C=O and shortening of the C–N distance. These bond-distance effects are not fully verified by more extended analyses of crystal data because they are too small and at the borderline of crystal-structure accuracy. In fact, C–N and C=O distances are, on average, 1.346(11) and 1.229(9) Å in tertiary amides (where the H-bond is forbidden) and 1.329(10) and 1.236(9) Å in secondary and primary amides (where the H-bond is always observed) (Allen *et al.*, 1987). It can be concluded that amides are barely susceptible to RAHB shortening because the O=C–NH system is too delocalized by itself to become further π-polarized.

Factors able to increase π-polarizability can strengthen the H-bond and this may occur when multiple conjugated systems overlap. This effect is well illustrated by compounds (*e*) and (*f*) that can be classified as (R_1+R_5)-chains and (R_1+R_5)-rings and whose N···O distances are seen to decrease to 2.761 and 2.782 Å, respectively. Even more efficient are R_3-RAHBs formed by the ···O=C–C=C–NH··· β-enaminone fragment that may occur in the form of simple R_3-chains [pyrazoloquinolinone (*g*); N···O = 2.694 Å], 2 × (R_3-chains) [pyridinone (*h*); N···O = 2.753 Å], and more complex 3 × (R_3-chains) [6-methylcytosine (*i*); N···O = 2.716 Å]. Finally, of smaller but comparable efficiency are R_5-RAHBs that are associable with the longer ···O=C–C=C–C=C–NH··· δ-dienaminone fragment. Two typical examples are the R_5-chain of the 2-formylpyrazole (*j*) and the R_5-ring of 3-methylxantine (*k*), having respective N···O distances of 2.856 and 2.734 Å.

The metric aspects of the H-bonds considered are summarized in the *d*(N···O) versus *d*(C=O) scatter plot and in the bar chart of the N···O distances of Figs. 3.32(a) and (b). The former is intended to establish a correlation between N···O shortening and the lengthening of the C=O bond taken as a simplified indicator of π-delocalization in the resonant ···O=R_n–NH··· fragment, the latter to assess

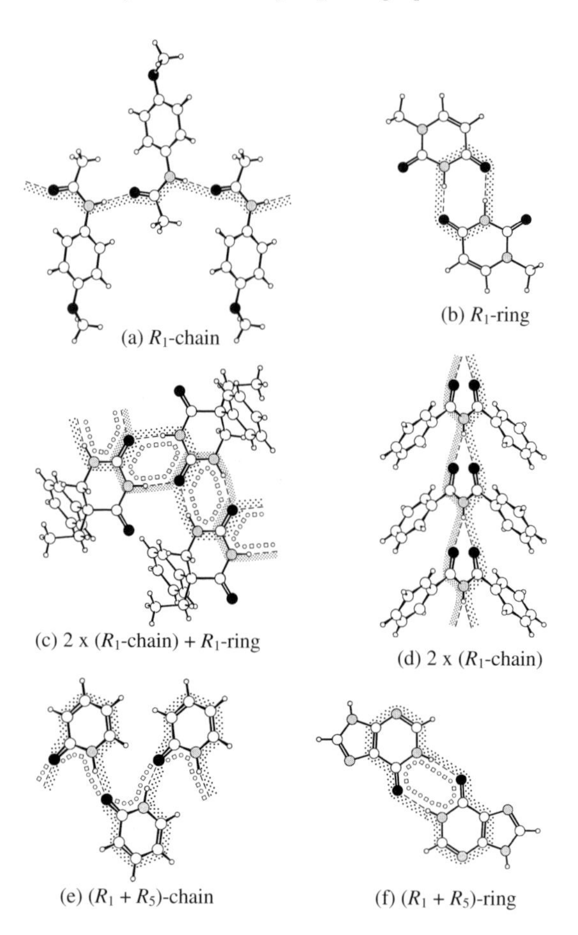

FIG. 3.31 Continues

the importance of the fragment length in RAHB strengthening. The order found is $R_1 \ll R_3 \cong R_5$, essentially the same order established for the O–H···O case (Fig. 3.21).

Substituent effects in intermolecular RAHBs are not easy to understand. Most probably, the rule that electron-attracting groups able to enhance the N–H acidity can also strengthen the H-bond is still valid but supported by very few data, such as the two *N*-substituted amides **3.2.XXVIa,b** whose N···O distances are, in fact, decreased to 2.76–2.78 Å for the effect of the two electron-attracting *N*-substituents. Also, simple *β*-enaminones **3.2.XXVII** display rather long N···O distances of 2.86±0.05 Å that are shortened down to 2.70–2.75 Å only in particular situations, such as cumulation of two, or more, resonance pathways in **3.2.XXVIIIa–c** or fusion with a pyrazolone ring in **3.2.XXVIIId,e** (Bertolasi *et al.*, 1995, 1998).

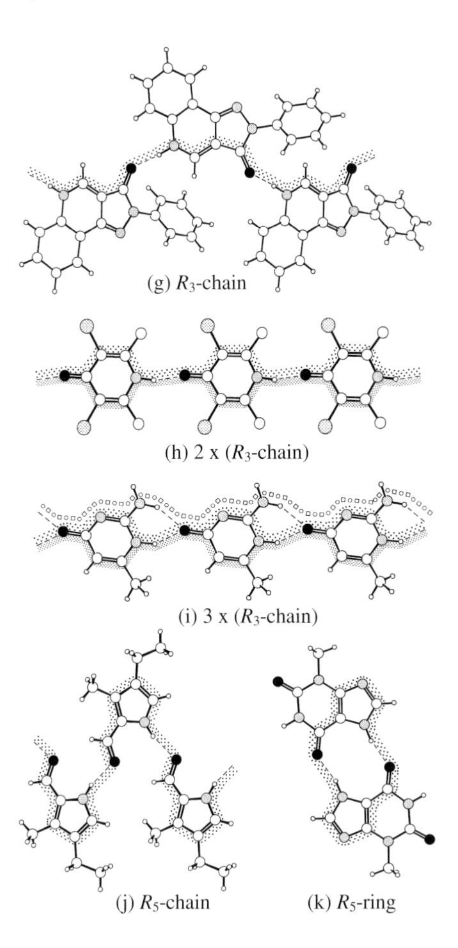

FIG. 3.31. The ten interaction patterns identified in a sample of 39 intermolecular N–H···O RAHBs. Resonant chains and rings are marked by darker or lighter shading or by lines of small open squares. References: (a) ACACTA, Haisa *et al.*, 1980; (b) METURA01, McMullan and Craven, 1989; (c) TAJXAD, Mastropaolo *et al.*, 1991; (d) CIHCUR01, Bertolasi *et al.*, 1995; (e) PYRIDO04, Ohms *et al.*, 1984; (f) GEBTUC, Schmalle *et al.*, 1988; (g) COVLII, Ferretti *et al.*, 1985; (h) MCHYPY10, Boer, 1972; (i) FETSEC, Lowe *et al.*, 1987; (j) DASROE, Smith *et al.*, 1985; (k) FADCUI, Low *et al.*, 1986. (Reproduced by permission from Bertolasi *et al.*, 1995)

d(N···O) = 2.762 Å
d(C=O) = 1.238 Å
(3.2.XXVIa) 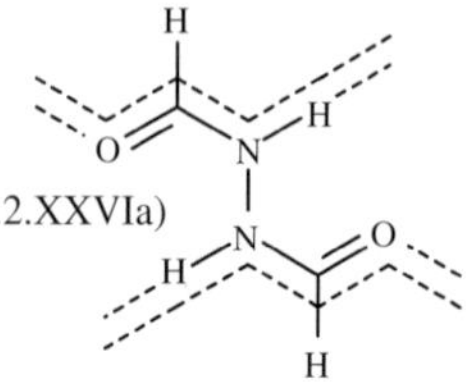
N,N'-Diformo hydrazide

d(N···O) = 2.784 Å
d(C=O) = 1.257 Å
(3.2.XXVIb) 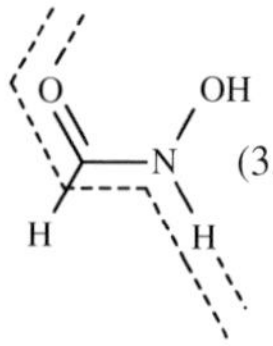
Formohydroxamic acid

d(N···O) = 2.86$_5$ Å
d(C=O) = 1.24$_1$ Å 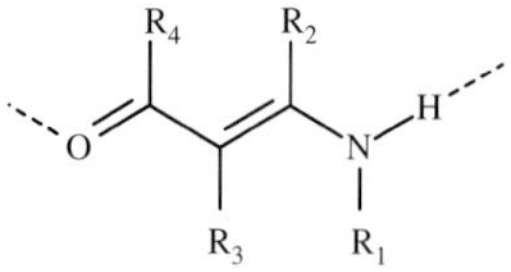
Simple enaminone (3.2.XXVII)

d(N···O) = 2.75$_4$ Å
d(C=O) = 1.27$_2$ Å
(3.2.XXVIIIa) 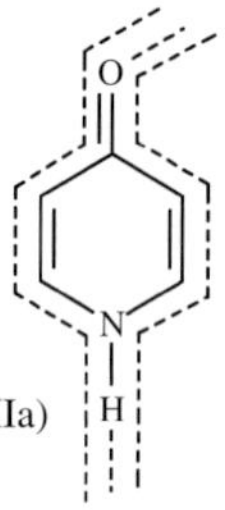
1H-Pyridin-4-one

d(N···O) = 2.71$_4$ Å
d(C=O) = 1.27$_2$ Å
(3.2.XXVIIIb) 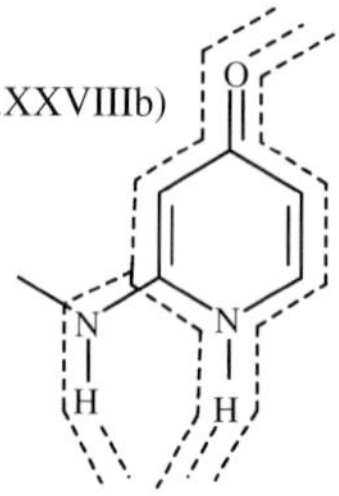
2-Amino-1H-pyridin-4-one

d(N···O) = 2.716 Å
d(C=O) = 1.256 Å
(3.2.XXVIIIc) 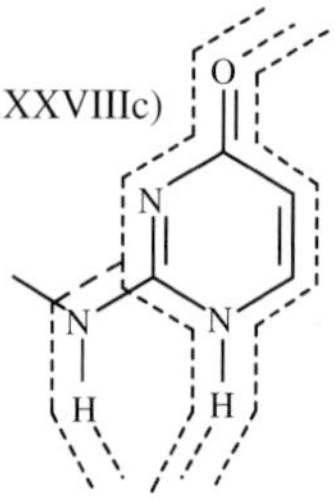
2-Amino-1H-pyrimidin-4-one

d(N···O) = 2.70$_6$ Å
d(C=O) = 1.26$_2$ Å 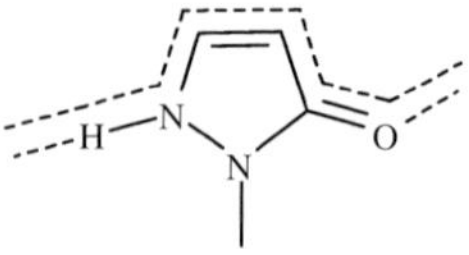
1,2-Dihydro-pyrazol-3-one (3.2.XXVIIId)

d(N···O) = 2.73$_3$ Å
d(C=O) = 1.25$_1$ Å 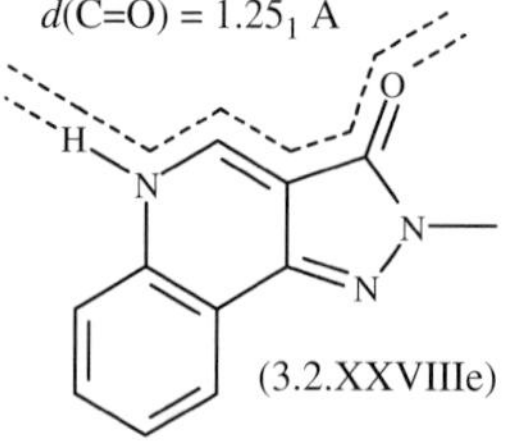
(3.2.XXVIIIe)
2,5-Dihydro-pyrazolo-[4,3]quinolin-3-one

Cooperative effects have been studied by Kobko and Dannenberg (2003) at the B3LYP/D95** level of theory in the H-bonded formamide chain of C_S symmetry **3.2.XXIX** containing from 2 to 15 monomeric units. Calculations show that the H-bond becomes stronger as the chain becomes longer and that bonds at the ends of the chain are weaker than those in its interior. Table 3.8 compares energies and geometries for the two extremes of this series, that is the formamide dimer, on the one hand, and the formamide unit at the center of a linear chain of 15 monomers, on the other. Cooperative effects are seen to increase the H-bond

TABLE 3.7. Structural data for the compounds forming intermolecular N–H···O bonds depicted in Fig. 3.31 (distances in Å and averaged values in brackets)

	Refcode	d(C=O)	d(N···O)	Type
(*a*)	ACACTA	1.228	2.905	R_1-chain
(*b*)	METURA01	1.241	2.810	R_1-ring
(*c*)	TAJXAD	<1.248>	<2.820>	$2\times(R_1$-chain$)+R_1$-ring
(*d*)	CIHCUR01	1.219	3.025	$2\times(R_1$-chain$)$
(*e*)	PYRIDO04	1.251	2.761	(R_1+R_5)-chain
(*f*)	GEBTUC	<1.246>	<2.782>	(R_1+R_5)-ring
(*g*)	COVLII	1.252	2.694	R_3-chain
(*h*)	MCHYPY10	1.252	2.753	$2\times(R_3$-chain$)$
(*i*)	FETSEC	1.256	2.716	$3\times(R_3$-chain$)$
(*j*)	DASROE	1.240	2.856	R_5-chain
(*k*)	FADCUI	1.227	2.734	R_5-ring

energy, E_{HB}, from 4.49 in the dimer to 12.99 kcal mol^{-1} in the middle of the chain. This large energy enhancement is associated with a decrease of 7.3% of the NH···O distance but with a much smaller C=O lengthening and C–N shortening of only 1.1–1.2%. Such small geometry variations raise the problem of whether the ···O=C–NH··· ↔ ···HO–C=N··· resonance is still the main determinant of H-bond strengthening, as it was in O–H···O RAHBs, or whether the observed effects are due to mutual electron-polarization. The authors maintain that the effect is not purely electrostatic but most probably due to a combination, in unknown ratios, of polarization, mutual polarization, and covalent (or RAHB) interactions. There are, in fact, several arguments supporting the role of resonance in intermolecular N–H···O bond formation: (*i*) variations of bond distances appear to be strictly coordinated, the d(NH···O) decrease being parallelled by the d(C=O) increase and d(C–N) decrease; (*ii*) the small C=O and C–N variations of Table 3.8 actually cause a consistent increase of π-delocalization within the O=C–N moiety, λ going from 0.23 to 0.34 with an increase of 48%; (*iii*) small geometrical changes can actually produce large energy variations because they are related as $E_{\mathrm{HB}} \propto n \propto \exp(-\Delta d)$, where n is the usual Pauling bond number and d is the d(NH···O) distance (Section 3.1.3.1); (*iv*) the dipole moment *per* residue increases from $\mu = 4.6$ Debye for the dimer to the asymptotic value of 5.9 Debye for the 15-mer, in agreement with previous observations that the dipole moment *per* amino acid tends to increase with the length of α-helices (Applequist and Mahr, 1966); this increase seems too large to be produced by simple polarization without a considerable charge-transfer contribution due to resonance.

A summary of the experimental data considered in this paragraph is shown in Fig. 3.33. Compounds are divided into five groups: **1.** tertiary amides (not forming H-bond); **2.** secondary and primary amides; **3.** shortened secondary

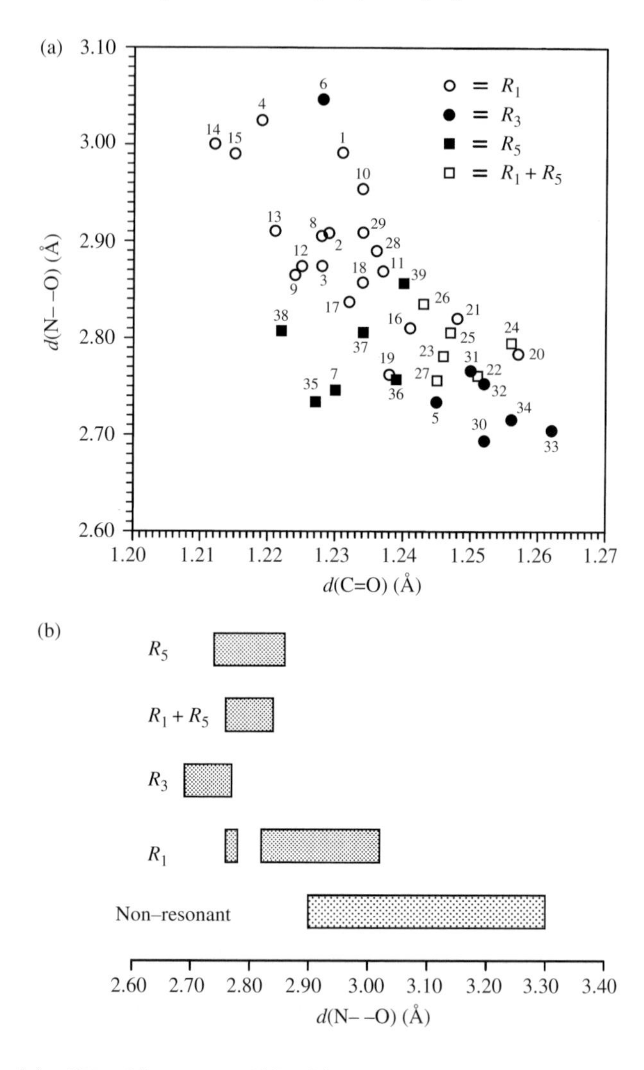

FIG. 3.32. (a) d(N···O) versus d(C=O) scatter plot for the 39 intermolecular N–H···O RAHBs discussed in the text, classified according to the length of the resonant spacer R_n. (b) Bar chart of the N···O distances in the N–H···O bonds formed by the same set of 39 structures. The small interval (2.76–2.78 Å) on the R_1 line refers to the two amides (**3.2.XXVIa,b**) whose nitrogen is linked to a hereoatom. (Reproduced by permission from Bertolasi *et al.*, 1995)

amides (**3.2.XXVIa,b**); **4.** simple enaminones (**3.2.XXVII**); and **5.** shortened enaminones (**3.2.XXVIIId,e**). The lower part of the figure reports the bar chart of the C–O and C–N bond distances together with d(N···O) distances and average π-delocalization indices, λ and Del%, of the O=C–N resonant fragment,

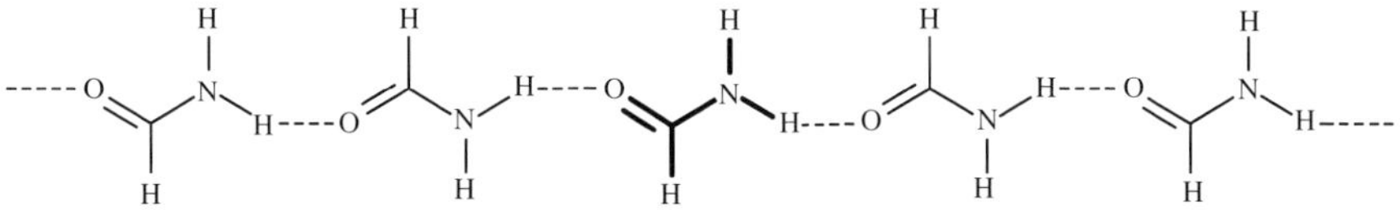

(3.2.XXIX)

TABLE 3.8. DFT emulation of the properties of H-bonded formamide chains (**3.2.XXIX**) containing from 2 to 15 monomer units at the B3LYP/D95** level of theory (Kobko and Dannenberg, 2003). The comparison is between the H-bond of the formamide dimer and the two H-bonds formed by the formamide at the center of a linear chain of 15 monomers. $\langle\mu\rangle$ is the mean dipole moment *per* formamide monomer within the chain.

	E_{HB} kcal/mol	d(NH···O) Å	d(C=O) Å	d(C−N) Å	λ	$\langle\mu\rangle$ Debye
Dimer	4.49	1.92	1.225	1.360	0.23	4.6
Center of 15	12.99	1.78	1.239	1.344	0.34	5.9
Δ	8.5	−0.14	0.014	−0.016	0.11	1.3
Δ%	189	−7.3	1.1	−1.2	48	28

while the upper part shows how C–O and C–N distances are related to the Pauling bond numbers, n, and to λ. As already remarked, amides are strongly delocalized even without forming any H-bond as, for instance, tertiary amides **1**, which show average delocalization of 58%. Formation of H-bonded amide (**2**) or enaminone (**4**) chains can only produce a slight increment of this delocalization to 62–64%. Only the presence of chemical substituents able to enhance the N–H acidity (**3** and **5**) can eventually raise the resonance mixing up to a π-bond delocalization of 90% causing, at the same time, a significant shortening of the H-bond formed. Hence, H-bond strengthening and increased π-delocalization of the O=C−N resonant fragment are correlated, as (negatively) intercorrelated are also the variations of C–O and C–N distances that make this delocalization possible. It seems inevitable to conclude that RAHB must be an important (though not necessarily the only) contributor to the cooperativity-induced strengthening of the intermolecular N–H···O bond.

3.3 Completing the H-bond classification: The chemical leitmotifs

3.3.1 *A full H-bond classification from the systematic analysis of the O–H···O system*

3.3.1.1 *A full CSD analysis of the O–H···O system*

A. *Introduction* The discovery of RAHB (Gilli *et al.*, 1989; Bertolasi *et al.*, 1991) and the finding that its strength can be easily rationalized in terms of Coulson's

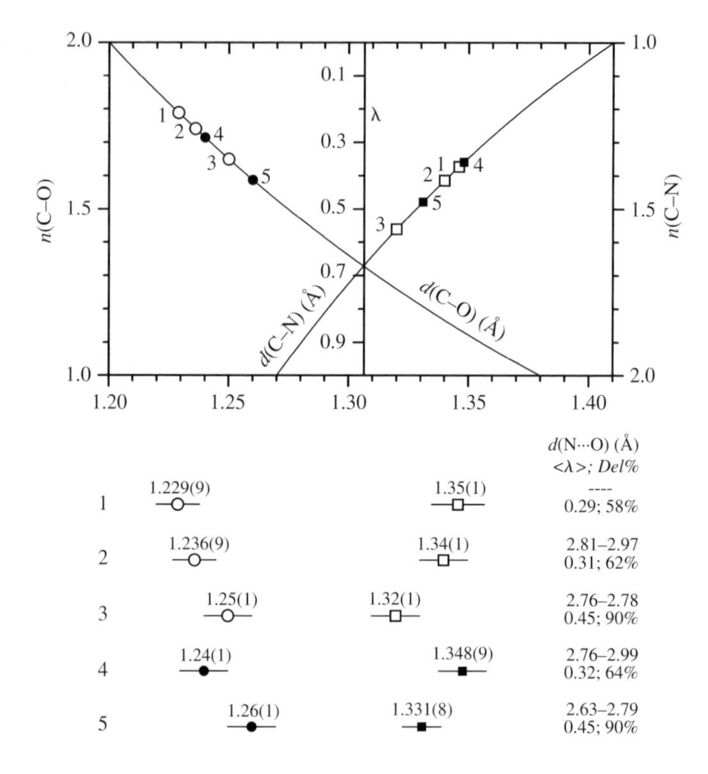

FIG. 3.33. (Upper part) Curves of the C–O and C–N distances of the O=C–N fragment versus the corresponding bond numbers, n, calculated by the Pauling eqn 3.2. (Lower part) Bar chart of the C–O and C–N bond distances, d(N···O) contact distances and values of the average π-delocalization parameters, λ and Del%, of the O=C–N resonant fragment for all classes of compounds considered in this section. **1.** Tertiary amides (without H-bond); **2.** Secondary and primary amides; **3.** Shortened secondary amides **3.2.XXVIa,b**; **4.** Simple enaminones **3.2.XXVII**; **5.** Shortened enaminones **3.2.XXVIIId,e**

formalism (Coulson, 1959; Coulson and Danielsson, 1954a,b) as a VB mixing of the two D–H···:A $\leftrightarrow$ $^-$D···H–A$^+$ resonance forms aroused a renewed interest for the role played by covalency in strong H-bonds after some thirty years of uncontested predominance of the electrostatic model. Since the debate between the supporters of the two models was at a deadlock, in 1993 we decided to reinvestigate the problem starting from a new systematic analysis of the O–H···O bond (the bond having the maximum amount of data) and adopting a purely empirical strategy: (*i*) suspend any previous ideas on its electrostatic or covalent nature; (*ii*) define the O–H···O bond as a simple topological structure where a hydrogen is connected to two, or more, oxygen atoms; (*iii*) collect and analyze all crystal structures having O–H···O bonds with d(O···O) $\leq$ 2.70 Å; (*iv*) collect all available IR ν(O–H) stretching frequencies, NMR δ(OH) chemical shifts, and gas-phase enthalpies $\Delta H°_{DIS}$ of the H-bond dissociation equilibrium; and (*v*) try to infer a conclusion on the very nature of the O–H···O bond from the ensemble of the data collected.

The project was carried out in successive steps (Gilli *et al.*, 1994a, 1996a,b; Gilli and Gilli, 2000; Gilli *et al.*, 2004), making it eventually possible to classify all conventional O–H···O bonds in the *six chemical leitmotifs (CLs)* summarized in Table 2.2 and Schemes **2.1.I–VI** above (Section 2.4.1) and to show that they can be interpreted only by adopting a twofold (electrostatic-covalent) H-bond model. There are only four possible classes of strong and covalent O–H···O bonds, called here the *four strong CLs*. One is assisted by resonance (RAHB = resonance-assisted H-bond) and has already been treated above. The other three are assisted by charge (CAHBs = charge-assisted H-bonds) and will be the subject of the present section.

All these strong bonds make reference to a well-defined chemical situation and, for this reason, were often considered in the past to be 3c–4e covalent interactions (i.e. true covalent bonds) different in nature from the true H-bonds that, according to a well-established tradition, were believed to be weak $D^{\delta-}–H^{\delta+}···A^{\delta-}$ electrostatic interactions with bonding energies not exceeding 4–6 kcal mol^{-1}. The latter, which constitute the great majority of the H-bonds ever observed, are called now *OHBs* or *ordinary H-bonds* and are intended to collect, without any internal subdivision, all bonds not belonging to one of the four strong CLs. Only a subclass of OHBs, called *polarization-assisted H-bonds (PAHBs)* or *σ-bond cooperative H-bonds* (Jeffrey and Saenger, 1991; Gilli and Gilli, 2000), has been differentiated in view of its theoretical and practical importance in determining proton transmission along cooperative chains of water molecules.

In analogy with what was previously done for RAHBs, in this section the geometrical features of CAHBs will be investigated by searching the CSD (version 2007) for all H-bonds with O···O distances $\leq$ 2.500 Å. For simplicity, only nearly linear bonds are considered (O–H–O angle $\geq$ 160°) and particular attention has been given to the exclusion of H-bonds that are possibly perturbed by other H-bonds or short contacts, particularly with ions. Fig. 3.34 reports the histograms of the $D = d$(O···O) distances and Charts 3.3–3.5 summarize the most typical chemical situations associated with the three classes of CAHBs considered.

B. *[O···H···O]$^-$. Negative charge-assisted H-bond or (–)CAHB* All (–)CAHBs have the form $[R_1–O···H···O–R_2]^-$ and are therefore *H-bonds formed by R–O–H acids with their own conjugated bases.* They belong to two different subclasses according to whether they are homomolecular ($R_1 = R_2$) or heteromolecular ($R_1 \neq R_2$) and, for the reasons to be discussed in the next section, the former are systematically stronger (shorter) than the latter that, accordingly, have been excluded from the CSD search. The histogram of the $D = d$(O···O) distances of Fig. 3.34(a) shows a peaked distribution with distances D(min) = 2.402 and

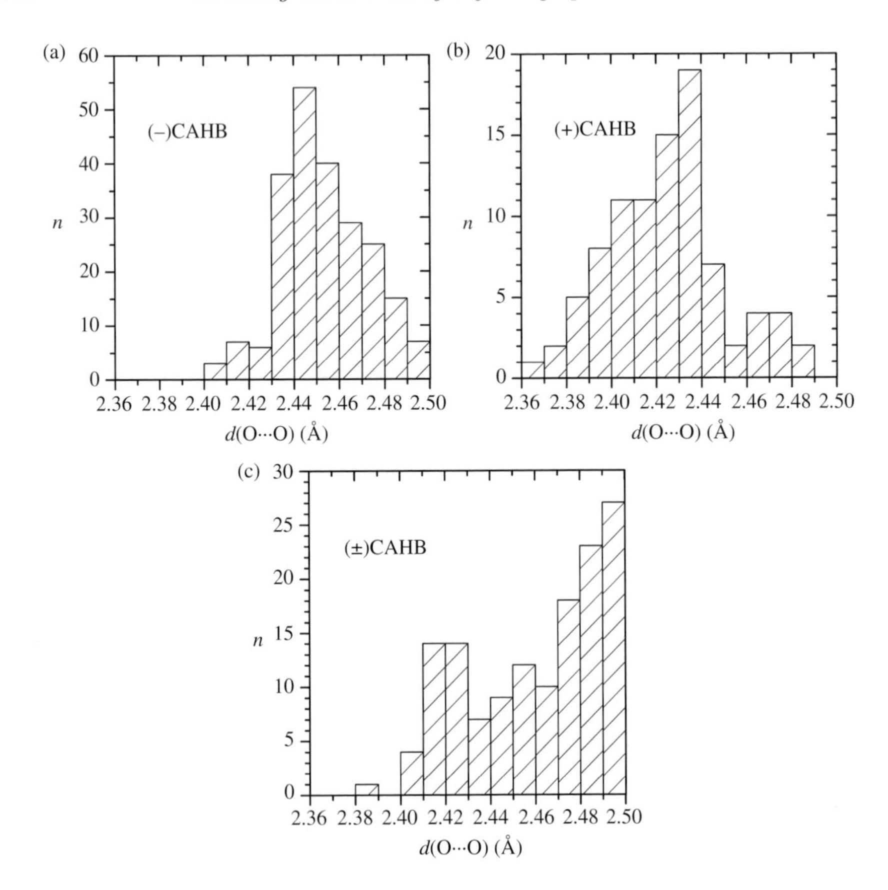

FIG. 3.34. Histograms of the $D = d(\text{O}\cdots\text{O})$ distances derived from the CSD analysis for the three CAHB classes considered: (a) homomolecular (−)CAHB; (b) homomolecular (+)CAHB; and (c) (±)CAHB. (Data from Gilli *et al.*, 2009).

$D(\text{mean}) = 2.45(2)$ Å ($n = 225$) that represent an impressive shortening with respect to the shortest distances in ordinary O−H···O bonds (some 2.70 Å). To notice, moreover, that most of these strong bonds are more or less perfectly proton centered.

The most characteristic chemical situations able to generate homomolecular (−)CAHBs are displayed in Chart 3.3. Only a few cases of intramolecular bonds are known (C3.3.a1–a3), out of which the H-maleate anion, $D(\text{mean}) = 2.393(3)$ Å, has been the object of extended studies by both diffraction and quantum-mechanical methods. The most common intermolecular (−)CAHBs are formed, in the order, by H-bis(carboxylates) (C3.3.b1,b2,b4), benzoates (C3.3.b3), phosphonates (C3.3.c1,c2), phenolates (C3.3.d1,d2), and inorganic acid

CHART 3.3. HOMONUCLEAR AND HOMOMOLECULAR O–H···O NEGATIVE CHARGE-ASSISTED HB, (–)CAHB

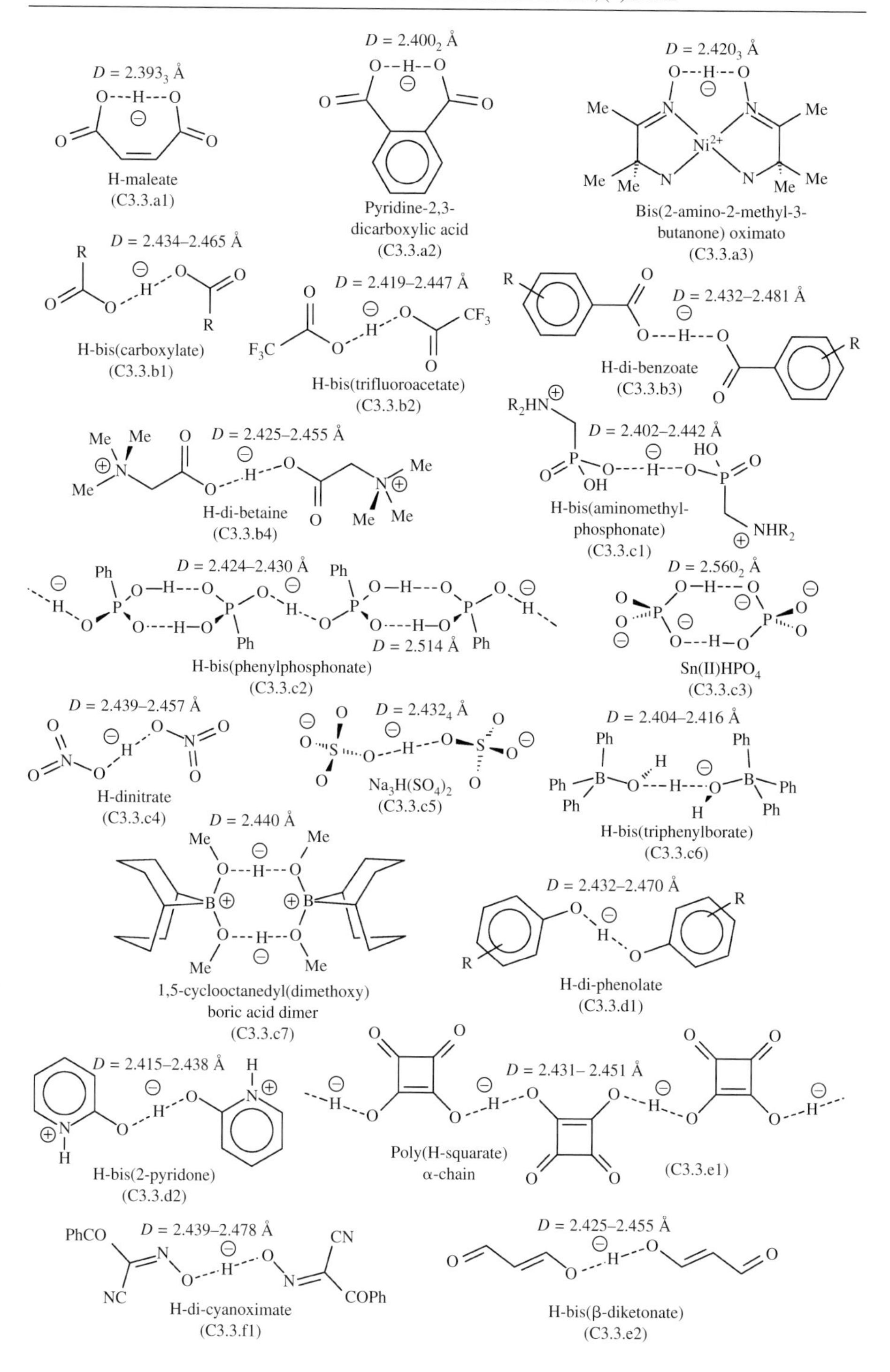

D = 2.393₃ Å
H-maleate
(C3.3.a1)
D = 2.400₂ Å
Pyridine-2,3-dicarboxylic acid
(C3.3.a2)
D = 2.420₃ Å
Bis(2-amino-2-methyl-3-butanone) oximato
(C3.3.a3)
D = 2.434–2.465 Å
H-bis(carboxylate)
(C3.3.b1)
D = 2.419–2.447 Å
H-bis(trifluoroacetate)
(C3.3.b2)
D = 2.432–2.481 Å
H-di-benzoate
(C3.3.b3)
D = 2.425–2.455 Å
H-di-betaine
(C3.3.b4)
D = 2.402–2.442 Å
H-bis(aminomethyl-phosphonate)
(C3.3.c1)
D = 2.424–2.430 Å
D = 2.514 Å
H-bis(phenylphosphonate)
(C3.3.c2)
D = 2.560₂ Å
Sn(II)HPO₄
(C3.3.c3)
D = 2.439–2.457 Å
H-dinitrate
(C3.3.c4)
D = 2.432₄ Å
Na₃H(SO₄)₂
(C3.3.c5)
D = 2.404–2.416 Å
H-bis(triphenylborate)
(C3.3.c6)
D = 2.440 Å
1,5-cyclooctanedyl(dimethoxy) boric acid dimer
(C3.3.c7)
D = 2.432–2.470 Å
H-di-phenolate
(C3.3.d1)
D = 2.415–2.438 Å
H-bis(2-pyridone)
(C3.3.d2)
D = 2.431– 2.451 Å
Poly(H-squarate) α-chain
(C3.3.e1)
D = 2.439–2.478 Å
H-di-cyanoximate
(C3.3.f1)
D = 2.425–2.455 Å
H-bis(β-diketonate)
(C3.3.e2)

salts (C3.3.c3–c6). Less frequent are H-bis(squarates), oximates, β-diketonates (C3.3.e1,f1,e2), and boric acid derivatives (C3.3.c7).

As for the strength of the H-bonds formed, there are indications that they are the shorter the more acidic are the molecules involved. For example, trifluoroacetates (C3.3.b2) are seen to give basically stronger bonds than carboxylates (C3.3.b1) and the bonds formed by both carboxylates and phosphonates are made stronger when a protonated aminomethyl group is acting as an electron-attracting substituent (C3.3.b4,c1).

C. *$[O\cdots H\cdots O]^+$. Positive charge-assisted H-bond or (+)CAHB* (+)CAHBs have the typical form $[R_1{=}O\cdots H\cdots O{=}R_2]^+$ and are therefore *H-bonds formed by two R=O bases that have captured a proton* from an external strong acid. Also in this case they can be homomolecular ($R_1 = R_2$) or heteromolecular ($R_1 \neq R_2$), out of which only the former ones, which are systematically stronger (shorter), have been considered in the CSD search. The histogram of their $D = d(O\cdots O)$ distances (Fig. 3.34(b)) shows a peaked distribution with distances $D(\text{min}) =$ 2.360 and $D(\text{mean}) = 2.42(2)$ Å ($n = 91$) that are considerably shorter than the corresponding (−)CAHB distances. Also in this case, most of the bonds formed are proton centered.

Very few examples of intramolecular (+)CAHB are known (Jaskólski *et al.*, 1982) and Chart 3.4 summarizes the most typical molecular situations producing intermolecular homomolecular bonds. More than 40 per cent of these bonds are associated with the $[H_2O\cdots H\cdots OH_2]^+$ hydron-bis(water) complex (C3.4.a1–a3). Other typical bases are phosphine oxides (C3.4.b1; P=O acceptors), amides (C3.4.c1), urea (C3.4.c2), carboxylic acids (C3.4.c3), and benzophenone (C3.4.c4) (all C=O acceptors), alcohols and ethers (C3.4.d1,d2; >O acceptors), pyridine- and pyrrole-*N*-oxides (C3.4.e1,e2; >N−O acceptors), sulphoxides (C3.4.f1,f2; S=O acceptors), and nitroderivates (C3.4.g1; NO_2 acceptors).

D. *$[^{1/2-}O\cdots H^+\cdots O^{1/2-}]$. Double charge-assisted H-bond or (±)CAHB* This last class of strong H-bonds can be related to the acid–base equilibrium

$$R_1{-}O{-}H\cdots{:}O{-}R_2 \leftrightarrows R_1{-}{}^{1/2-}O\cdots H^+\cdots O^{1/2-}{-}R_2 \leftrightarrows R_1{-}{}^{-}O{:}\cdots H{-}O^+{-}R_2$$

between an acid, $R_1{-}O{-}H$, and a base, $:O{-}R_2$, whose conjugated acid is $H{-}O^+{-}R_2$. As discussed in detail later, this equilibrium is known to give three extreme outcomes (plus an infinite number of intermediate ones) that are modulated by the difference of the acid–base parameters of the H-bond donor and acceptor, that is $\Delta PA = PA(R_1{-}O^-) - PA(O{-}R_2)$ or $\Delta pK_a = pK_a(R_1{-}O{-}H) - pK_a(H{-}O^+{-}R_2)$. When $\Delta PA/\Delta pK_a$ is large and positive, weak and neutral O−H···O bonds are formed, while when this difference is large and negative always weak but charged $^-O\cdots H{-}O^+$ bonds are obtained. When, however, the difference tends to vanish, very strong and proton-centered H-bonds develop that could be represented as $R_1{-}{}^{1/2-}O\cdots H^+\cdots O^{1/2-}{-}R_2$ as well as $R_1{-}{}^{1/2-}O\cdots H\cdots O^{1/2+}{-}R_2$ and that, for the presence of a double charge inside the bond, have been called *double charge-assisted H-bonds or (±)CAHBs* (Gilli

CHART 3.4. HOMONUCLEAR AND HOMOMOLECULAR O–H···O POSITIVE CHARGE-ASSISTED HB, (+)CAHB

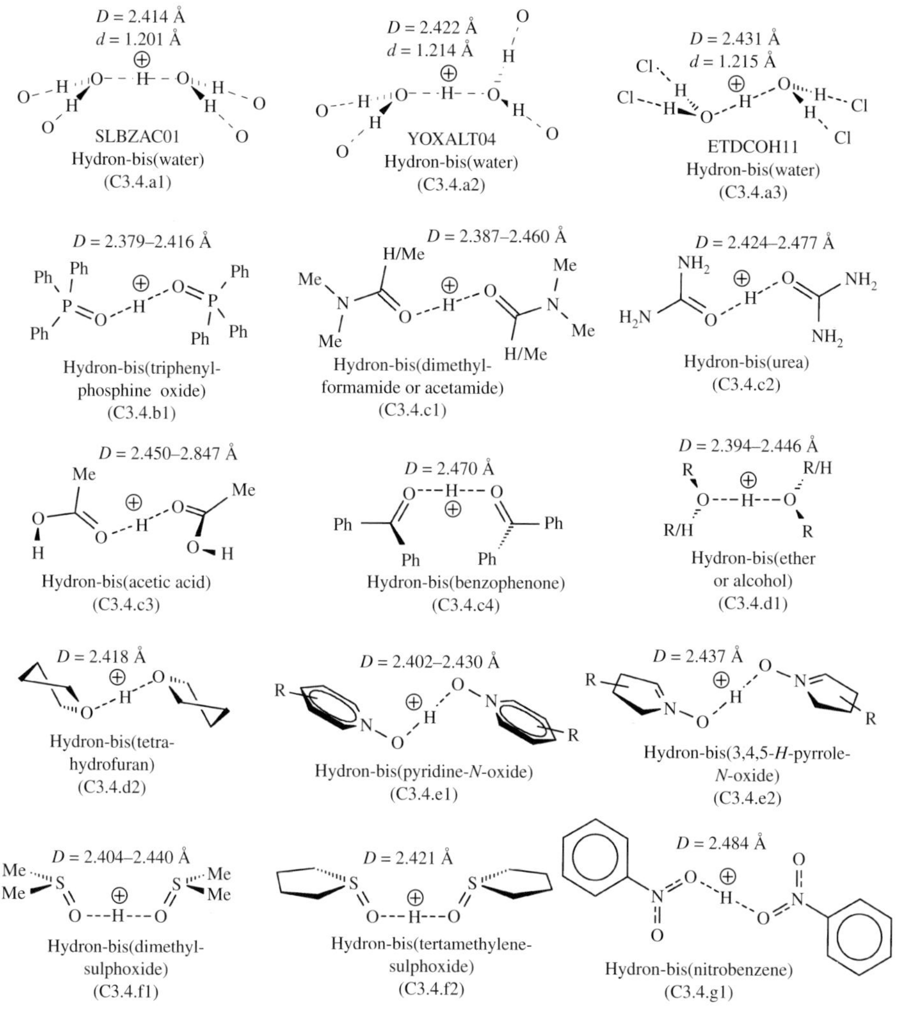

$D = 2.414$ Å
$d = 1.201$ Å
SLBZAC01
Hydron-bis(water)
(C3.4.a1)

$D = 2.422$ Å
$d = 1.214$ Å
YOXALT04
Hydron-bis(water)
(C3.4.a2)

$D = 2.431$ Å
$d = 1.215$ Å
ETDCOH11
Hydron-bis(water)
(C3.4.a3)

$D = 2.379$–2.416 Å
Hydron-bis(triphenyl-phosphine oxide)
(C3.4.b1)

$D = 2.387$–2.460 Å
Hydron-bis(dimethyl-formamide or acetamide)
(C3.4.c1)

$D = 2.424$–2.477 Å
Hydron-bis(urea)
(C3.4.c2)

$D = 2.450$–2.847 Å
Hydron-bis(acetic acid)
(C3.4.c3)

$D = 2.470$ Å
Hydron-bis(benzophenone)
(C3.4.c4)

$D = 2.394$–2.446 Å
Hydron-bis(ether or alcohol)
(C3.4.d1)

$D = 2.418$ Å
Hydron-bis(tetra-hydrofuran)
(C3.4.d2)

$D = 2.402$–2.430 Å
Hydron-bis(pyridine-*N*-oxide)
(C3.4.e1)

$D = 2.437$ Å
Hydron-bis(3,4,5-*H*-pyrrole-*N*-oxide)
(C3.4.e2)

$D = 2.404$–2.440 Å
Hydron-bis(dimethyl-sulphoxide)
(C3.4.f1)

$D = 2.421$ Å
Hydron-bis(tertamethylene-sulphoxide)
(C3.4.f2)

$D = 2.484$ Å
Hydron-bis(nitrobenzene)
(C3.4.g1)

and Gilli, 2000). The condition by which these strong bonds are generated is often referred to as *PA/p*K_a *matching* or *PA/p*K_a *equalization*.

Ordinary O–H···O bonds have very large ΔpK_a values, for instance 21–25 for the alcohol–ketone couple, and only very recently a number of (±)*CAHB*s have been successfully studied by diffraction methods. Some of the most accurate low-temperature neutron structures (Chart 3.5) show very short $D = d(\mathrm{O}\cdots\mathrm{O})$ distances in the range 2.41–2.44 Å for a ΔpK_a interval from −1.3 to 3.3 pK_a units. To notice that the shape of the histogram of distances (Fig. 3.34(c)) is not peaked but increases monotonically with the distance itself in relation to

CHART 3.5. HOMONUCLEAR O–H···O DOUBLE CHARGE-ASSISTED HB, (±)CAHB

$D = 2.409$ Å, $d = 1.157$ Å

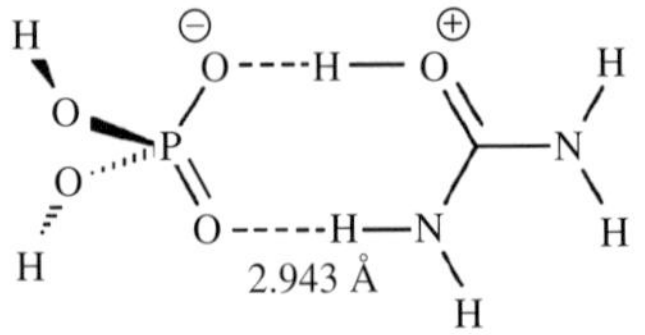

2.943 Å

CRBAMP01, N-100K, $\Delta pK_a = 2.1$
Phosphoric acid–urea
(C3.5.a1)

$D = 2.428$ Å, $d = 1.211$ Å

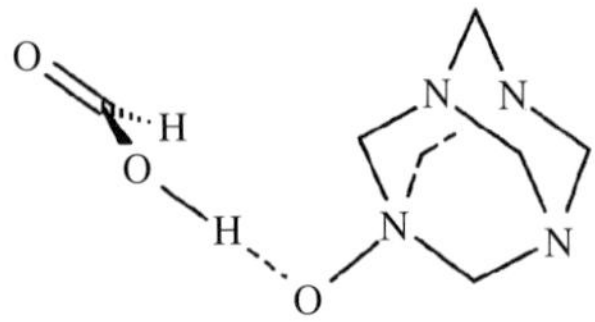

HMTOFA07, N-123K, $\Delta pK_a = -0.9$
Formic acid–HMTA-*N*-oxide
(C3.5.a2)

$D = 2.430$ Å, $d = 1.148$ Å

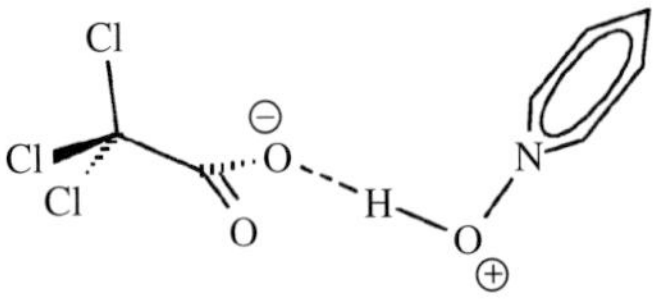

PYOTCA01, N-120K, $\Delta pK_a = -1.3$
Trichloroacetic acid–pyridine-*N*-oxide
(C3.5.a3)

$D = 2.437$ Å, $d = 1.123$ Å

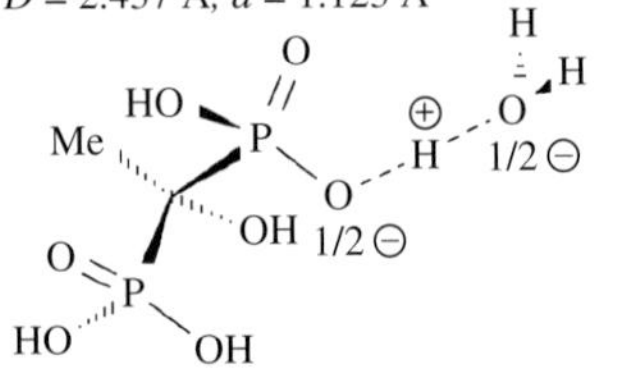

ETHDPH01, N-293K, $\Delta pK_a = 3.3$
(OH)-Ethane-1,1-diphosphoric acid–water
(C3.5.a4)

the ever-increasing number of H-bonds endowed with ever-larger ΔpK_a values, so showing that the transition from (±)CAHBs to ordinary O–H···O bonds is continuous.

E. *O–H···O ordinary H-bond or OHB* Having defined four classes of potentially strong bonds [RAHB, (−)CAHB, (+)CAHB, and (±)CAHB], it is useful to define a class of bonds that can never become really strong because they lack the conditions that make the other four classes strong. These *ordinary H-bonds or OHBs* can therefore be defined as the bonds that *are not assisted by either charge or resonance.*

F. *O–H···O σ-bond cooperative or polarization-assisted H-bond (PAHB)* The concept of σ-bond cooperativity has been already introduced in Section 2.4.1. It was firstly suggested by Jeffrey and Saenger (1991) in relation to the hypothesis that H-bonds within long chains of waters (or alcohols or phenols) should be stronger than ordinary O–H···O bonds because of the increased O–H bond polarization induced by the chain. Later, these bonds have been added to the CL list with the name *polarization-assisted H-bond or PAHB* (Gilli and Gilli, 2000). Chart 3.6 reports a few typical cases of PAHBs that clearly show that the bond shortening is small, most probably because of the too small polarizability of the σ-bond, but sufficient to keep the O···O distances slightly below the

CHART 3.6. HOMONUCLEAR AND HOMOMOLECULAR O–H···O σ-COOPERATIVE OR POLARIZATION-ASSISTED HB, PAHB

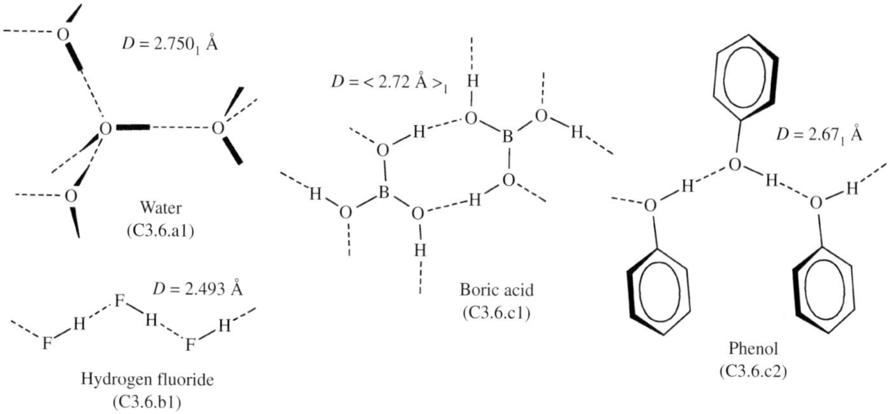

$D = 2.750_1$ Å

Water (C3.6.a1)

$D = 2.493$ Å

Hydrogen fluoride (C3.6.b1)

$D = <2.72\text{ Å}>_1$

Boric acid (C3.6.c1)

$D = 2.67_1$ Å

Phenol (C3.6.c2)

shortest distances normally associated with the ordinary O–H···O bond (some 2.70–2.75 Å).

3.3.1.2 *Interpreting the O–H···O system: The electrostatic-covalent H-bond model*

A. *Symmetry and covalency of the strong O–H···O bond* Having classified all plausible weak, strong, and moderate O–H···O bonds, we can now try to infer a conclusion on the very nature of this bond from the ensemble of the data collected. This analysis will employ the correlation methods already used for studying RAHB in β-diketone enols (Section 3.2.2) focusing the attention on the relationships between H-bond strength and H-bond symmetry, a property that directly connects experimental data to the H-bond Coulson formalism (Chart 2.2; Coulson, 1959; Coulson and Danielsson, 1954a,b) and can provide a simple rational base for interpreting the H-bond in terms of simplified VB theory.

The fact that the O–H···O bond becomes increasingly symmetric with its increasing strength can be substantiated in several ways. For instance, the already discussed Fig. 3.3(a) has shown, for a subset of accurate neutron diffraction data, that d(O–H) and d(H···O) distances are strictly correlated and strong O–H···O bonds continuously tend to the limit $d(\text{O–H}) = d(\text{H}\cdots\text{O}) \simeq 1.20$ Å for a $d(\text{O}\cdots\text{O})$ of nearly 2.40 Å. This result is corroborated by the scatter plot of Fig. 3.35(a), reporting the correlation between IR ν(O–H) stretching frequencies and O···O distances obtained by extending the original compilation by Novak (1974) (Bertolasi *et al.*, 1996). It well illustrates the continuous weakening (lengthening) of the O–H covalent bond that is associated with the shortening of the O···O, and then H···O, distances.

The increase of symmetry within the O–H···O moiety is parallelled by that of the two adjacent X–O distances in the extended X_1–O–H···O–X_2 molecular fragment, as shown by the $\Delta d/\Delta d_0$ versus d(O···O) scatterplot of Fig. 3.35(b), where Δd is the difference between the actual X_1–O and O–X_2 distances, and Δd_0 the same difference in the absence of H-bonding. Finally, Fig. 3.35(c) shows that the full symmetry achieved by the fragment in very strong O–H···O bonds

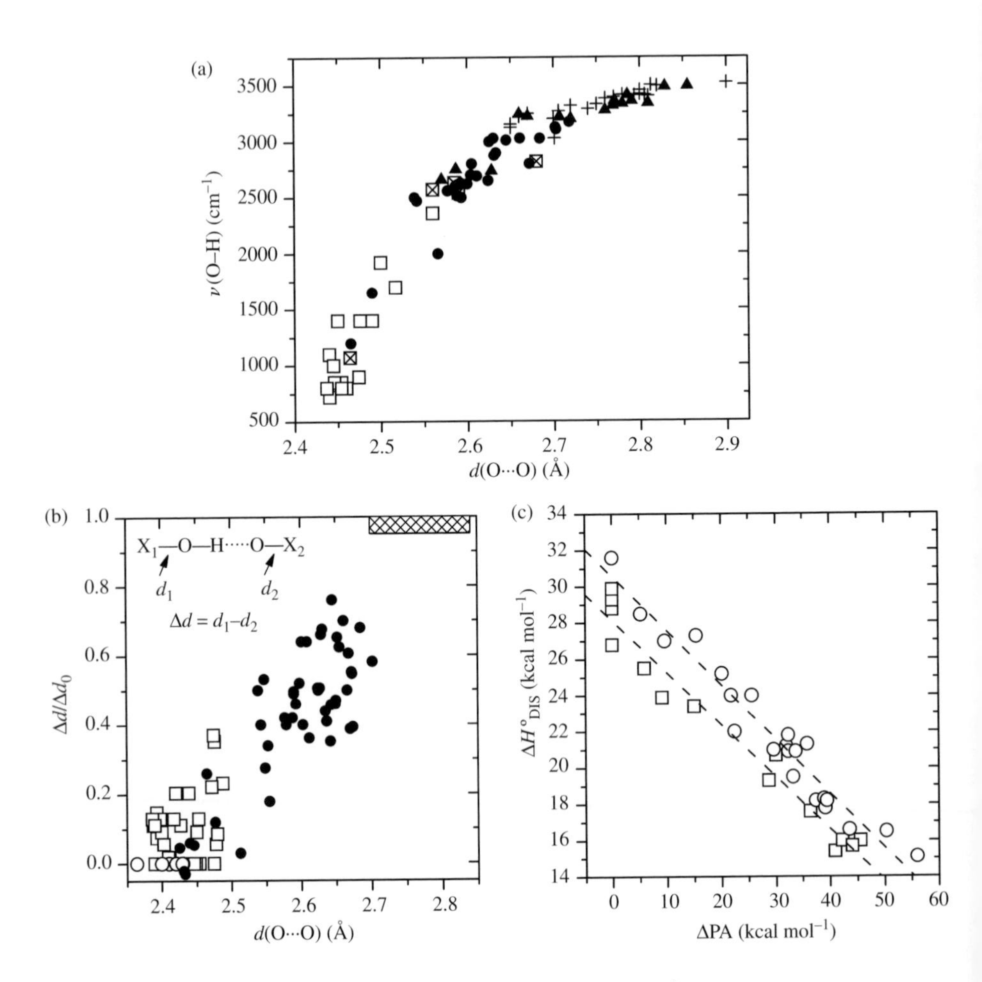

FIG. 3.35. (a) IR ν(O–H) (cm^{-1}) versus d(O···O) (Å) scatterplot for intermolecular O–H···O bonds; (b) $\Delta d/\Delta d_0$ versus d(O···O) (Å) scatterplot for intermolecular O–H···O bonds in organic compounds; (c) Gas-phase O–H···O bond dissociation enthalpies $\Delta H°_{\mathrm{DIS}}$ (kcal mol^{-1}) versus ΔPA (kcal mol^{-1}), the difference of proton affinities of the two interacting molecules. Symbols: □ = organic (−)CAHB; ⊠ = inorganic (−)CAHB; ○ = (+)CAHB; ● = RAHB; ▲ = PAHB; + and top-right rectangle in (b) = ordinary H-bonds. (Reproduced by permission from Gilli and Gilli, 2000)

is not simply geometrical but physicochemical in nature (Gilli *et al.*, 1996a). It reports the values of the gas-phase dissociation enthalpies, $\Delta H^{\circ}_{\text{DIS}}$, measured by pulsed-electron-beam mass spectroscopy (Meot-Ner (Mautner), 1984; Meot-Ner (Mautner) and Sieck, 1986) for a number of charged H-bonded complexes belonging to the classes (−)CAHB (open squares) and (+)CAHB (open circles) as a function of ΔPA, the difference of proton affinities of the two interacting molecules. It is seen that the greatest H-bond energies occur for homomolecular complexes whose ΔPA is identically zero, while heteromolecular associations are the weaker the larger the PA difference is.

In conclusion, the transition of the homonuclear O−H···O bond from weak (long) to strong (short) goes with a parallel transition from dissymmetric to symmetric of all the properties of the X_1−O−H···O−X_2 fragment, the condition

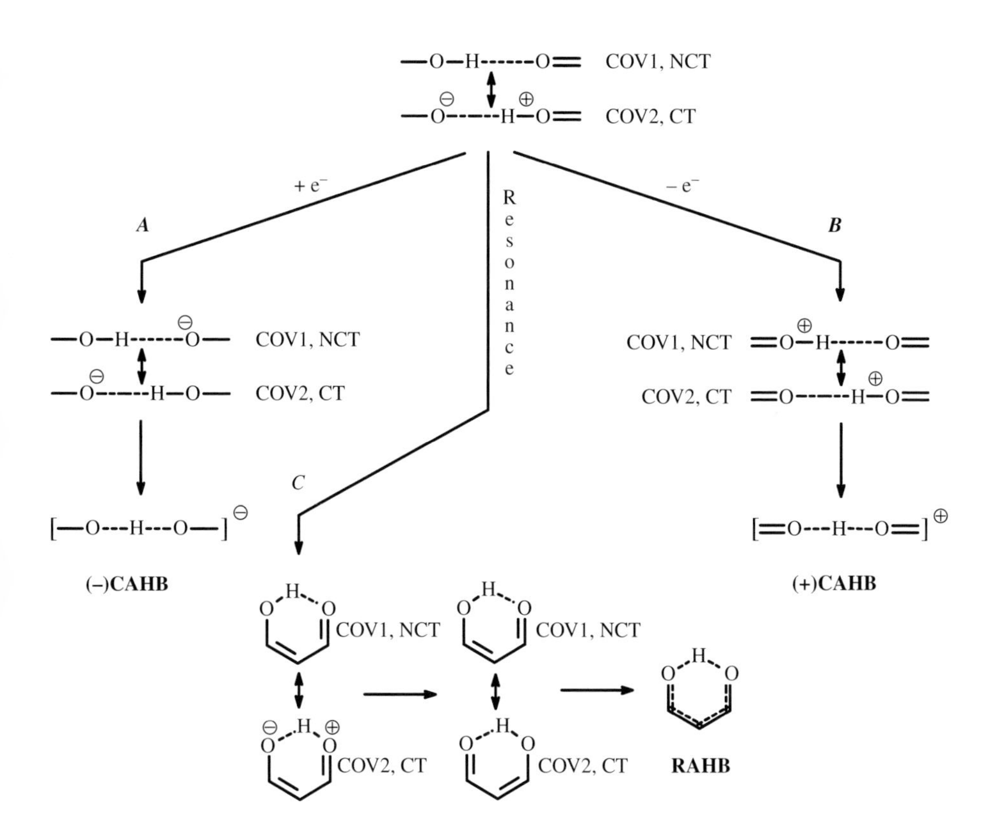

FIG. 3.36. Three ways for making isoenergetic the two VB resonance forms COV1,NCT and COV2,CT: (a) by adding an electron; (b) by removing an electron; (c) by connecting the two oxygens by a π-conjugated chain. (Reproduced by permission from Gilli and Gilli, 2000)

$X_1 = X_2$ included. Hence, the progressive equalization of both O–X and O–H distances indicates that *very strong O–H···O bonds are to be considered as fully delocalized 3c–4e covalent bonds*, a situation that can be expressed in VB terms (Chart 2.2 of Section 2.2) by the linear combination

$$\Psi = a_1\ \Psi_{\text{COV1, NCT}}\,(-\text{O}-\text{H}\cdots\text{O}=) + a_2\ \Psi_{\text{COV2, CT}}\,(-\text{O}^{-}\cdots\text{H}-{}^{+}\text{O}=), \qquad (3.32)$$

of two VB resonant wavefunctions which are isoenergetic (Scheme (b) of Chart 2.2) and then can mix with identical *a* coefficients. This simple rule makes it possible to predict when chemical substances are able to form strong H-bonds: *only chemical situations for which the resonance forms –O–H···O= and –O⁻···H–⁺O= may become energetically (and then chemically) equivalent can lead to strong or very strong H-bonds.*

It can be easily shown that such chemical equivalence can be achieved for the O–H···O bond and, in general, for any homonuclear X–H···X bond whose X is electronegative enough, in four ways (Fig. 3.36):

1) by reacting a couple of H-bond donors and acceptors that happen to have identical PA/pK_a values;
2) by adding an electron and so giving rise to symmetric $[-\text{O}\cdots\text{H}\cdots\text{O}-]^-$ bonds (pathway A);
3) by removing an electron and so producing the symmetric situation for which a proton is shared by two oxygens, that is $[=\text{O}\cdots\text{H}\cdots\text{O}=]^+$ (pathway B);
4) by connecting the two oxygens by a chain of conjugated double bonds (pathway C).

Clearly, these four cases correspond to the four strong CLs already identified by screening structural data, that is, in the order, (±)CAHB, (–)CAHB, (+) CAHB, and RAHB.

B. *The electrostatic-covalent H-bond model (ECHBM)* Previous considerations can be summarized in the *electrostatic-covalent H-bond model (ECHBM)* (Gilli *et al.*, 1994a; Gilli and Gilli, 2000) which was worked out for the O–H···O bond but should hold for any other X–H···X homonuclear bond, that is:

a) Weak X–H···X bonds are *dissymmetric* but become increasingly *symmetric* with their increasing strength;
b) Weak X–H···X bonds are *electrostatic* in nature but become increasingly *covalent* with their increasing strength; at the limit, very strong bonds are to be regarded as essentially *3c–4e covalent interactions* where the two X–H bonds have identical bond number of ½;
c) The degree of covalency depends on the degree of mixing of the two wavefunctions of eqn 3.32, reaching the maximum of 1:1 when $\Delta E = E(\Psi_{\text{COV2, CT}}) - E(\Psi_{\text{COV1, NCT}}) = 0$.
d) The four strong CLs are essentially molecular devices able to annihilate the normally very large ΔE associated with weak electrostatic H-bonds.

This annihilation may occur in two ways:

d′) in X–H···X (±)CAHB, by accidental mixing of H-bond donors and acceptors having identical, or very similar, PA/pK_a values;

d″) in X–H···X (–)CAHB, (+)CAHB and RAHB, by mixing H-bond donors and acceptors that are *symmetrical* as far as the distribution of chemical groups on the two sides of the H-bond is concerned; this guarantees that the two VB wavefunctions are *isoenergetic* because they are identical by symmetry.

3.3.1.3 *Interpreting the O–H···O system: The PA/pK_a equalization principle*

The VB treatment given above is not entirely satisfactory for two reasons. The first is that its conclusions do not directly emerge from experiments but are mediated by the interleaving VB theory, a fact to be avoided whenever possible. The second, and more important reason, is that the model proposed does not identify the physical variable (or combination of variables) that drives the transition from weak (electrostatic) to strong (covalent) H-bond.

An indication for solving these problems comes from (±)CAHBs, whose strengths appear to be largely determined by a difference of acid–base indicators expressed in the form ΔpK_a = pK_a(D–H) – pK_a(A–H^+) or ΔPA = PA(D^-) – PA(:A), where pK_a is the co-logarithm of the acid–base dissociation constant (normally in water) and PA the proton affinity in the gas phase. To become really of general use, however, this concept needs to be extended to all other classes of strong H-bonds and Chart 3.7 shows how this result can be achieved (Gilli *et al.*, 2007).

Scheme 1 of the chart illustrates the mechanism of ΔPA/ΔpK_a annihilation for the two charge-assisted leitmotifs (–)CAHB and (+)CAHB (for simplicity, only the pK_a notation is used). The water dimer C3.7.1.I is well known to form a rather weak bond with energy of 4–5 kcal mol^{-1} and O···O distance of some 2.75 Å, this weakness being the result of the large acidity difference between the donor (HO–H) and the conjugated acid of the acceptor (H_2O–H^+) and of the consequent large ΔpK_a of 15.7–(–1.7) = 17.4 pK_a units. However, the water dimer is transformed by *proton removal* into the [H–O···H···O–H]$^-$ hydroxyl–water complex C3.7.1.II where *the proton is shared by two identical acids* and, accordingly, its ΔpK_a = pK_a(HO–H) – pK_a(H–OH) drops to zero, causing a dramatic increase of the binding energy (E_{HB} = 25–30 kcal mol^{-1}) and an equally dramatic fall of the O···O contact distance to *d*(O···O) = 2.45(2) Å. An even more spectacular effect is produced by *proton addition* that can transform the water dimer into the [H_2O···H···OH_2]$^+$ hydronium–water complex C3.7.1.III where *the proton is shared by two identical bases* causing again the ΔpK_a = pK_a(H_2O–H^+) – pK_a(H^+–OH_2) to drop to zero while E_{HB} is increased to 25–31 kcal mol^{-1} and *d*(O···O) decreased to 2.42(2) Å, so providing one of the shortest H-bonds of all the O–H···O system. A similar sort of reasoning can be applied to the resonant-assisted leitmotif, RAHB, of Scheme 2. The RO–H···O=CR_2 complex C3.7.2.I is known to form rather weak H-bonds [E_{HB} = 4–5 kcal mol^{-1}; *d*(O···O, *inter*) = 2.70–2.95 Å; *d*(O···O, *intra*) = 2.65–2.90 Å] because of the very large ΔpK_a

(21–25 pK_a units) associated with the alcohol–ketone couple. This bond can be, however, strengthened *by connecting the two oxygens by a π-conjugated group* transforming it into the β-diketone enol C3.7.2.II whose two equally probable ···O=C–C=C–OH··· and ···HO–C=C–C=O··· resonance forms (C3.7.2.IIa,b) naturally lead to a ΔpK_a of zero whenever the full π-delocalized structure C3.7.2.IIc is reached, leading at the same time to much stronger (E_{HB} = 15–25 kcal mol^{-1}) and shorter [d(O···O, *intra*) = 2.39–2.55 Å; d(O···O, *inter*) = 2.46–2.65 Å] bonds.

Hence, the four strong CLs can be rationalized in terms of the unifying concept that all strong H-bonds must be associated with the condition $\Delta PA/\Delta pK_a \cong 0$ or, in other words, that *the driving force able to transform weak, long, asymmetric*

CHART 3.7. CHEMICAL LEITMOTIFS AND PA/pK_a EQUALIZATION PRINCIPLE

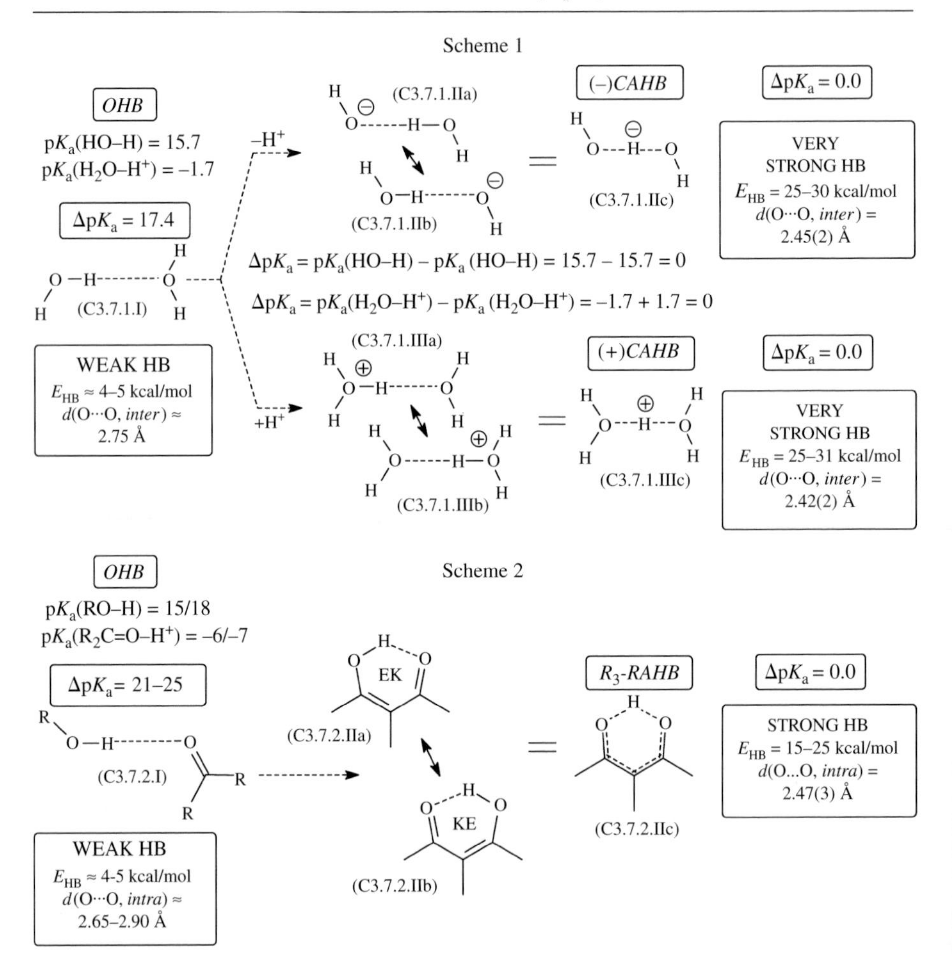

and electrostatic H-bonds into strong, short, symmetric and covalent ones is to be sought in the annihilation of the difference between the proton affinities, or related acid–base dissociation constants, of the H-bond donor and acceptor moieties. In agreement with our previous work (Gilli *et al.*, 2004, 2005a, 2007), this basic concept will be indicated in the following as the *PA/pK_a equalization principle* though the more colloquial expression *PA/pK_a matching* can be used as well.

This analysis indicates that CLs are not only a useful taxonomic criterion but can also be considered as a kind of *artifice*, not to say *trick*, that molecules may adopt to wipe out the normally very large ΔpK_a (or ΔPA) between the H-bond donor and acceptor moieties as well as *recipes* to tune the molecular autoassociation properties by changing the p*H* of the environment (Scheme 1) or by connecting, or not, the donor and the acceptor by a short resonant fragment (Scheme 2).

A last consideration is that the PA/pK_a matching, besides being a way for *understanding* the H-bond, can become a method for *predicting* how strong the H-bonds to be formed will be (see below). It may be noticed from now that these predictions, easy to realize for CAHBs, do not work for RAHBs because the π-delocalization occurring during H-bond formation upsets the acid–base properties of the interacting molecules.

3.3.2 *CAHB generalization to other homonuclear X–H···X bonds*

The considerations made for the O–H···O bond can be directly extended to all other homonuclear X–H···X bonds. *Strong* homonuclear bonds must fulfil three conditions: (*i*) X is sufficiently electronegative; (*ii*) the bond formed belongs to one of the four classes previously identified, i.e. (±)CAHB, (–)CAHB, (+)CAHB or RAHB; (*iii*) in the case of RAHB, the X atom must be able to form double bonds. Homonuclear RAHBs have been illustrated in Chart 3.1 (Section 3.2.3.1) and the most significant homonuclear CAHBs not involving O atoms retrieved from CSD and, in a few cases, ICSD are summarized in Chart 3.8. All X···X distances are, according to expectation, abnormally short. Their comparison with weak OHBs is delayed to the discussion of Table 3.9 below.

Typical (–)CAHBs include the [Hal···H···Hal]$^-$ bonds (Hal = F, Cl, Br; c3.8.a1–a3) and a single case of [S···H···S]$^-$ (c3.8.a4) together with a limited number of [N···H···N]$^-$ bonds (c3.8.b1,b2). The [F···H···F]$^+$ bond (c3.8.c1) is found in a small number of inorganic structures and is the only (+)CAHB known to involve halogens. Conversely, [N···H···N]$^+$ bonds (c3.8.d1–e4) are quite frequent because they constitute the class of intramolecular and intermolecular *proton sponges*, compounds with a variety of chemical applications that, for this reason, have been extensively studied from a structural point of view (Alder, 1990; Staab and Saupe, 1988; Llamas-Saiz *et al.*, 1994).

Finally, about thirty N–H···N (±)CAHBs short enough to suggest a strict PA/pK_a matching have been retrieved that would deserve a more detailed thermodynamic study of their acid–base properties. The examples reported in Chart 3.8 concern the pyridine–bis(methylsulphonyl)amide (C3.8.f1; ΔpK_a = –2.4) and 1,1-dimethylhydrazine–hydrazoic acid (C3.8.f2; ΔpK_a = –3.4) complexes.

CHART 3.8. HOMONUCLEAR X–H···X (X = F, Cl, Br, S, N)
CHARGE-ASSISTED HBs, CAHBs

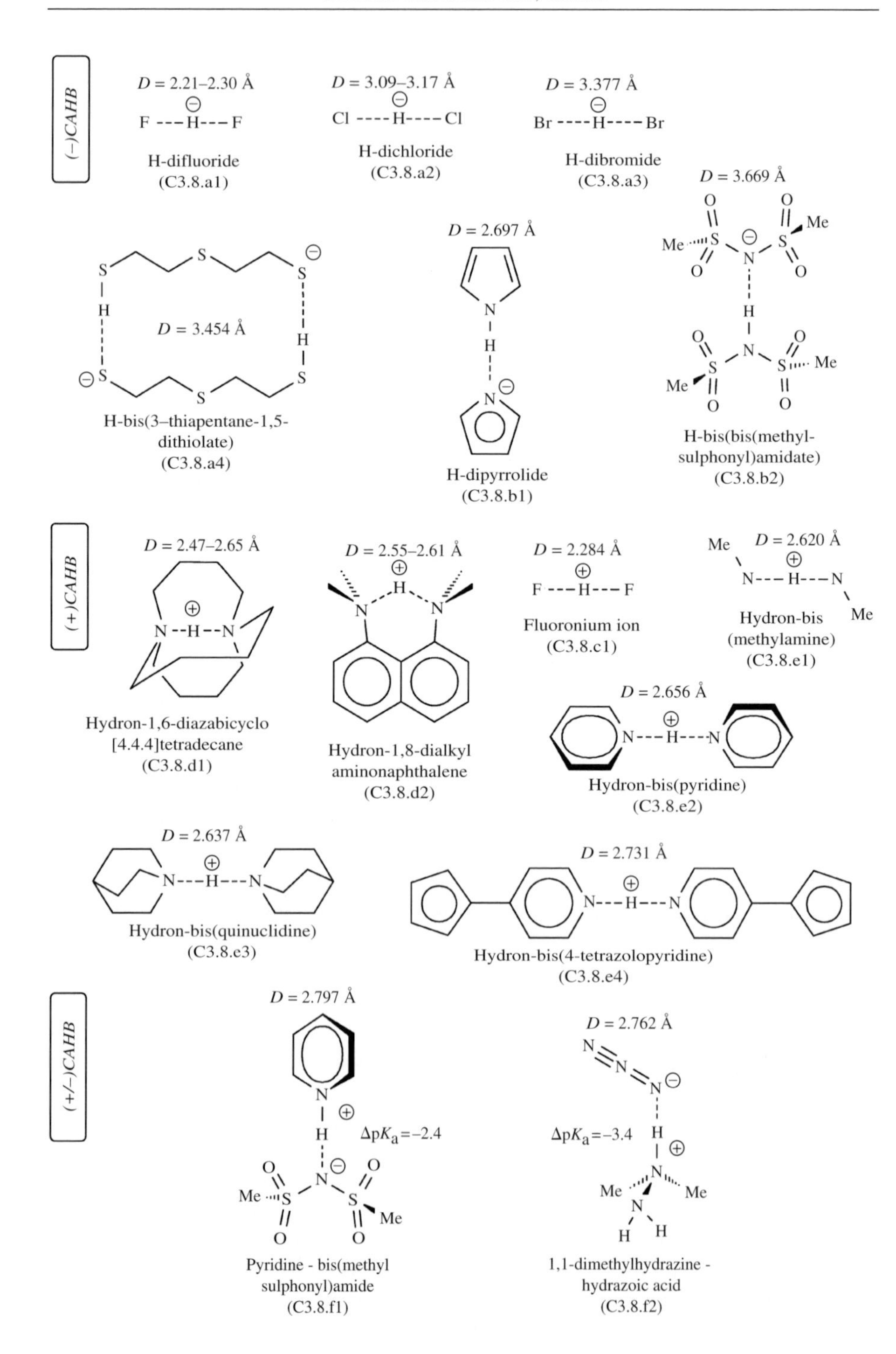

(–)CAHB
D = 2.21–2.30 Å
F ---H--- F
H-difluoride
(C3.8.a1)
D = 3.09–3.17 Å
Cl ----H---- Cl
H-dichloride
(C3.8.a2)
D = 3.377 Å
Br ----H---- Br
H-dibromide
(C3.8.a3)
D = 3.454 Å
H-bis(3–thiapentane-1,5-dithiolate)
(C3.8.a4)
D = 2.697 Å
H-dipyrrolide
(C3.8.b1)
D = 3.669 Å
H-bis(bis(methyl-sulphonyl)amidate)
(C3.8.b2)
(+)CAHB
D = 2.47–2.65 Å
Hydron-1,6-diazabicyclo [4.4.4]tetradecane
(C3.8.d1)
D = 2.55–2.61 Å
Hydron-1,8-dialkyl aminonaphthalene
(C3.8.d2)
D = 2.284 Å
F ---H--- F
Fluoronium ion
(C3.8.c1)
D = 2.620 Å
Hydron-bis (methylamine)
(C3.8.e1)
D = 2.656 Å
Hydron-bis(pyridine)
(C3.8.e2)
D = 2.637 Å
Hydron-bis(quinuclidine)
(C3.8.e3)
D = 2.731 Å
Hydron-bis(4-tetrazolopyridine)
(C3.8.e4)
(+/–)CAHB
D = 2.797 Å
ΔpKa = –2.4
Pyridine - bis(methyl sulphonyl)amide
(C3.8.f1)
D = 2.762 Å
ΔpKa = –3.4
1,1-dimethylhydrazine - hydrazoic acid
(C3.8.f2)

3.3.3 *CAHB generalization to heteronuclear X—H···Y bonds*

ECHBM, though specifically developed for the homonuclear X—H···X case, can give us some insight also into the charged *X—H···Y heteronuclear bond.* (±)CAHBs, whose strength simply depends on the more or less accidental PA/pK_a matching of the H-bond donor and acceptor, are not expected to be seriously affected by the heteronuclearity of the bond. Conversely (−)CAHBs, (+)CAHBs and RAHBs, which base their strengthening on the perfect homonuclear and homomolecular symmetry around the H-bond, are easily predicted to become much weaker in the heteronuclear case unless made stronger by introducing appropriate *chemical substituents* that are able to reduce the electronegativity difference between the H-bond donor and acceptor atoms. These order of problems has been discussed in detail for RAHB in Section 3.2.3.4. As for CAHBs, a systematic CSD study (Gilli and Gilli, 2000) carried out on 318 strong N—H···O/O—H···N bonds with d(N···O) $\leq$ 2.66 Å has singled out 265 (±)CAHBs (192 N—H$^+$···O$^-$ and 73 O—H···N), 7 (−)CAHBs (5 N—H···O$^-$ and 2 O—H···N$^-$), and 29 (+)CAHBs (29 N—H$^+$···O bonds), showing that strong heteromolecular (−)CAHBs and (+)CAHBs are quite uncommon and that the analysis can be limited, at a first level of approximation, to the (±)CAHB bonds only. A small collection of high-quality crystal structures of these very short bonds is shown in Chart 3.9.

3.3.4 *CAHB geometry–energy relationships*

According to the PA/pK_a equalization principle proposed above, H-bonds whose PA/pK_a differences are not far from zero should be much *shorter* and *more energetic* than those produced by OHBs, for which such differences are normally much larger (for instance, 21–25 and 14–19 pK_a units for common alcohol–ketone and amide–amide bonds). Verification of this hypothesis requires the simultaneous knowledge of acid–base indicators, H-bond geometries, and H-bond energies of a large number of bonds, data that are accessible only for (−)CAHBs and (+)CAHBs because their ΔPA/ΔpK_a = 0 by definition, their geometries are fairly well known, and their H-bond energies, E_{HB}, can be derived from the gas-phase dissociation enthalpies, $\Delta H°_{DIS}$, of the [X···H···X]$^-$ and [X···H···X]$^+$ bonds collected in the NIST database (Meot-Ner (Mautner) and Lias, 2005; Bartmess, 2005). Conversely, these enthalpies are not accessible for (±)CAHBs or OHBs because their measure requires globally charged species.

Table 3.9 summarizes all data retrieved. Minimum and average CAHB distances, $d_{D···A}$(CAHB, min) and $d_{D···A}$(CAHB, mean), are compared with minimum OHB distances, $d_{D···A}$(OHB, min), and van der Waals distances, $d_{D···A}$(vdW). Bond shortening is expressed as *per cent* shrinking with respect to $d_{D···A}$(vdW), SHR%. E_{HB}s for OHBs are averaged from a variety of sources (Pimentel and McClellan, 1971; Joesten and Schaad, 1974); E_{HB}s for (−) and (+)CAHBs are from the NIST database, those for (±)CAHBs are interpolated by eqn 3.33 (in italics) or evaluated by the LS-HB method (in parentheses). Data analysis shows that the transition from OHBs to (−) or (+)CAHBs (that is from large to null

CHART 3.9. HETERONUCLEAR
N–H···O/O–H···N DOUBLE CHARGE-ASSISTED HB, (±)CAHB

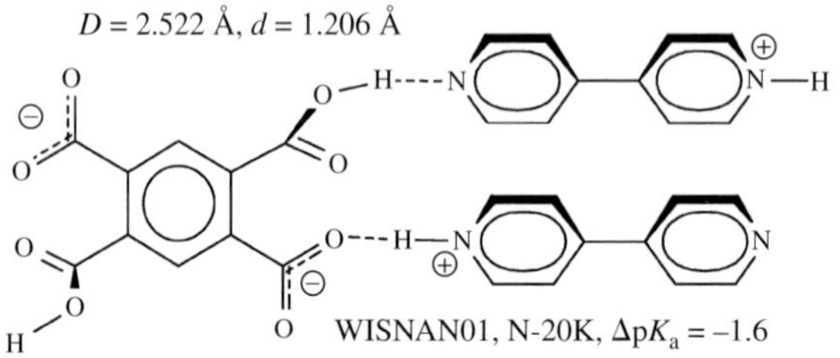

$D = 2.522$ Å, $d = 1.206$ Å

WISNAN01, N-20K, $\Delta pK_a = -1.6$
4,4'-Bypiridine - benzene-1,2,4,5-tetracarboxylic acid
(C3.9.a1)

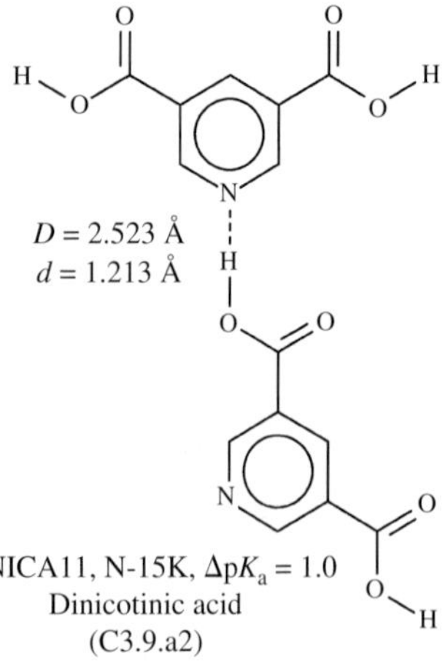

$D = 2.523$ Å
$d = 1.213$ Å

DINICA11, N-15K, $\Delta pK_a = 1.0$
Dinicotinic acid
(C3.9.a2)

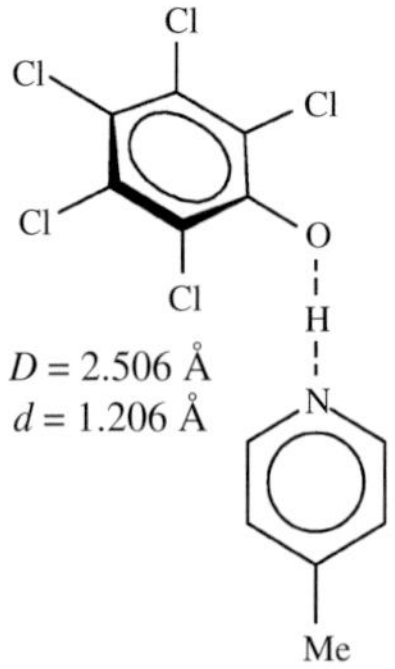

$D = 2.506$ Å
$d = 1.206$ Å

RAKQAV, N-20K, $\Delta pK_a = -0.7$
4-Methylpyridine-
pentachlorophenol
(C3.9.a3)

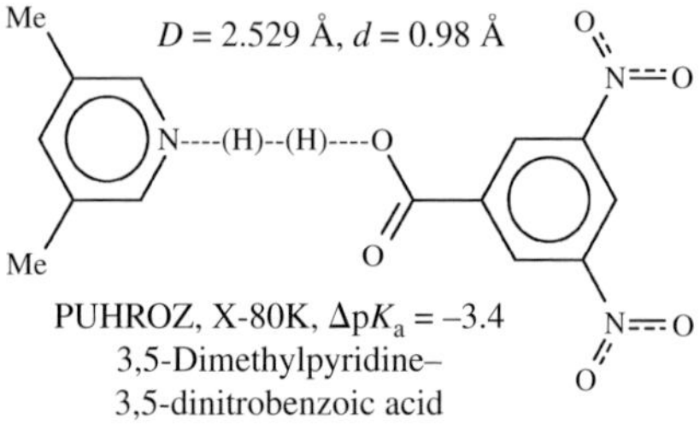

$D = 2.529$ Å, $d = 0.98$ Å

PUHROZ, X-80K, $\Delta pK_a = -3.4$
3,5-Dimethylpyridine–
3,5-dinitrobenzoic acid
(C3.9.a4)

$\Delta PA/\Delta pK_a$ values) causes a remarkable shrinking of the D···A distances (0.21–0.28 for F, 0.30–0.46 for N and O, and 0.53–0.60 Å for Cl, Br, and S) together with a much more important increase of the H-bond energies (from 9 to 36 with an average of 21.3 kcal mol^{-1}) so confirming beyond any reasonable doubt the validity of the pK_a equalization principle.

The data of Table 3.9 can be summarized in the form of a bar chart (Fig. 3.37) where the H-bond energies of OHBs, marked as (•), and of CAHBs, marked as (—) or (+), are plotted as vertical bars occurring at their proper D···A distances, whose full variation range, starting from $d_{D\cdots A}(\mathrm{vdW})$, is also indicated by the horizontal lines on the bottom. Energies are seen to increase exponentially with the D···A shrinking and the exact functional form of this dependence can be appreciated following the Lippincott and Schroeder treatment for linear O–H···O bonds (eqn 3.13 of Section 3.1.3.3 with $\theta = 180°$ or $\theta' = 0°$), where energies and distances are linked by the regression equation

$$E_{HB} = E_{HB,\,MAX} \exp\left[-k\left(d_{D\cdots A} - d_{D\cdots A,\,\min}\right)\right], \tag{3.33}$$

TABLE 3.9. Comparison of D···A distances, $d_{D\cdots A}$ in Å, and H-bond energies (E_{HB} in kcal mol^{-1}) in CAHBs and in OHBs. χ_P = Pauling electronegativity; $d_{D\cdots A}$(vdW)= van der Waals D···A distance without H-bond; $d_{D\cdots A}$(OHB,min) and $d_{D\cdots A}$(CAHB,min)= minimum D···A distances observed in OHBs and CAHBs; $d_{(D\cdots A)}$(CAHB,mean) = average D···A distances for the current samples of n elements; SHR%= *per cent* $d_{D\cdots A}$ shrinking with respect to $d_{D\cdots A}$(vdW). E_{HB}s for OHBs are averages from a variety of sources; E_{HB}s for CAHBs are gas-phase dissociation enthalpies, $\Delta H°_{DIS}$, of the homomolecular $[X\cdots H\cdots X]^-$ and $[X\cdots H\cdots X]^+$ bonds derived from the NIST database; energy values in italics and in parentheses are respectively interpolated by eqn 3.33 or by the LS–HB method.

D–H···A	χ_P	$d_{D\cdots A}$ (vdW)	$d_{D\cdots A}$	SHR%	E_{HB}	$d_{D\cdots A}$	SHR%	E_{HB}	$d_{D\cdots A}$	n
			(OHB, min)			(CAHB, min)			(CAHB, mean)	
$[F\cdots H\cdots F]^-$	3.98	3.59	2.49	–31	≤6	2.207	–38	42(3)	2.27(3)	17
$[O\cdots H\cdots O]^-$	3.44	3.70	2.70	–27	≤5	2.402	–35	27(1)	2.45(2)	225
$[Cl\cdots H\cdots Cl]^-$	3.16	4.22	3.69	–13	≤2	3.093	–30	24(3)	3.13(3)	15
$[N\cdots H\cdots N]^-$	3.04	3.76	3.05	–19	≤3	2.669	–29	12(–)	2.72(3)	7
$[Br\cdots H\cdots Br]^-$	2.96	4.46	3.91	–12	≤2	3.377	–24	19(3)	–	1
$[S\cdots H\cdots S]^-$	2.58	4.24	4.00	–6	≤1	3.454	–18	13(1)	–	1
$[F\cdots H\cdots F]^+$	3.98	3.59	2.49	–31	≤6	2.284	–36	25(2)	2.292(2)	2
$[O\cdots H\cdots O]^+$	3.44	3.70	2.70	–27	≤5	2.360	–36	32(2)	2.42(2)	91
$[N\cdots H\cdots N]^+$	3.04	3.76	3.05	–19	≤3	2.592	–31	26(2)	2.70(5)	85
$[O\cdots H\cdots O]^\pm$	3.44	3.70	2.70	–27	≤5	2.381	–36	*28.7*		141
$[N\cdots H\cdots O]^\pm$	<3.24>	3.73	2.87	–23	≤4	2.506	–33	(15.2)		305
$[N\cdots H\cdots N]^\pm$	3.04	3.76	3.05	–19	≤3	2.682	–29	*16.4*		27

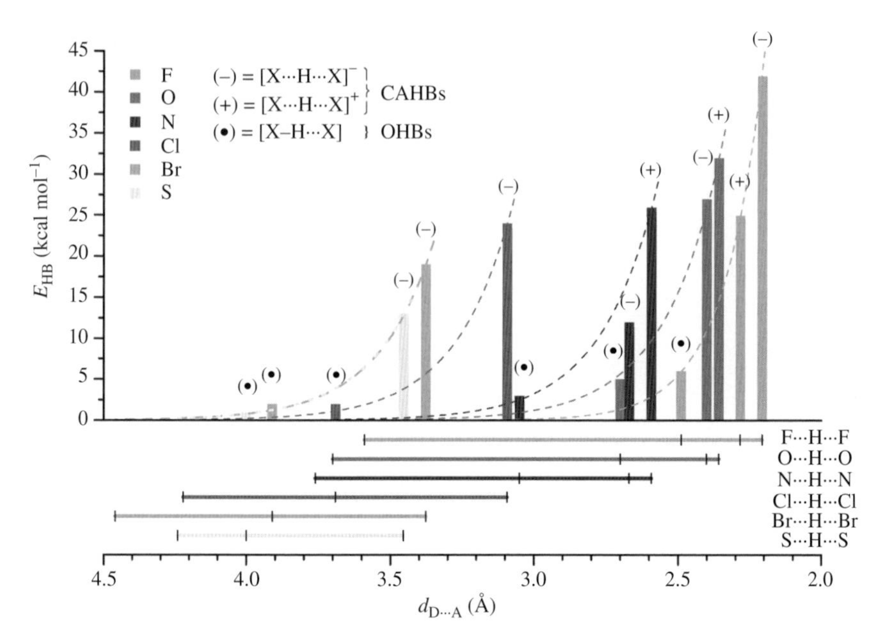

FIG. 3.37. Full range of H-bond energies and D···A distances spanned by the most common homonuclear bonds fulfilling, or not, the pK_a equalization condition (CAHBs and OHBs, respectively). Data from Table 3.9. H-bond energies, E_{HB}(OHB,min) or E_{HB}(CAHB,min) in kcal mol^{-1}, are marked as vertical bars occurring at their relative $d_{D\cdots A}$(OHB,min) or $d_{D\cdots A}$(CAHB,min) values in Å; horizontal lines on the bottom show the full ranges of the D···A distances involved from $d_{D\cdots A}$(vdW) to the shortest value $d_{D\cdots A} = d_{D\cdots A,min}$ having $E_{HB} = E_{HB,MAX}$. The dashed curves have equation $E_{HB} = E_{HB,MAX} \exp[-k(d_{D\cdots A} - d_{D\cdots A,min})]$ and represent the exponential dependence of E_{HB} on $d_{D\cdots A}$ according to eqn 3.33. (Reproduced by permission from Gilli *et al.*, 2009; see also Color Plate 1)

where $E_{HB,MAX}$ is the highest energy associated with the minimum $d_{D\cdots A,\,min}$ distance and k a least-squares constant. This equation can be used to interpolate the data using $k = 5.1$ for all bonds except F···H···F whose energy increase is steeper and a value of $k = 7.0$ is needed. The resulting functions, added to the figure as dashed curves, are seen to fit reasonably well all data except that for $[N\cdots H\cdots N]^-$, which is experimentally uncertain.

Table 3.9 and Fig. 3.37 represent the widest intercorrelation between H-bond distances and energies ever attempted. It has the obvious limit that experimental H–bond energies can be actually measured only for OHBs, (−)CAHBs, and (+)CAHBs, because no reliable experimental measures of energies are available for the other two strong CLs, namely RAHBs and (±)CAHBs. Nevertheless, the energies of *all homonuclear H-bonds* can be interpolated, in principle, from the dashed curves of Fig. 3.37 and, more precisely, by the use of eqn 3.33. For example, the shortest D···A distances of the $[O \cdots H \cdots O]^{\pm}$ and $[N \cdots H \cdots N]^{\pm}$ bonds are 2.381 and 2.682 Å, respectively, which allows estimatation of the corresponding maximum H-bond energies to be some 28.7 and 16.4 kcal mol^{-1} (values in italics).

Finally, no $E_{HB,MAX}$ value is experimentally accessible for the *heteronuclear N–H···O bond* though an approximate value of 15.2 kcal mol^{-1} (in parentheses) can be evaluated by the HB-LS method for the shortest (±)CAHB known having $d_{N \cdots O} = 2.506$ Å, so providing the tentative equation

$$E_{HB,N \cdots O} = 15.2 \exp[-5.1(d_{N \cdots O} - 2.506)] \text{ kcal mol}^{-1}.$$

A better approximation is probably achievable by computing more accurate $E_{HB,MAX}$ values by high-level quantum mechanical methods.

4

MODELLING THE H-BOND BY THERMODYNAMIC METHODS

4.1 Introduction

The PA/pK_a equalization, or PA/pK_a matching, principle is essentially the hypothesis that the strength of the H-bond can be accounted for in terms of the differences ΔPA = PA(D$^-$) – PA(:A) or ΔpK_a = pK_a(D–H) – pK_a(A–H$^+$). This hypothesis, already introduced to interpret the O–H···O bond in Section 3.3.1.3, is certainly not new, having been invoked several times in older H-bond studies performed in gas phase or solution by thermodynamic (Huyskens and Zeegers-Huyskens, 1964; Zeegers-Huyskens, 1986, 1988; Kebarle, 1977; Meot-Ner (Mautner), 1987), IR (Ault *et al.*, 1975; Malarski *et al.*, 1982; Barnes, 1983), and NMR (Cleland and Kreevoy, 1994; Frey *et al.*, 1994) methods, but the analysis performed has never been extensive enough to give to this basic idea the status of general rule (natural law) instead of that of plausible hypothesis. The problem can be now reconsidered by taking advantage of the ever more extended sources of thermodynamic acid–base indicators, including new pK_a compilations in aqueous solutions (Martell *et al.*, 2004), the NIST gas-phase PA database (Hunter and Lias, 2005), and two readily accessible and comprehensive sources of H-bond strengths: (*i*) the often-quoted CSD (Allen *et al.*, 1979; Allen, 2002), containing solid-state geometries of thousands of intra- and intermolecular H-bonded molecules; and (*ii*) the NIST database (Meot-Ner (Mautner) and Lias, 2005; Bartmess, 2005), collecting standard gas-phase dissociation enthalpies, $\Delta H°_{DIS}$, and entropies, $\Delta S°_{DIS}$, of hundreds of charged H-bonded complexes. This wealth of new data is not sufficient, however, to find a definitive solution of the problem by itself because it has been recently shown (Gilli *et al.*, 2007) that the combined application of PA/pK_a indicators, crystal geometries, and gas-phase dissociation enthalpies is not straightforward but implies a number of thermodynamic subtleties that need to be cleared up before developing an efficient method of analysis. This is what is attempted in the next sections.

4.2 The use of ΔPA and ΔpK_a indicators in H-bond studies

4.2.1 *PA and pK_a definitions*

The most common indicators of molecular acidity are PAs in the gas phase and pK_a(H_2O)s or, less often, pK_a(DMSO)s in solution. The thermodynamic cycles involved in their definition are given in eqns 4.1 and 4.2 distinguishing, both in the gas phase and in the solvent (S), the proton dissociation of a neutral

$$\mathrm{AH(g)} \underset{}{\overset{K^{\circ}_{\mathrm{AH}}}{\rightleftharpoons}} \mathrm{H^+(g) + A^-(g)} \quad (4.1a)$$

$$-\mathrm{S} \updownarrow +\mathrm{S} \qquad\qquad -\mathrm{S} \updownarrow +\mathrm{S}$$

$$\mathrm{AH(S) + S} \underset{}{\overset{K^{\ominus}_{\mathrm{AH}}}{\rightleftharpoons}} \mathrm{SH^+(S) + A^-(S)} \quad (4.1b)$$

$$\mathrm{BH^+(g)} \underset{}{\overset{K^{\circ}_{\mathrm{BH+}}}{\rightleftharpoons}} \mathrm{H^+(g) + B(g)} \quad (4.2a)$$

$$-\mathrm{S} \updownarrow +\mathrm{S} \qquad\qquad -\mathrm{S} \updownarrow +\mathrm{S}$$

$$\mathrm{BH^+(S) + S} \underset{}{\overset{K^{\ominus}_{\mathrm{BH+}}}{\rightleftharpoons}} \mathrm{SH^+(S) + B(S)} \quad (4.2b)$$

acid (AH) from that of a positively charged acid (BH^+) that is the Brønsted conjugate of the base (B).

This distinction is basic in H-bond studies where AH and B represent, respectively, the H-bond donor (D–H) and acceptor (:A). Thermodynamic parameters of the four reactions are:

$$\begin{aligned} \Delta_{\mathrm{AH}}G^{\circ} &= \Delta_{\mathrm{AH}}H^{\circ} - T\Delta_{\mathrm{AH}}S^{\circ} = -RT\ln K^{\circ}_{\mathrm{AH}} \\ &= \textit{gas-phase acidity of AH} = \textit{gas-phase basicity of } \mathrm{A^-}; \\ \Delta_{\mathrm{AH}}H^{\circ} &= \textit{proton affinity of } \mathrm{A^-} = \mathrm{PA(A^-)}; \end{aligned} \quad (4.3a)$$

$$\begin{aligned} \Delta_{\mathrm{AH}}G^{\ominus} &= \Delta_{\mathrm{AH}}H^{\ominus} - T\Delta_{\mathrm{AH}}S^{\ominus} = -RT\ln K^{\ominus}_{\mathrm{AH}} \\ &= 2.303\ RT\ \mathrm{p}K_{\mathrm{AH}} = 1.364\ \mathrm{p}K_{\mathrm{AH}}\ (\mathrm{kcal\ mol^{-1}\ at\ 25^{\circ}\ C}); \end{aligned} \quad (4.3b)$$

$$\begin{aligned} \Delta_{\mathrm{BH+}}G^{\circ} &= \Delta_{\mathrm{BH+}}H^{\circ} - T\Delta_{\mathrm{BH+}}S^{\circ} = -RT\ln K^{\circ}_{\mathrm{BH+}} \\ &= \textit{gas-phase basicity of B}; \\ \Delta_{\mathrm{BH+}}H^{\circ} &= \textit{proton affinity of B} = \mathrm{PA(B)}; \end{aligned} \quad (4.4a)$$

$$\begin{aligned} \Delta_{\mathrm{BH+}}G^{\ominus} &= \Delta_{\mathrm{BH+}}H^{\ominus} - T\Delta_{\mathrm{BH+}}S^{\ominus} = -RT\ln K^{\ominus}_{\mathrm{BH+}} \\ &= 2.303\ RT\ \mathrm{p}K_{\mathrm{BH+}} = 1.364\ \mathrm{p}K_{\mathrm{BH+}}\ (\mathrm{kcal\ mol^{-1}\ at\ 25^{\circ}\ C}). \end{aligned} \quad (4.4b)$$

Correct quantities for comparing acid–base properties of H-bond donors and acceptors should have dimensions of free enthalpies (i.e. Gibbs free energies), such as pK_a in solution and gas-phase acidity and basicity in the gas. Proton affinities, PA, having dimensions of enthalpies are often used in place of gas-phase basicities, so introducing an error believed to be small in view of the nearly constant value of ΔS° (18–27 cal K^{-1} mol^{-1}) typical of these gas-phase reactions. While PAs are accessible from a unique NIST database (Hunter and Lias, 2005), obtaining exhaustive lists of pK_a values is not an easy matter because of the extremely wide acidity range spanned by the H-bond phenomenon ($-15 \leq \mathrm{p}K_a \leq 53$) in respect to the autoprotolysis range of water ($0 \leq \mathrm{p}K_a \leq 14$), the interval in which pK_a(H_2O)s can easily be measured.

4.2.2. *Proton-transfer and proton-sharing H-bonds*

In principle, the application of PA/pK_a equalization methods seems the most practical and inexpensive way for the prediction of H-bond strengths. This is true provided a number of underlying thermodynamic problems are solved and perfectly understood. The first, discussed in this section, is that the different CLs are not equivalent in respect of their acid–base properties. The second, to be discussed in the next section, is that, while ΔpK_a values can be easily computed for all CLs, ΔPAs cannot for the two very important OHB and (±)CAHB CLs, thus strongly limiting the predictive ability of proton affinities.

As for the acid–base properties of CLs, it has been already shown (Section 3.3.1.3) that each one of the four strong CLs attains PA/pK_a equalization in a different way: (±)CAHBs by direct acid–base matching, (−)CAHBs and (+) CAHBs by matching induced by proton loss or addition, and RAHBs by matching induced by π-bond delocalization. These different mechanisms indicate that CLs are not chemically equivalent, a fact that can be rationalized in terms of the distinction between proton-transfer and proton-sharing H-bonds. Only (±) CAHBs can be considered to be true *proton-transfer H-bonds* where the proton is really *transferred* from an acid to a base

$$R_1{-}D{-}H\cdots{:}A{-}R_2 \rightleftarrows R_1{-}^{\frac{1}{2}-}D\cdots H^{+}\cdots A^{\frac{1}{2}-}{-}R_2 \rightleftarrows R_1{-}^{-}D{:}\cdots H{-}A^{+}{-}R_2 \qquad (4.5)$$

while (−)CAHBs, (+)CAHBs, and RAHBs are more properly classified as *proton-sharing H-bonds* where the proton is *shared* between two acids (4.6a), two bases (4.6b), or two groups that can be indifferently considered acids or bases (4.6c)

$$R_1{-}D_1{-}H\cdots{:}D_2^{-}{-}R_2 \rightleftarrows [R_1{-}D_1\cdots H\cdots D_2{-}R_2]^{-} \rightleftarrows R_1{-}^{-}D_1{:}\cdots H{-}D_2{-}R_2, \qquad (4.6a)$$

$$R_1{-}^{+}A_1{-}H\cdots{:}A_2{-}R_2 \rightleftarrows [R_1{-}A_1\cdots H\cdots A_2{-}R_2]^{+} \rightleftarrows R_1{-}A_1{:}\cdots H{-}A_2^{+}{-}R_2, \qquad (4.6b)$$

$$\cdots{:}A{=}R_n{-}D{-}H\cdots \rightleftarrows \cdots H\cdots A\overline{\cdots}R_n\overline{\cdots}D\cdots H\cdots \rightleftarrows \cdots H{-}A{-}R_n{=}D{:}\cdots. \qquad (4.6c)$$

Hence, proton-transfer H-bonds include what are normally presumed to be true acid–base or donor–acceptor H-bonds, that is all (±)CAHBs (where the matching occurs) or OHBs (where it does not), so forming a continuous scale of strengths going from weak $R_1{-}D{-}H\cdots{:}A{-}R_2$ to very strong $R_1{-}^{\frac{1}{2}-}D\cdots H^{+}\cdots A^{\frac{1}{2}-}{-}R_2$ and again to weak $R_1{-}^{-}D{:}\cdots H{-}A^{+}{-}R_2$ ionized bonds according to whether the $\Delta pK_a = pK_{AH}(R_1{-}D{-}H) - pK_{BH+}(R_2{-}A{-}H^{+})$ is large and positive, nearly zero, or large and negative, respectively.

Proton-sharing H-bonds are of two subtypes according to whether their sharing mechanism is of the charge-assisted (eqns 4.6a,b) or resonance-assisted type (eqn 4.6c). The latter (RAHB) represents the only class of strong H-bonds whose strength cannot be predicted by either ΔpK_a or ΔPA criteria because RAHB formation heavily affects the delocalization of the interleaving π-conjugated fragment with consequent large perturbation of the tabulated pK_a and PA values of the two interacting groups, which stresses the point that not even PA/pK_a

matching is the universal method for predicting H-bond strengths: the RAHB class of strong bonds remains, however, excluded.

The other two classes (eqns 4.6a,b: (−)CAHB and (+)CAHB) are now *acid–acid* (or *donor–donor*) and *base–base* (or *acceptor–acceptor*) *H-bonds*, respectively, whose ΔpK_a and ΔPA values are the differences of two acids [$\Delta pK_a = pK_{AH}(R_1{-}D_1{-}H) - pK_{AH}(R_2{-}D_2{-}H)$ and $\Delta PA = PA(R_1{-}D_1^-) - PA(R_2{-}D_2^-)$] or bases [$\Delta pK_a = pK_{BH+}(R_1{-}A_1{-}H^+) - pK_{BH+}(R_2{-}A_2{-}H^+)$ and $\Delta PA = PA(R_1{-}A_1) - PA(R_2{-}A_2)$], where the positional order between R_1 and R_2 is arbitrary. This leads to an important consequence that deeply affects the distribution of very strong bonds in nature: *whenever the H-bond is both homonuclear ($D_1 = D_2$ or $:A_1 = :A_2$) and homomolecular ($R_1 = R_2$) the special matching condition $\Delta PA = \Delta pK_a = 0$ will hold irrespective of the actual pK_as of the two interacting moieties.*

4.2.3 *Computing ΔPA and ΔpK_a values: The problem of ΔPA evaluation*

Evaluation of ΔPA poses some intricate thermodynamic problems that severely limit the scope of PA-matching methods. Let's call $\Delta PA^{D\cdots A}$ the PA difference between the H-bond donor, D, and acceptor, :A, experienced by the proton *inside the H-bond,* and ΔPA the usual difference between tabulated PA values. Clearly, $\Delta PA^{D\cdots A} = 0$ (and not $\Delta PA = 0$) is the condition that guarantees the formation of the proton-centered 3c–4e covalent interaction D···H···A, which is the real 1:1 mixture of the two D−H···:A and D:···H−A VB resonance forms. Hence, what is to be ascertained is whether $\Delta PA = 0$ actually verifies the required $\Delta PA^{D\cdots A} = 0$ condition for both proton-transfer and proton-sharing H-bonds.

By definition, ionization in the gas phase is a process for which the proton is extracted from the molecule and carried at an infinite distance from it. Accordingly, the difference $\Delta PA = PA(A^-) - PA(B) \equiv PA(D^-) - PA(A)$ is the enthalpy needed to transfer the proton from the acid to the base (or from the H-bond donor to the H-bond acceptor) only when the protonated (p) and deprotonated (dep) forms, with respective charges z_p and z_{dep}, are at infinite distance from each other. Hence, an additional term (Perchard, 1981; Maes, 1991) must be included in the expression for $\Delta PA^{D\cdots A}$ to account for the work possibly made by the electrical forces to take back the H-bond donor and acceptor from infinity to the H-bond distance, R, that is

$$\begin{aligned}\Delta PA^{D\cdots A} &= \Delta PA - z_p z_{dep}\int_{\infty}^{R}(e^2/r^2)\mathrm{d}r = \Delta PA + z_p z_{dep}(e^2/R)\\ &= \Delta PA + z_p z_{dep}\,332.15/R\ (\mathrm{kcal\ mol^{-1}\ \AA^{-1}}).\end{aligned} \quad (4.7)$$

This electrical term is controlled by the product $z_p z_{dep}$ that can only assume the values $(1)\cdot(-1) = -1$, or $(\pm 1)\cdot(0) = 0$. In proton-transfer H-bonds (eqn 4.5: (±)CAHBs and OHBs) such a product is equal to −1, so that

$$\begin{aligned}\Delta PA^{D\cdots A} &= PA(D^-) - PA(A) + z_p z_{dep}(e^2/R)\\ &= \Delta PA - 332.15/R\ (\mathrm{kcal\ mol^{-1}\ \AA^{-1}}),\end{aligned} \quad (4.8a)$$

while in proton-sharing H-bonds (eqns 4.6a,b: (−)CAHBs and (+)CAHBs) this product is null and the two definitions of ΔPA become identical

$$\Delta PA^{D\cdots A} = PA(D_1^-) - PA(D_2^-) = \Delta PA, \tag{4.8b}$$

$$\Delta PA^{D\cdots A} = PA(A_1) - PA(A_2) = \Delta PA. \tag{4.8c}$$

In conclusion, ΔPA is an excellent indicator of $\Delta PA^{D\cdots A}$ only for proton-sharing H-bonds where the electrostatic term vanishes and, as such, has been extensively used to interpret the (−)CAHB and (+)CAHB standard dissociation enthalpies, $\Delta H°_{DIS}$, obtained from gas-phase thermodynamic measurements.

Conversely, ΔPA is not a proficient indicator of $\Delta PA^{D\cdots A}$ for proton-transfer H-bonds because of the indetermination of the value of R, a quantity unclear in its very definition of 'distance between the plus and minus charges in the ion pair'. A coarse estimate of R can be obtained starting from the consideration that, just because of this electric term, the PA scale of acids (or H-bond donors) is shifted upwards with respect to that of bases (or H-bond acceptors) by some 120 kcal mol^{-1}, which obviously represents the work needed to take the two charges generated from the ionization reaction from R to infinity. This gives $\langle R \rangle \approx 332.15/120 \approx 2.227$ Å, a value quite reasonable from a physical point of view but too inexact to be of practical use. To overcome this difficulty Pimentel and coworkers (Ault *et al.*, 1975) proposed a semiempirical parameter called the 'normalized proton affinity difference' $\Delta = [PA(A) - PA(D^-)] \;/\; [PA(A) + PA(D^-)]$ and used it to correlate H-bond IR stretching-frequency data. It is evident that this artifice, though moderately efficient, does not have an unequivocal thermodynamic meaning and then cannot be considered as a reliable basis for an accurate H-bond analysis.

Such difficulties do not subsist in solution where no electric work is involved because the proton is exchanged within the close environment of the solvated ions and, accordingly, ΔpK_a turns out to be a reasonably good indicator of pK_a equalization both for proton-transfer and proton-sharing H-bonds, though two minor aspects connected with its use need to be briefly considered. The first is how strictly strong H-bonds have to fulfil the matching condition $\Delta pK_a = 0$. In practice, it is found that the pK_a matching does not need to be so perfect and a widespread analysis of experimental data may suggest that misfits of ± 2.5 pK_a units are still perfectly compatible with strong H-bond formation. The second arises from the circumstance that pK_as are normally measured in water, a solvent of high dielectric constant that tends to stabilize ionic against neutral H-bonds, while bond strengths are normally measured in a non-polar environment, solids and gas phase (Huyskens and Zeegers-Huyskens, 1964; Malarski *et al.*, 1982; Huyskens *et al.*, 2002; Pearce and Simkins, 1968; Johnson and Rumon, 1965; Sobczyk, 1998). Calling ΔpK_a(critical) the ΔpK_a for which the proton transfer amounts to 50%, there are indications that this quantity is not necessarily zero but may assume small negative values depending on the particular experimental environment. For the solid state, in particular, it has been reported that the ΔpK_a(critical) for both ^{35}Cl NQR and diffraction experiments is nearly −1.5 pK_a units, suggesting that, in crystal structures, the proton transfer may occur only when ΔpK_a is smaller than −1.5 and that the best interval of close pK_a matching is to be shifted in the range $-4.0 \leq \Delta pK_a \leq 1.0$.

4.2.4 *The use of* PA *and* pK_a *as predictors of the H-bond strength: A summary*

Previous discussion leads to the practical rules summarized in Table 4.1. The table can be considered to define the intersection domain between two great areas of chemistry, acid–base theory and the H-bond, that, though apparently distinct, actually deal with the same phenomenon: the sharing and the transfer of *protons* (or, more properly, *hydrons*) among molecules. It makes clear that the comparison between PA/pK_a acid–base indicators and H-bond strength estimates derived from crystal geometries, say $d(\text{D}\cdots\text{A})$ or $d(\text{H}\cdots\text{A})$, or gas-phase dissociation enthalpies, $\Delta H°_{\text{DIS}}$, is in no way a method of general applicability because of the many restrictions subsisting. Only (−)CAHBs and (+)CAHBs do not have constraints because both crystal and gas-phase data are accessible and interpretable through either ΔpK_a or ΔPA parameters. The worst happens with RAHBs where only crystal data are obtainable, but cannot be interpreted in terms of any acid–base indicator. The proton-transfer leitmotifs, (±)CAHB and OHB, represent an intermediate case where crystal data can be obtained but interpreted only in terms of ΔpK_a.

In conclusion, two combinations appear more promising for predicting H-bond strengths from acid–base parameters:

(*i*) the *gas-phase enthalpies* – ΔPA *combination*, which has the great advantage of providing a direct knowledge of the H-bond energies but permits treatment of only the two proton-sharing CLs, (−)CAHB and (+) CAHB;

TABLE 4.1. Summary of thermodynamic acid–base indicators (ΔpK_a and ΔPA) that can be used to interpret crystal and gas-phase H-bond strength data, subdivided for chemical leitmotifs. (From Gilli *et al.*, 2007)

Experimental method	Measured quantity	Chemical leitmotif (CL)			
		(±)CAHB and OHB	(−)CAHB	(+)CAHB	RAHB
		Proton transfer	Proton sharing		
X-ray and neutron crystallography	$d(\text{D}\cdots\text{A})$ or $d(\text{H}\cdots\text{A})$	Both data and analysis (ΔpK_a)	Both data and analysis (ΔpK_a and ΔPA)	Both data and analysis (ΔpK_a and ΔPA)	Data but no analysis
Gas-phase thermodynamics	$\Delta H°_{\text{DIS}}$	No data and no analysis	Both data and analysis (ΔpK_a and ΔPA)	Both data and analysis (ΔpK_a and ΔPA)	No data and no analysis

(*ii*) the *crystal data* – ΔpK_a *combination*, which allows treatment of four CLs, the two proton-sharing (–)CAHB and (+)CAHB and the two proton-transfer (±)CAHB and OHB leitmotifs, the last of which includes the largest fraction of the H-bonds existing in nature.

Applications of these two methods will be discussed in the following Sections 4.3 and 4.4. To notice that, in principle, acid-base indicators can also be used in combination with many other physicochemical techniques (such as IR, NMR, NQR spectroscopies, inelastic neutron scattering, dipole moments, etc.) obtaining results of great potential interest for H-bond studies. Their analysis has not even been attempted here because the data needed have never been organized in databases and then cannot be analyzed in a systematic way.

4.3 Predicting (–)CAHB and (+)CAHB strengths from enthalpy versus proton affinity correlations

4.3.1 $\Delta H°_{DIS}$ *against* ΔPA *correlations*

Though crystal geometries can also be used in principle, the method of choice for interpreting and predicting (–)CAHB and (+)CAHB strengths is based on $\Delta H°_{DIS}$ versus ΔPA correlations, a method having the great advantage that both quantities come from a common source of thermodynamic gas-phase data (Linstrom and Mallard, 2005). In consequence, a large number of these correlations have been published starting from the 1970s for which the reader is referred to a recent review on the subject (Meot-Ner (Mautner), 2005). Most of these correlations are linear

$$\Delta H°_{DIS} = a - b\,\Delta \mathrm{PA}, \tag{4.9}$$

and include only a small number of H-bonds whose donor or acceptor groups are limited by chemical similarities (Meot-Ner (Mautner), 1984, 1987, 1988; Meot-Ner (Mautner) and Sieck, 1985, 1986; Larson and McMahon, 1983, 1984; Davidson *et al.*, 1979; Speller and Meot-Ner (Mautner), 1985; Humbel, 2002). The constant a is generally interpreted in terms of PA matching as the $\Delta H°_{DIS}$ value of the strongest and most symmetric bond of the series. The constant b varies from 0.15 to 0.59 depending on the system and its physical meaning has not been clearly explained so far. More complex correlations have also been proposed, such as the three-parameter equation

$$\Delta H°_{DIS} = a - b\,\mathrm{PA(D)} + b'\,\mathrm{PA(A)}, \tag{4.10}$$

where D is the donor and A the acceptor (Caldwell *et al.*, 1984), the exponential equation

$$\Delta H°_{DIS} = C_1 \exp(-C_2\,\Delta \mathrm{PA}), \tag{4.11}$$

which reportedly performs better than the linear ones in fitting a larger spectrum of H-bond strengths (Zeegers-Huyskens, 1986, 1991, 1999) and, more recently, the relation

$$\Delta H°_{DIS} = k\,x - \alpha, \tag{4.12}$$

where α and k are constants to be evaluated and x = PA(A)/PA(D) (Bian, 2003).

In spite of the many efforts made, it cannot be said that the search for a 'universal equation' to be used for most (if not all) different H-bonded systems has been so far successful and we can only be confident that it will be found in the nearest future.

4.3.2 *A verification of the* PA *equalization principle*

$\Delta H°_{DIS}$ versus ΔPA correlations can, in theory, provide one of the most striking confirmations of the PA equalization principle. This result cannot be achieved, however, by a straightforward analysis of raw data because the use of tabulated $\Delta H°_{DIS}$ values raises some tricky problems that need to be previously considered.

$\Delta H°_{DIS}$ is defined as the experimental standard enthalpy of the dissociation equilibrium of the H-bonded complex *as written* from left to right but does not necessarily represent the H-bond enthalpy, $\Delta H°_{HB}$, a quantity that is to be defined as the enthalpy gap leading from the complex to the lower (i.e. more stable) of the two dissociated forms. This fact is illustrated in Chart 4.1 for the particular case of the $[D_1\cdots H\cdots D_2]^- \rightleftharpoons D_1{-}H + D_2^-$ equilibrium (but the scheme would be identical for the $[A_1\cdots H\cdots A_2]^+ \rightleftharpoons A_1{-}H^+ + A_2$ case). It is seen that the evaluation of $\Delta H°_{HB}$ can be reduced to a simple algorithm: when the $D_1{-}H + D_2^-$ form is really the more stable (C4.1.a), the value of $\Delta PA = PA(D_1^-) - PA(D_2^-)$ is positive and $\Delta H°_{HB} = \Delta H°_{DIS}$. When, on the contrary, it is the less stable one, ΔPA will be negative and $\Delta H°_{HB} = \Delta H°_{DIS} + \Delta PA$ (C4.1.b).

CHART 4.1. RELATIONSHIP BETWEEN THE H-BOND ENTHALPY, $\Delta H°_{HB}$, AND THE STANDARD ENTHALPY OF THE GAS-PHASE DISSOCIATION EQUILIBRIUM OF THE H-BONDED COMPLEX, $\Delta H°_{DIS}$

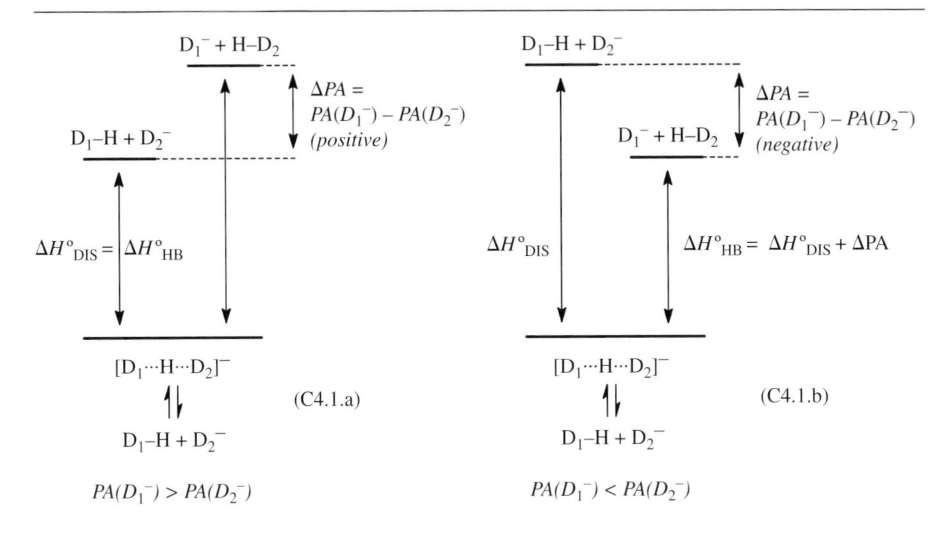

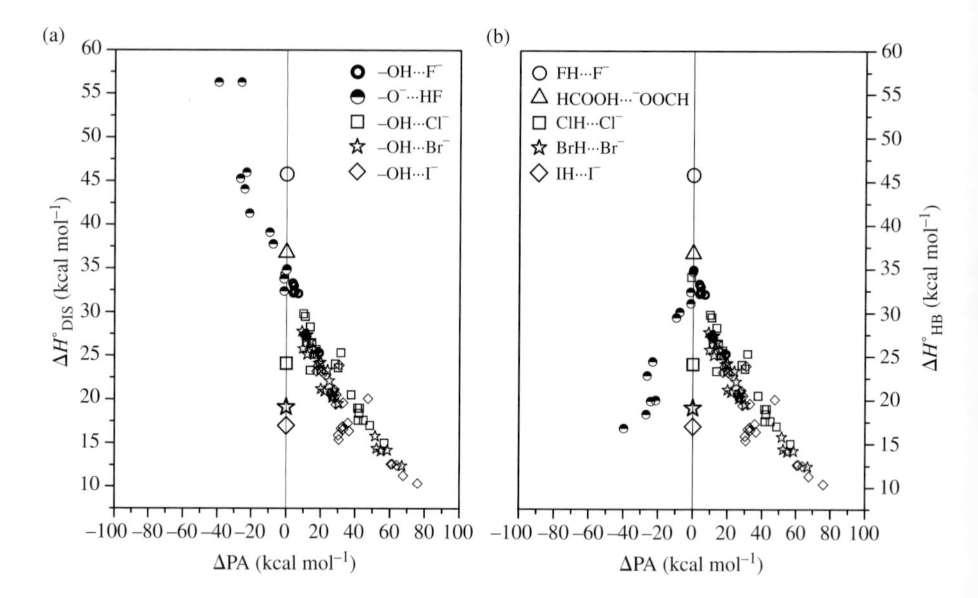

FIG. 4.1. (a) ΔH°_{DIS} versus ΔPA plot for the gas-phase $[-O\cdots H\cdots X]^- \rightleftharpoons -O-H+X^-$ dissociation equilibrium where X^- are halide ions (F^-, Cl^-, Br^-, and I^-) and $-O-H$ a series of organic alcohols and acids. (b) The corresponding ΔH°_{HB} versus ΔPA plot for the same set of H-bonded complexes. The ΔH°_{DIS} to ΔH°_{HB} conversion is accomplished by the method outlined in Chart 4.1. (Reproduced by permission from Gilli *et al.*, 2007)

The importance of the conversion from ΔH°_{DIS} to ΔH°_{HB} is illustrated in Fig. 4.1 for the well-known gas-phase dissociation equilibrium $[-O\cdots H\cdots X]^- \rightleftharpoons -O-H + X^-$ (Yamdagni and Kebarle, 1971; Larson and McMahon, 1983; Caldwell and Kebarle, 1984), where X^- indicates the halide ions (F^-, Cl^-, Br^-, and I^- with respective PAs of 371.3, 333.4, 323.5, and 314.3 kcal mol^{-1}) and $-O-H$ a series of organic alcohols ($373 \leq$ PA ≤ 391 kcal mol^{-1}) and acids ($345 \leq$ PA ≤ 371 kcal mol^{-1}). These PA values indicate that the $-O-H + X^-$ dissociated form is the physically correct one for all H-bonded complexes except those formed by the organic acids with fluorides where, because of the lower acidity of HF, the stable form is $-O^- +$ HF. Because of this, these latter complexes have negative ΔPAs and their ΔH°_{DIS} values increase systematically while moving away from the vertical line of $\Delta PA = 0$, as shown in the ΔH°_{DIS} versus ΔPA plot of Fig. 4.1(a). This representation is formally correct from a thermodynamic point of view and, as such, has been adopted in other papers (Zeegers-Huyskens, 1986, 1991). Its main drawback is that it seems to violate the PA equalization rule for which the strongest H-bonds are *always* to be associated with the condition of ΔPA annihilation. This violation is shown to be much more apparent than real by the parallel ΔH°_{HB} versus ΔPA plot of Fig. 4.1(b), where, after the ΔH°_{DIS} to ΔH°_{HB} conversion according to the rules of Chart 4.1, the graph is reduced to the shape of a cusp centered around the zero of the ΔPA scale that would be expected in terms of the PA/pK_a equalization principle.

4.4 Predicting H-bond strengths from crystal geometry versus pK_a correlations

4.4.1 pK_a *tables for the most common H-bond donors and acceptors*

There are two main series of acidity constants measured in water, $pK_a(H_2O)$, and in dimethyl-sulphoxide, pK_a(DMSO), which could constitute a valid basis for the prediction of H-bond strengths. However, the number of values for the latter (Bordwell, 1988; Reich, 2008) is still considerably limited so that the only really accessible source of acid–base indicators in solution remains limited to the values measured in (or reduced to) water. The main difficulty arises from the fact that the acidity range of common H-bond donors and acceptors is extremely wide ($-15 \leq pK_a \leq 53$) while the normal water scale is rigorously defined only within its autoprotolysis range ($0 \leq pK_a \leq 14$). Values must therefore be expanded by measurements carried out in solvents more acidic or more basic than water and then rescaled to water by various methods though, of course, the final values will become the less accurate the farther from the water range they are.

Present values were mostly taken from previous compilations (Maskill, 1985; Cookson, 1974; Smith and March, 2001; Lide, 2006; Martell *et al.*, 2004) and complemented by a number of single values from the literature. Since all data refer to water, the short notation pK_a is used instead of $pK_a(H_2O)$ while the symbols pK_{AH} and pK_{BH+} are used to distinguish between the dissociation of acids and protonated bases. Final data are arranged for chemical functionality (inorganic acids, C–H acids, organic acids as H-bond donors; organic oxygen, nitrogen and phosphorous bases as H-bond acceptors) in Tables 4.2 and 4.3 given in the Appendix of this chapter. The assignment of the correct pK_a to some chemical groups, particularly bases, has raised several non-simple problems. As an example, acetamide ($CH_3C(=O)NH_2$) is certainly a N–H acid but could be either an =O or ≡N base. Here, a relatively wide literature (Smith and March, 2001) suggests that amides, as well as carboxylic acids and esters, are normally protonated at the carbonyl oxygen but, in other instances, no such information was available. In these cases, protonation sites were attributed by comparison with real H-bonds retrieved from crystallographic sources. Data of Tables 4.2 and 4.3 have been arranged for classes of chemical functionalities in a unique bar chart called *the* pK_a *slide rule* (Gilli *et al.*, 2009) which provides a simple visual method for the quick determination of the ΔpK_a values (Fig. 4.2 and Color Plates 3 and 4).

4.4.2 *The* pK_a *slide rule*

The pK_a slide rule summarizes, in a pictorial form, the complex relationships between H-bond and acid–base properties. Being D–H donors (or A–H acids) on the right and :A acceptors (or B bases) on the left, three H-bond types remain defined: right-right, left-left, and left-right bonds corresponding, respectively, to proton-sharing acid–acid (–)CAHBs and base–base (+)CAHBs and to

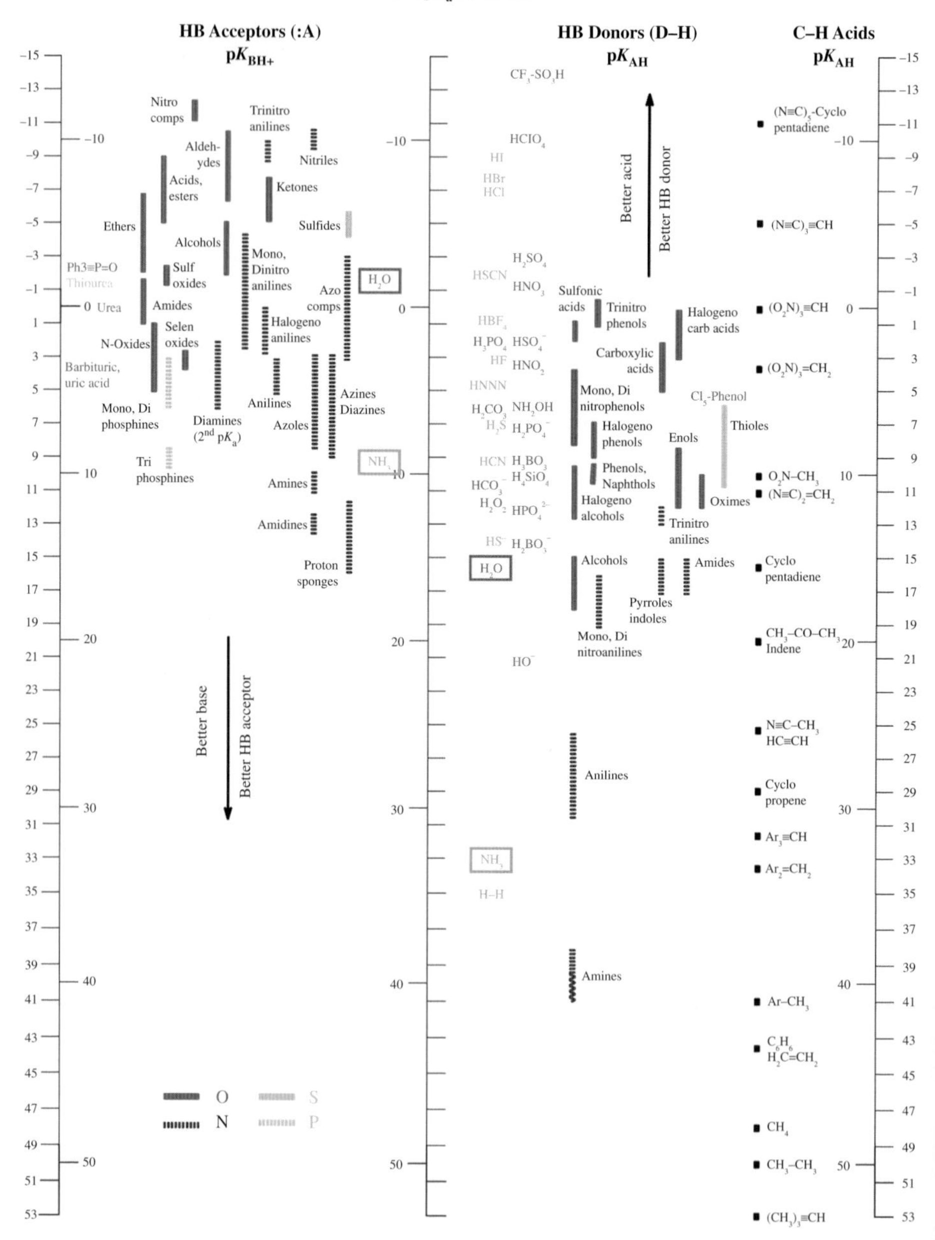

FIG. 4.2. The pK_a slide rule. The figure represents, in the form of a bar chart, the pK_a values of the most important H-bond donor (right) and acceptor (left) molecules. Strong H-bonds with strict pK_a matching are to be formed by donors and acceptors lying on a same horizontal line. Donor and acceptor atoms as coded in the lower-left corner. (Reproduced by permission from Gilli *et al.*, 2009; see also Color Plates 3 and 4)

proton-transfer acid–base (±)CAHBs and OHBs. For the two proton-transfer H-bonds, the vertical direction of the plot is the coordinate of proton transfer, the positive or negative sign of $\Delta pK_a = pK_a(\text{right}) - pK_a(\text{left}) = pK_a(\text{D–H}) - pK_a(\text{A–H}^+)$ distinguishing between neutral D–H···:A and ionized $^-$D:···H–A$^+$ bonds. Therefore, OHBs will be neutral or doubly charged according to whether the acid (on the right side) is lower or higher than the base (on the left side) while, of course, strong (±)CAHBs remain defined as the bonds where donor and acceptor face each other.

The slide rule allows us to have, at a glance, a comprehensive view of the most common types of H-bond donors and acceptors as well as of the features of the bonds they are going to form.

C–H acids display the greatest spread of values, $-11 \leq pK_a \leq 53$, practically covering the full range of the H-bond phenomenon. It was reported in Section 2.4.3.1 that the pK_a(DMSO)s of C–H donors are correlated with crystal C···O distances in a wide set of C–H···:O bonds with carbonyl acceptors (Pedireddi and Desiraju, 1992). Unfortunately, there are few data for very acidic C–H donors that could match the pK_a of the carbonyl acceptor and then form strong C–H···:O bonds. The only known structure (Bock *et al.*, 1993) concerns the adduct of trinitromethane ($pK_a = 0.0$) with dioxane ($pK_a \approx 2.1$) that, being not far from full matching, displays a C···O distance of 2.936 Å, probably the shortest ever observed.

Other organic H-bond donors lie in the interval $-1 \leq pK_a \leq 40$, the strongest acids being trinitrophenols ($-0.7 \leq pK_a \leq 0.33$) and the weakest ones organic amines (pK_a around 39–40). The order of decreasing acidity is carboxylic acids > phenols > alcohols and amides > anilines > amines and, in each class, the acidity is seen to be enhanced by halogenation or nitration. Thioles ($6.5 \leq pK_a \leq 11$), enols ($8.5 \leq pK_a \leq 12$), and oximes ($10 \leq pK_a \leq 12$) are significantly more acidic than the corresponding alcohols ($15 \leq pK_a \leq 18$).

The full range of H-bond acceptors ($-12 \leq pK_a \leq 16$) is shifted upwards with respect to H-bond donors. In consequence, a large group of acceptors (nitro and carbonyl compounds, nitriles, ethers, alcohols, and sulphoxides) falls in a region facing only a few inorganic acids but no organic donors with which they are then expected to form only rather weak H-bonds. The same happens for the weakest H-bond donors (amines, anilines, and alcohols) that do not face any known acceptor and can then form only weak and neutral D–H···:A bonds. The H-bond is therefore a dissymmetric phenomenon where neutral are favored over ionic H-bonds and this reflects the basic dissymmetry between proton donors, which cover the full range of accessible pK_a values, and proton acceptors, which are seen to cover less than one half of it.

The maximum overlap between donors and acceptors occurs in the interval $0 \leq pK_a \leq 14$ (corresponding, perhaps not by chance, to the autoprotolysis range of water) where the greatest part of strong H-bonds are then expected to occur. It has been remarked above that misfits of ±2.5 pK_a units can still be compatible with strong H-bond formation and that this interval has to be downshifted to

$-4.0 \leq pK_a \leq 1.0$ to compensate for the ionizing properties of water. As a consequence, H-bonds are predicted to be in the $^-$D:···H–A$^+$ ionized form only when the acceptor on the left is at least some 4 pK_a units lower than the donor on the right.

Finally, the slide rule permits appreciation of the enormous differences of properties that there are among the most common and, we could say, familiar types of H-bonds. For instance, we speak without distinction of amine–nitroderivative, aniline–ketone, water–water, carboxylic acid–pyridine, and perchloric acid–aniline H-bonds, without considering that their ΔpK_as are as different as 50, 36, 17.4, −1, and −15 pK_a units or, being 1 pK_a unit = 1.364 kcal mol^{-1} at 25 °C (eqns 4.3,4), as 68, 49, 24, −1.4, and −21 kcal mol^{-1} and that this impressive series of numbers actually represent the energy difference between their tautomeric $^-$D:···H–A$^+$ and D–H···:A forms.

4.4.3 *Two projects for validating the* pK_a *equalization principle*

Full verification of the pK_a equalization rule is clearly impracticable in front of the tens of thousands of H-bonded crystal structures and thousands of combining molecules with often unknown or uncertain pK_a values. We'd better try to develop smaller projects that, though retaining reasonably good diagnostic capability, need to handle more limited sets of structural data and can circumvent, in some way, the chronic lack of accurate pK_a values (Gilli *et al.*, 2009).

4.4.3.1 *First project.* pK_a *equalization in CAHBs with* ΔpK_a *near zero*

This project was intended to verify that all CAHBs having small or null ΔpK_a values are *always* associated with very short (strong) H-bonds. A first analysis was performed on homomolecular (−) and (+)CAHBs that have ΔpK_a= 0 by definition. A systematic search of the CSD and ICSD databases, carried out paying particular attention to data quality, has led to the identification of 266 homomolecular [D···H···D]$^-$ bonds with D = F, O, Cl, N, Br, S and 178 homomolecular [A···H···A]$^+$ bonds with A = F, O, N. These data, summarized in Table 3.9 and already discussed in Section 3.3.4, have shown that the transition from OHBs to the corresponding (−) and (+)CAHBs is associated with a shortening of the $d_{D\cdots A}$ contact distances of such entities to fully confirm the validity of the equalization principle.

The following analysis was aimed to verify the same fact for (±)CAHBs. In this case ΔpK_a= pK_a(D–H) − pK_a(A–H$^+$) is changing with the nature of the interacting molecules and the procedure adopted was to search the database for the shortest D···A distances up to a cutoff and then to check whether they were really associated with small ΔpK_a. Not more than one third of the pK_a values could be retrieved from tabular sources and the others were interpolated on a chemical-class criterion, obtaining final estimates in the plausible $-4 \leq \Delta pK_a \leq 4$ interval. At the end, 141 O–H···O, 305 N–H···O, and 27 N–H···N bonds were retained that are summarized in Table 3.9 and confirm beyond doubt the equalization rule.

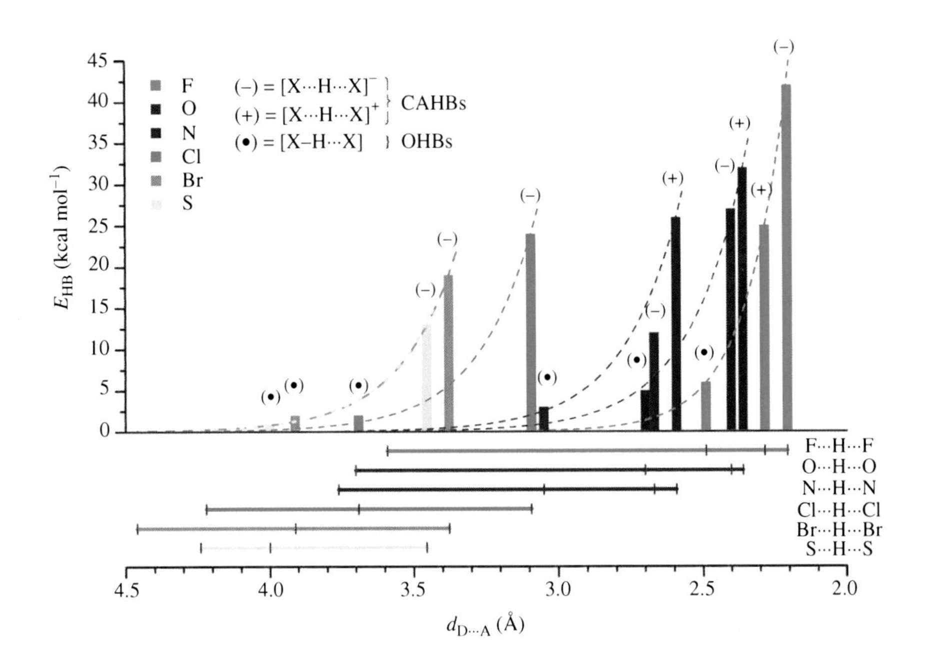

PLATE 1. Correlation between H-bond energies and D···A distances in the most common homonuclear bonds fulfilling, or not, the pK_a equalization condition, (–) and (+)CAHBs or OHBs, respectively. Data from Table 3.9 and Figure 3.37. H-bond energies, E_{HB} in kcal mol^{-1}, are marked as vertical bars corresponding to the shortest D···A distances (Å) observed for charge-assisted and ordinary bonds; horizontal lines on the bottom show the full ranges of the D···A distances involved from $d_{D...A}$(vdW) to the shortest value $d_{D...A} = d_{D...A,min}$ having $E_{HB} = E_{HB,MAX}$. The dashed curves represent the exponential dependence of E_{HB} on $d_{D...A}$ according to eqn 3.3.3: $E_{HB} = E_{HB,MAX} \exp[-k\,(d_{D...A} - d_{D...A,min})]$. (Reproduced by permission from Gilli *et al.*, 2009)

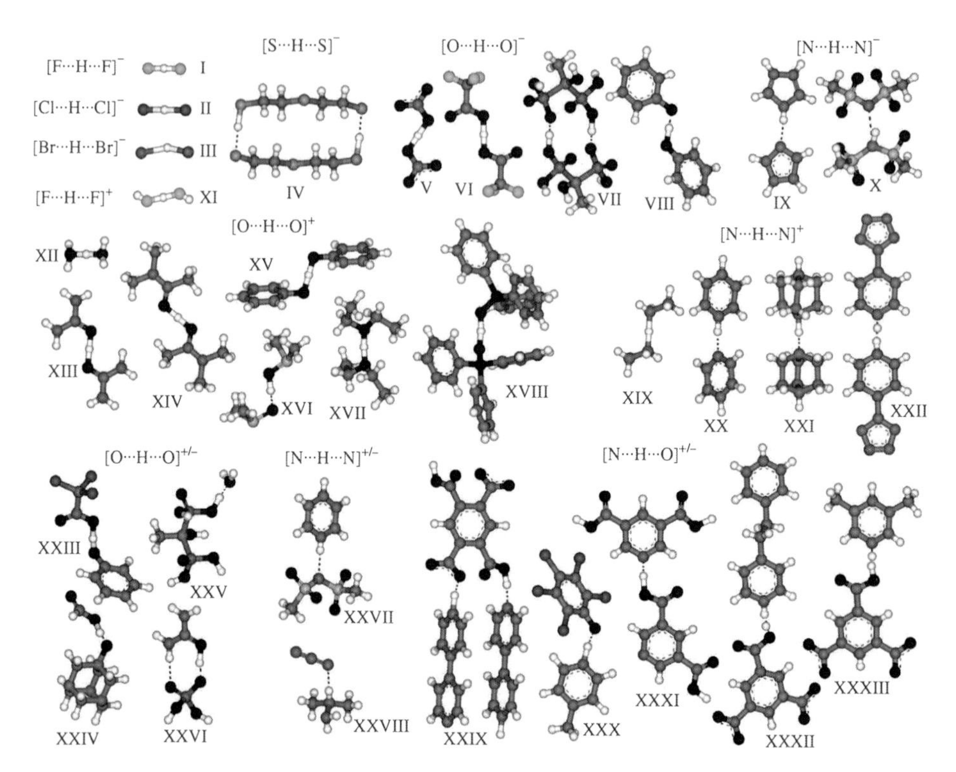

PLATE 2. Famous H-bonds: Some of the most characteristic examples of molecules forming strong homomolecular (−)CAHBs and (+)CAHBs with $\Delta pK_a = 0$ and (±)CAHBs selected for their small ΔpK_a values. Each structure is identified by the string: CSD/ICSD Refcode, X/Neutron-Temperature; $d_{D\cdots A}$ (Å), ΔpK_a. **[D···H···D]⁻**. **I.** Hydrogen-difluoride (FEDDOH, X-295K; 2.238, 0.0) (Farnham *et al.*, 1987); **II.** Hydrogen-dichloride (CERVAW, X-295K; 3.108, 0.0) (Swann *et al.*, 1984); **III.** Hydrogen-dibromide (KEGXUP, X-295K; 3.377, 0.0) (Atwood *et al.*, 1990); **IV.** 3-Thiapentane-1-thiol-5-thiolate (YAJHEW, X-171K; 3.454, 0.0) (Boorman *et al.*, 1992); **V.** Hydrogen-dinitrate (AFUZAC, X-295K; 2.439, 0.0) (Huta *et al.*, 2002); **VI.** Hydrogen-bis(trifluoroacetate) (KHFLAC01, N-295K; 2.419, 0.0) (Macdonald *et al.*, 1972); **VII.** Trihydrogen-1-hydroxyethane-1,1-diphosphate (SATHIE01, N-295K; 2.433, 0.0 (Silvestre *et al.*, 1990); **VIII.** Hydrogen-bis(pyridinium-2-hydroxylate) (GOHHOA, X-143K; 2.418, 0.0) (Wijaya *et al.*, 1999); **IX.** Hydrogen-dipyrrolide (YAKXIS, X-153K; 2.697, 0.0) (Heldt and Behrens, 2005); **X.** Hydrogen-bis(bis(methylsulphonyl) amidate (RUJDAB, X-173K; 2.669, 0.0) (Wijaya *et al.*, 1997); **[A···H···A]⁺**. **XI.** Hydron-bis(hydrogen fluoride) (71453-ICSD, X-295K; 2.284, 0.0) (Mootz and Bartmann, 1991); **XII.** Hydron-bis(water) (COLNUM01, N-20K; 2.430, 0.0) (Cotton *et al.*, 1984); **XIII.** Hydron-bis(urea) (BADCIS10, X-295K; 2.424, 0.0) (Zhang *et al.*, 1981); **XIV.** Hydron-bis(*N,N*-dimethylacetamide)

(HDMAAU01, N-295K; 2.432, 0.0) (Hussain and Schlemper, 1980); **XV.** Hydron-bis(pyridine-*N*-oxide) (BALGUQ10, X-295K; 2.410, 0.0) (Hussain and Schlemper, 1982); **XVI.** Hydron-bis(dimethylsulphoxide) (JIMKAR01, X-295K, 2.414, 0.0) (Jaswal *et al.*, 1990); **XVII.** Hydron-bis(diethylether) (DIDTIT, X-203K; 2.394, 0.0) (Kolesnikov *et al.*, 1985); **XVIII.** Hydron-bis(triphenylphosphineoxide) (TPOPHX01, X-153K; 2.416, 0.0) (Antipin *et al.*, 1980); **XIX.** Hydron-bis(methylamine) (ROHTIR, X-200K; 2.620, 0.0) (Bock *et al.*, 1997); **XX.** Hydron-bis(pyridine) (PYDMPS, X-295K; 2.656, 0.0) (Minshall and Sheldrick, 1978); **XXI.** Hydron-bis(quinuclidine) (YERLOW, X-200K; 2.637, 0.0) (Bock *et al.*, 1994); **XXII.** Hydron-bis(4-tetrazolopyridine) (AYADOT, X-90K; 2.631, 0.0) (Mautner *et al.*, 2004); $[\mathbf{D}\cdots\mathbf{H}\cdots\mathbf{A}]^{\pm}$ **. XXIII.** Pyridine-*N*-oxide–Trichloroacetic acid (PYOTCA01, N-120K; 2.430, −1.3) (Eichhorn, 1991); **XXIV.** HMTA-*N*-oxide–Formic acid ((HMTOFA07, N-123K; 2.428, −0.9) (Nygren *et al.*, 2005); **XXV.** 1-Hydroxyethane-1,1-diphosphoric acid–Water (ETHDPH01, N-293K; 2.437, 3.3) (Silvestre *et al.*, 2002); **XXVI.** Urea–Phosphoric acid (CRBAMP01, N-100K; 2.409, 2.1) (Rodrigues *et al.*, 2001); **XXVII.** Pyridine–Bis(methylsulphonyl)amide (QUSLEV, X-143K; 2.797, −2.4) (Moers *et al.*, 2001); **XXVIII.** 1,1-Dimethylhydrazine–Hydrazoic acid (CORRUW, X-173K; 2.762, −3.4) (Klapotke *et al.*, 1999); **XXIX.** 4,4'-Bipyridine–Benzene-1,2,4,5-tetracarboxylic acid (WISNAN01, N-20K; 2.522, −1.6) (Cowan *et al.*, 2003); **XXX.** 4-Methypyridine–Pentachlorophenol (RAKQAV, N-20K; 2.506, −0.7) (Steiner *et al.*, 2001); **XXXI.** Dinicotinic acid (DINICA11, N-15K; 2.523, 1.0) (Cowan *et al.*, 2005); **XXXII.** 1,2-Dipyrid-4-ylethane–3,5-Dinitrobenzoic acid (QIBSOJ, X-150K; 2.521, −2.2) (Burchell *et al.*, 2001); **XXXIII.** 3,5-Dimethylpyridine–3,5-Dinitrobenzoic acid (PUHROZ, X-80K; 2.529, −3.4) (Jerzykiewicz *et al.*, 1998). (Reproduced by permission from Gilli *et al.*, 2009)

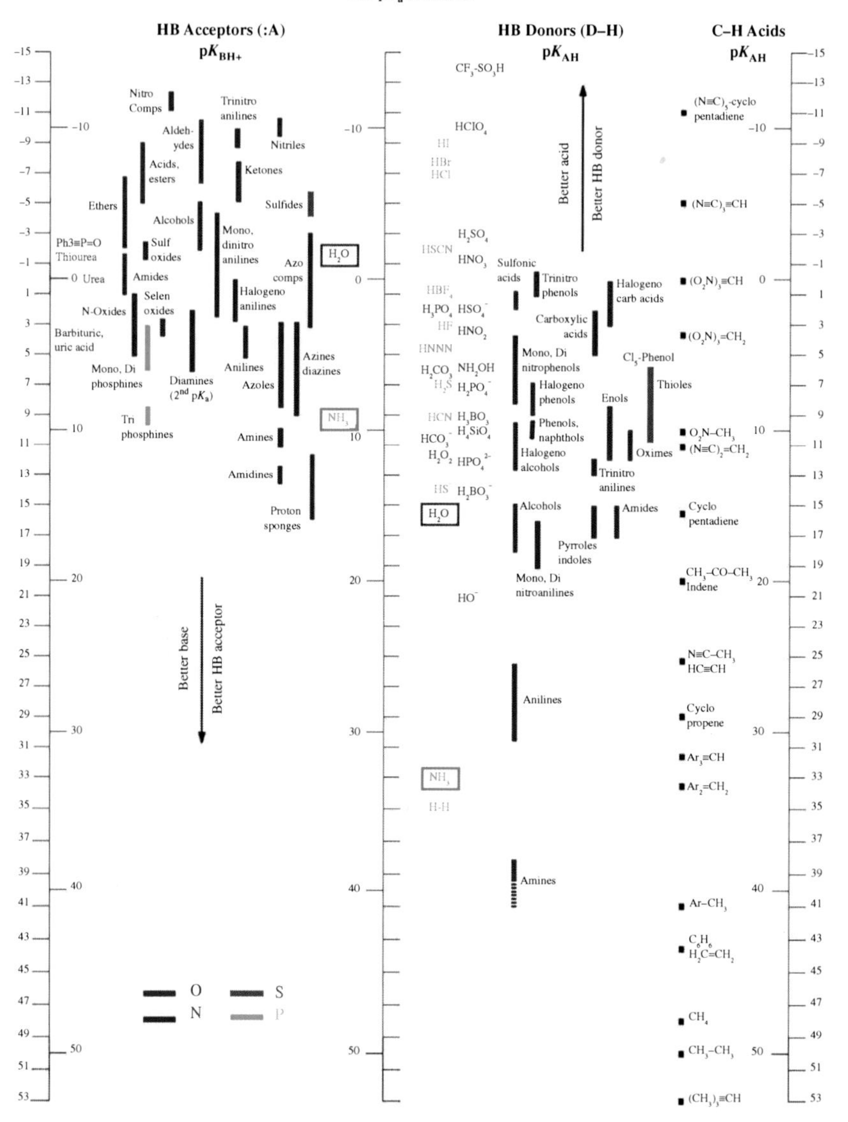

PLATE 3. A comprehensive representation of the pK_a slide rule. It summarizes, in the form of a bar chart, the pK_a values of the most important H-bond donor (right) and acceptor (left) molecules. Data from Tables 4.2 and 4.3 and Figure 4.2. According to the pK_a equalization principle, the strongest H-bonds are to be formed by donors and acceptors with strict pK_a matching lying on a same horizontal line of the ruler. Donor and acceptor atoms are coded in the lower-left corner and inorganic hydracids, oxyacids, and C–H acids are respectively represented in green, red, and black. (Reproduced by permission from Gilli *et al.*, 2009)

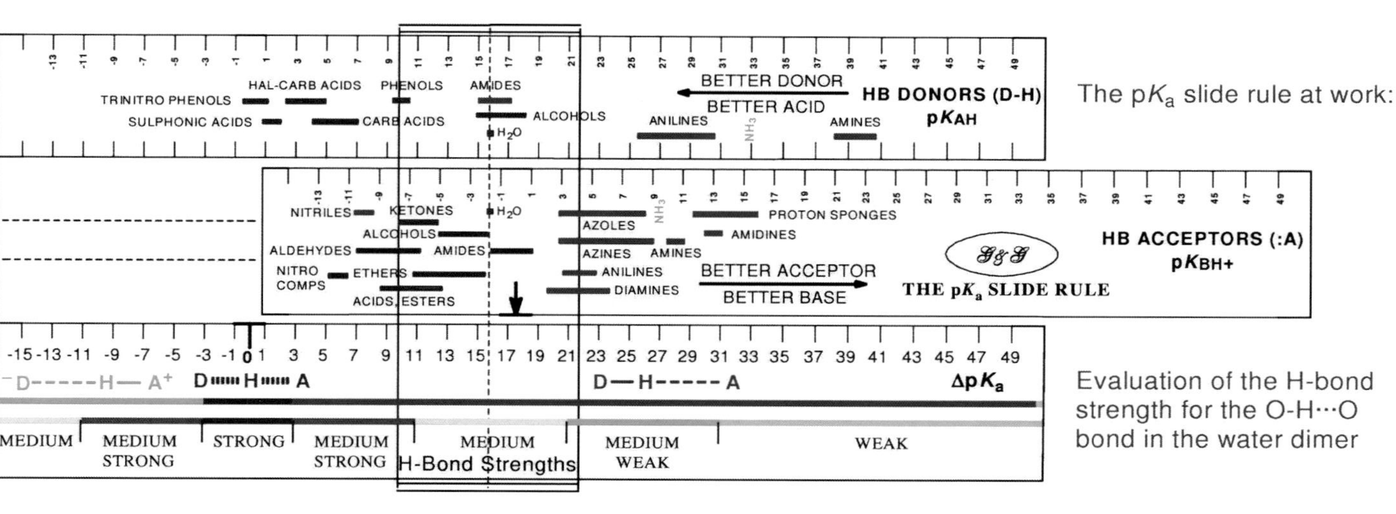

PLATE 4. An operational version of the pK_a slide rule. Allowing the two scales to shift reciprocally so to bring into coincidence the donor and acceptor molecules, the ruler permits graphical evaluation of the ΔpK_a and then empirical appreciation of the D–H···:A bond strength according to the pK_a equalization principle. In the figure above the slide rule is set in such a way to evaluate the strength of the O–H···:O bond in the water dimer. To this aim, water as a donor (upper scale) is put in vertical correspondence with water as an acceptor (central scale) allowing to predict (lower scale) that the ΔpK_a will be 17.4 and that the H-bond formed will be of medium strength. (Reproduced by permission from Gilli *et al.*, 2009)

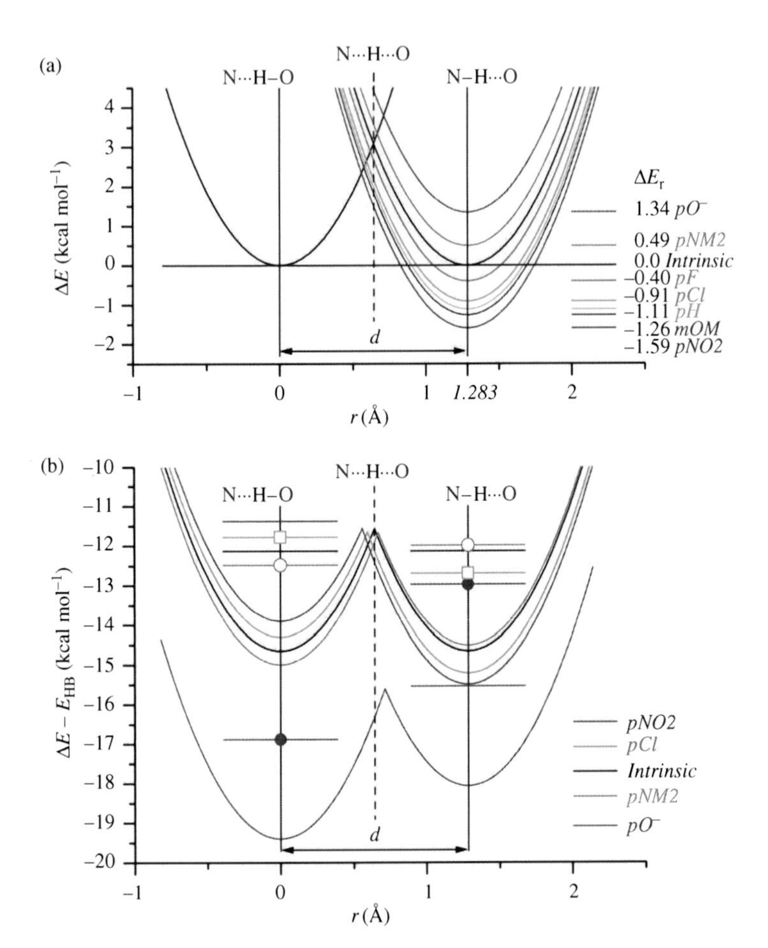

PLATE 5. (a) Marcus modelling of the N···H–O → N–H···O PT reaction for the compounds of Table 6.2 (without ZPC). The two symmetric parabolas (in black) crossing at the intrinsic barrier $\Delta^{\ddagger}E_{o} = 3.08$ kcal mol^{-1} for $r^{\ddagger}/d = 0.5$ represent the intrinsic PT profile. Curves for the other compounds (in color) are shifted upwards or downwards by their respective reaction energies, ΔE_{r}, and cross at $r^{\ddagger}/d$ values larger or smaller than 0.5 with PT barriers $\Delta^{\ddagger}E$ higher or lower than $\Delta^{\ddagger}E_{o}$, respectively. (b) Marcus modelling of figure (a) as modified by subtracting the values of the N···H–O bond energy, E_{HB}, to each individual curve. The curve in black corresponds to the *intrinsic H-bond* with $\Delta E_{r} = 0$ and $\Delta^{\ddagger}E = \Delta^{\ddagger}E_{o}$. Horizontal lines mark the approximate vibrational levels of the proton; full points indicate single-well and open symbols double-well H-bonds. (Reproduced by permission from Gilli *et al.*, 2006)

It is interesting that the data collected clearly indicate that H-bond strengths depend on two different factors, the first being the ΔpK_a and the second the electronegativity of the atoms involved. This is particularly evident from an analysis of the parameter $d_{D\cdots A}$(CAHB, min), the shortest D···A distance observed for CAHBs. For instance, $[F\cdots H\cdots F]^-$, $[O\cdots H\cdots O]^-$, and $[N\cdots H\cdots N]^-$ bonds have increasing distances of 2.207, 2.402, and 2.669 Å and $[F\cdots H\cdots F]^+$, $[O\cdots H\cdots O]^+$, and $[N\cdots H\cdots N]^+$ of 2.284, 2.360, and 2.592 Å against decreasing values of F, O, and N electronegativities of 3.98, 3.44, and 3.04. In a similar way, $[O\cdots H\cdots O]^{\pm}$, $[N\cdots H\cdots O]^{\pm}$, and $[N\cdots H\cdots N]^{\pm}$ bonds have increasing distances of 2.381, 2.506, and 2.682 Å against decreasing O, N/O, and N electronegativities of 3.44, <3.24>, and 3.04. To our knowledge, these are the first estimates of H-bond lengths (strengths) carried out *at constant* ΔpK_a and are seen to fully confirm the traditional ideas on the role played by the increasing electronegativity in strengthening the H-bond. This problem will be further discussed in terms of H-bond energies in Section 5.3.

4.4.3.2 *Second project.* pK_a *equalization in the N–H···O/O–H···N system*
The previous project has shown that the pK_a equalization rule positively holds in the restricted pK_a interval around zero. The problem is now if and how such a rule applies to the entire ΔpK_a range where the H-bond is defined, which covers the respectable $-30 \leq \Delta pK_a \leq 65$ interval according to the pK_a slide rule of Fig. 4.2. To that aim, this last project tries to ascertain whether, in any selected D–H···:A system (in this example the N–H···O/O–H···N one), the H-bond lengths (strengths) are actually ΔpK_a driven over the full ΔpK_a range.

Crystal data were retrieved from the CSD database. Firstly, all functional groups of known pK_a range and more frequently implied in N–H···O/O–H···N bonds were identified. Next, 10 classes of donors and 11 of acceptors with the greatest number of occurrences were selected and the search restarted for each separate donor–acceptor couple. Altogether, 8681 bonds were analyzed (3968 N–H···O, 2295 O–H···N and 2418 $^-$O···H–N$^+$). For each compound, the N–H or O–H distances were normalized by the method described in Section 2.4.2.1 and then the N···O distances evaluated as $d'_{N\cdots O} = d_{N-H} + d_{H-O}$ to account for the variability of the N–H–O angle. Finally, the shortest and average distances, $d_{N\cdots O}$(min) and $d_{N\cdots O}$(mean), were registered for each donor–acceptor group. These values are compared in Fig. 4.3 with the acid–base features of the donors (pK_{AH} range), acceptors (pK_{BH+} range), and their combinations (ΔpK_a range). Each box of the table is divided into two parts to account for neutral (N–H···O or O–H···N bonds, on the left) or charged H-bonds ($^+$N–H···O$^-$ bonds, dashed on the right). Bond lengths are divided into the five grey-tone-coded groups shown in the right top of the figure. For each group a tentative range of H-bond energies derived from the LS–HB method is also given.

Note that the information obtained from Fig. 4.3 is statistical in nature because individual pK_as are unknown and *average D···A distances* can only be compared with *average* ΔpK_a *ranges* for each donor–acceptor group. Nevertheless, a number of systematic regularities are observed, all of which can be shown to conform

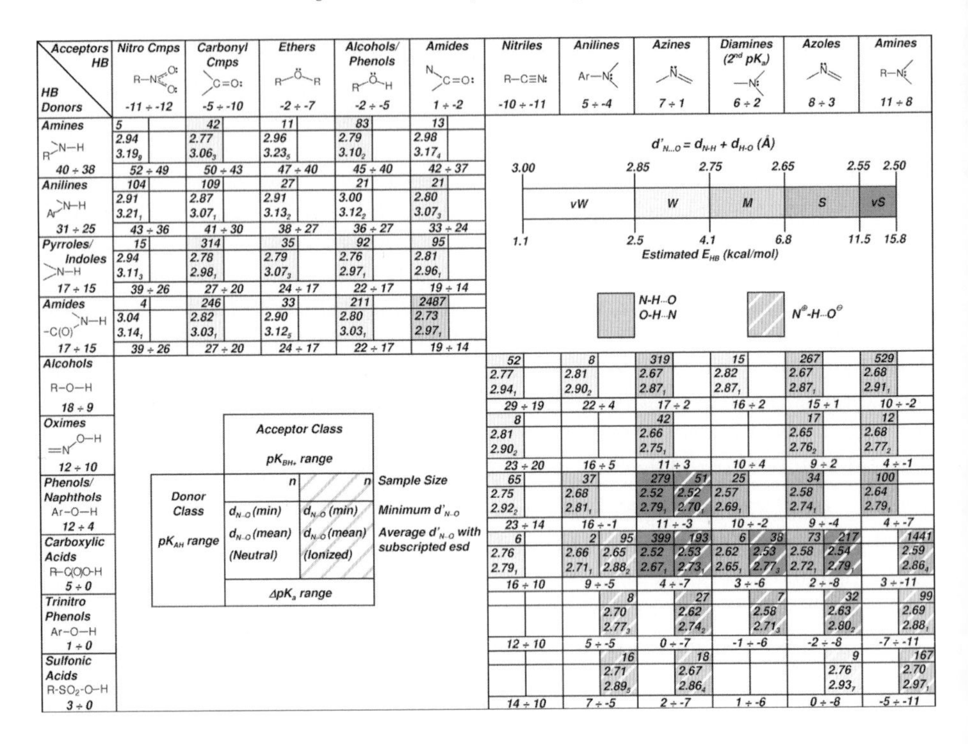

FIG. 4.3. Minimum and average N···O distances [$d_{N\cdots O}$(min) and $d_{N\cdots O}$(mean), in Å] obtained from a CSD search of the N–H···O/O–H···N bonds formed between common O–H and N–H donors and :O and :N acceptors. $d_{N\cdots O}$ is calculated as $d'_{N\cdots O} = d_{N-H} + d_{H-O}$ to account for the variability of the N–H–O angle. Observed values are compared with the acid–base properties of donors (pK_{AH} range) and acceptors (pK_{BH+} range) and of their combinations (ΔpK_a range). Each box is divided in two parts to account for neutral (N–H···O or O–H···N; on the left) or charged ($^+$N–H···O$^-$; dashed on the right) H-bonds. (Reproduced by permission from Gilli *et al.*, 2009)

with the basic hypothesis that H-bond strengths are essentially ΔpK_a-driven in the full range of ΔpK_a values.

The N–H···O bonds are concentrated in the upper left rectangle of the figure. According to expectation, all of them are relatively weak because the two main donors (amines and anilines) are exceedingly weak acids ($25 \le pK_a \le 40$) that do not face any acceptors of comparable acidity in the pK_a slide rule. Only pyrroles and amides ($15 \le pK_a \le 17$) could start to be acidic enough to make moderately strong bonds with less acidic protonated acceptors, such as amides themselves ($-2 \le pK_a \le 1$). The resulting amide–amide ΔpK_a range (14–19) remains too far from zero, however, to allow formation of really strong H-bonds.

H-bonds formed by O–H acids with N bases cluster in the lower right rectangle. They include all potentially strong H-bonds because their global ΔpK_a range ($-11 \leq \Delta pK_a \leq 29$) encompasses positive, zero and negative values and make it possible to produce, respectively, weak O–H···N, strong $^{½-}$N···H$^+$···O$^{½-}$, and weak $^+$N–H···O$^-$ bonds. Since donors and acceptors have been plotted in decreasing and increasing order of pK_a, respectively, the ΔpK_a intervals for the different classes are decreasing from positive to negative along the main diagonal of the rectangle, leaving nearly zero values in the proximity of the center. Along this diagonal we move, in fact, from weak O–H···N alcohol–nitrile bonds ($\Delta pK_a = 29$–19; $d_{N \cdots O}$(min) = 2.77 Å; $d_{N \cdots O}$(mean) = 2.94(1) Å; $n = 52$) to equally weak $^+$N–H···O$^-$ sulfonic acid–amine bonds ($\Delta pK_a = -5$ to -11; $d_{N \cdots O}$(min) = 2.70 Å; $d_{N \cdots O}$(mean) = 2.97(1) Å; $n = 167$), while the center is occupied by a block of bonds (marked in dark grey) associated with the complexes of phenols and carboxylic acids with azines, azoles, and second aminic groups of monoprotonated diamines that are much stronger (average $d_{N \cdots O}$(min) = 2.53(1) Å) because their global ΔpK_a range of -8 to 11 encompasses the zero and a relevant part of them (nearly one third) can be supposed to fall within the interval of true pK_a matching.

A last property that appears to change with regularity along the main diagonal is the ratio between ionized and neutral H-bonds. Out of the 56 donor–acceptor couples of Fig. 4.3 for which sufficient data could be retrieved, only 15 located in the lower right angle are seen to form ionized bonds, confirming our previous considerations on the fact that the N–H···O/O–H···N system (as any other H-bonded system) is strongly dissymmetric as far as proton transfer is concerned.

4.4.3.3 *Conclusions*

In summary, the two partial projects developed in this section lead to similar conclusions that strongly support the validity of the pK_a equalization principle for the four CLs OHB, (−)CAHB, (+)CAHB, and (±)CAHB, that is for the great majority of the H-bonds. In this list PAHBs can also be included because, owing to the small changes induced by the weak σ-bond polarizability, they do not substantially differ from OHBs of which they are only slightly shorter. Hence, the only bonds that remain excluded are RAHBs for the practical impossibility of evaluating the proper ΔpK_a value.

This makes the pK_a equalization rule the method of choice or, at least, the most promising empirical method for quick and efficient H-bond-strength prediction that can escape the inevitable complications of extended quantum-mechanical calculations. Equalization methods can become a quite efficient way for the rational design of complex H-bonded interaction patterns in crystals and mesophases provided the slide rule will be complemented by inclusion of the many chemical compounds whose acid-base constants are still unknown.

4.5 Appendix. pK_a tables arranged for chemical functionality

TABLE 4.2. pK_a lists of the most common H-bond donors

H-BOND DONORS: 1. INORGANIC HYDRACIDS AND OXYACIDS

Inorganic acid	Step	pK_{AH} (H_2O)	pK_{AH} (DMSO)	PA(A^-) kcal mol^{-1}	Type
$F_3C\text{-}SO_2OH$		−14	0.3	305.4	XOH
$HClO_4$		−10		288	XOH
HI		−9.5		314.3	XH
HBr		−8.5	0.9	323.5	XH
HCl		−7	1.8	333.4	XH
H_2SO_4	1	−3		306.3	XOH
	2	2.0			XOH
HSCN		−1.8		325.3	XH
HNO_3		−1.4		324.5	XOH
$Me\text{-}SO_2OH$		−1.3	1.6	320.9	XOH
HBF_4		0.5			XFH
$Ph\text{-}SO_2OH$		0.70			XOH
H_2CrO_4	1	0.74			XOH
	2	6.49			XOH
HIO_3		0.78			XOH
$H_4P_2O_7$	1	0.91			XOH
	2	2.10			XOH
	3	6.70			XOH
	4	9.32			XOH
H_2NSO_2OH		0.99			XOH
H_3PO_3	1	1.3			XOH
	2	6.70			XOH
Me_2AsOOH, Cacodylic acid		6.27			XOH
HIO_4		1.64			XOH
H_2SeO_4	1	1.7			XOH
H_2SO_3	1	1.85			XOH
	2	7.2			XOH
$HClO_2$		1.94			XOH
H_3PO_4	1	2.16		330.5	XOH
	2	7.21			XOH
	3	12.32			XOH
$MeO\text{-}PO(OH)_2$	1	1.1			XOH
	2	6.3			XOH
$PhO\text{-}PO(OH)_2$	1	0.9			XOH
	2	5.8			XOH

TABLE 4.2. *Continued*

H-BOND DONORS: 1. INORGANIC HYDRACIDS AND OXYACIDS

Inorganic acid	Step	pK_{AH} (H_2O)	pK_{AH} (DMSO)	PA(A^-) kcal mol^{-1}	Type
Me-PO$(OH)_2$	1	2.10			XOH
	2	7.5			XOH
R-PO$(OH)_2$	1	2.1/2.6			XOH
	2	7.5/8.5			XOH
Ph-PO$(OH)_2$	1	1.7			XOH
	2	7.0			XOH
Ar-PO$(OH)_2$	1	1.0/2.6			XOH
	2	3.5/7.7			XOH
H_3AsO_4	1	2.26			XOH
	2	6.76			XOH
	3	11.29			XOH
H_2TeO_3	1	2.48			XOH
	2	7.7			XOH
H_2SeO_3	1	2.62			XOH
	2	8.32			XOH
H_2Te	1	2.6		331.1	XH
	2	11			XH
HF		3.18	15	371.3	XH
HNO_2		3.3		340.2	XOH
HCNO		3.46			XH
H_2Se	1	3.89		341.5	XH
	2	11.0			XH
HNNN		4.6	7.9	343.9	XH
NH_2OH		5.94			XOH
H_2CO_3	1	6.35			XOH
	2	10.33			XOH
H_2S	1	7.0		351.3	XH
	2	14.0			XH
HClO		7.40		355.6	XOH
H_2TeO_4	1	7.68			XOH
	2	11.0			XOH
HBrO		8.55		353.5	XOH
H_2GeO_3	1	9.01			XOH
	2	12.13			XOH
HCN		9.21	12.9	350.9	XH
H_3BO_3	1	9.27			XOH
	2	>14			XOH
H_2AsO_3	1	9.29			XOH

TABLE 4.2. *Continued*

H-BOND DONORS: 1. INORGANIC HYDRACIDS AND OXYACIDS

Inorganic acid	Step	pK_{AH} (H_2O)	pK_{AH} (DMSO)	PA(A^-) kcal mol^{-1}	Type
H_4SiO_4	1	9.9			XOH
	2	11.8			XOH
	3	12			XOH
	4	12			XOH
HIO		10.15			XOH
H_2O_2	1	11.7		353.8	XOH
H_2O	1	15.7	32	390.3	XOH
	2	≈21			$^-$OH
NH_3		≈33	41	403.4	XH

H-BOND DONORS: 2. ORGANIC CARBON ACIDS (C–H ACIDS)

Carbon acid (C–H)	pK_{AH} (H_2O)	pK_{AH} (DMSO)	PA(A^-) kcal mol^{-1}
Cyano compounds			
$(N{\equiv}C)_5$Cyclopentadiene	–11		
$(N{\equiv}C)_3\underline{CH}$	–5		
$(N{\equiv}C)_2\underline{CH_2}$	11	11–12.5	335
$N{\equiv}C\text{–}\underline{CH_2}\text{–}R/Ar$	21–25	22	370
Halogeno compounds			
$(F_3C)_3\underline{CH}$	11		334.3
$F_3\underline{CH}$	32		378.0
Nitro compounds			
$(NO_2)_3\underline{CH}$	0		
$(NO_2)_2\underline{CH_2}$	3.6		
$NO_2\text{–}\underline{CH_2}\text{–}R$	10		355
Sulfones, Sulfoxides			
$(Me\text{-}SO_2)_2\underline{CH_2}$	12.5		
$Me\text{-}SO_2\text{-}\underline{Me}$	33	31.1	366.2
Me-SO-$\underline{Me}$	33	35.1	374.7
***β*-Dicarbonyl compounds**			
Malondialdehyde	5		
1,3-*cyclo*-Pentanedione	5.35		
1,3-*cyclo*-Hexanedione	4.8	11.2	
R-CO-$\underline{CH_2}$-CO-R	9–11	13–14	345
RO-CO-$\underline{CH_2}$-CO-OR	11–13	16	350
Carbonyl compounds			
R/Ar-$\underline{CH_2}$-CHO	16–19	25	360–365

TABLE 4.2. *Continued*

H-BOND DONORS: 2. ORGANIC CARBON ACIDS (C–H ACIDS)

Carbon acid (C–H)	pK_{AH} (H_2O)	pK_{AH} (DMSO)	PA(A^-) kcal mol^{-1}
R-$\underline{CH_2}$-CO-R	19–20	27	370
R-$\underline{CH_2}$-CO-OR	24	30	370
π-Anions			
$(N{\equiv}C)_5$Cyclopentadiene	−11		
Cyclopentadiene	15.5	18.0	353
Indene	20	20.1	354.5
9*H*-Fluorene	23		350.5
cyclo-Propene	29		
$CH_2{=}CH$-$\underline{Me}$	43	44	390
Alkines			
Ph-C≡$\underline{CH}$	21–23	28.8	370.6
HC≡$\underline{CH}$	25–26		379
Aromatic compounds			
$Ar_3{\equiv}\underline{CH}$	31.5	31	260
$Ar_2\underline{CH_2}$	33.5	32	364
Ar-$\underline{Me}$	41	43	380
Benzene	≈44		401.7
Alkenes			
$CH_2{=}\underline{CH_2}$	44		407.5
$CH_2{=}\underline{CH}$-Me	43	44	390
Alkanes			
Cyclopropane	≈47	56	
Methane	≈48	≈55	417.3
Me-$\underline{Me}$	50		420.5
$Me_2\underline{CH_2}$	51		419.2
$Me_3\underline{CH}$	53		413.5
Cyclohexane	≈51		404.0

H-BOND DONORS: 3. ORGANIC ACIDS

Organic acid	pK_{AH} (H_2O)	pK_{AH} (DMSO)	PA(A^-) kcal mol^{-1}	Type
Sulfonic Acids				
Ar-Sulfonic Acids	1–3			Ar-SO_2-OH
R-Sulfonic Acids	0–2	1.5	320	R-SO_2-OH
Ar-Carboxylic Acids				Ar-CO-OH
Ar-Nitroacids	2–3		330	
Ar-Halogenoacids	2–4		335	

TABLE 4.2. *Continued*

H-BOND DONORS: 3. ORGANIC ACIDS

Organic acid	pK_{AH} (H_2O)	pK_{AH} (DMSO)	PA(A^-) kcal mol^{-1}	Type
Ar-Hydroxyacids	3–5		325–335	
Ar-Aminoacids	2–2.5		340	
Ph-COOH	4.2	11.1	340.1	
R-Carboxylic Acids				R-CO-OH
R-Halogenoacids	−0.5–3		325–335	
R-Carboxylic acids	3–5	12	345–350	
Phenols				Ar-OH
Tri-, Dinitrophenols	−1–4			
2,4,6-Trinitrophenol	0.33		310.3	
2,4-Dinitrophenol	4.09		316.1	
3,5-Dichloro-2,4,6-trinitrophenol	−0.7			
Mononitrophenols	7.1–8.4	11	328–334	
Halogenophenols	7–9		342–343	
Pentachlorophenol	5.25			
Phenols	10–10.4	18–19	350	
Phenol	10	18	350.4	
Naphtols		16–17	345	
Thiols				
Ph-SH	6.5	10.3	340.4	Ar-SH
R-SH	10–11	17	354–355	R-SH
Cisteine	8.37			
Alcohols				
Halogenoalcohols	9–13			R-OH
$(F_3C)_3C$-OH	9.3		331.6	
F_3C-CH_2-OH	12.4	23.5	361.7	
Alcohols	15–18	29–32	374–382	
Water	15.74	31.3	390.3	H-OH
Hydroxyl anion	≈21			$^-$OH
Oximes	10–12	20	365	>C=N-OH
Anilines				Ar-NRH
Nitroanilines	12–18			
2,4,6-Trinitroaniline	12.2			
2,4-Dinitroaniline	15			
4-Nitroaniline	18.4	20.9	343.5	
Anilines	25–31			
Ph_2NH	24.95	25	350.8	
Aniline	30.6/≈27	30.6	366.4	

TABLE 4.2. *Continued*

H-BOND DONORS: 3. ORGANIC ACIDS

Organic acid	pK_{AH} (H_2O)	pK_{AH} (DMSO)	PA(A^-) kcal mol^{-1}	Type
Amides, Amines, Imines				
Amides	15–17			R-CO-NRH
Amines	>38			R_2NH
Ammonia	38/≈33	41	403.5	H_2NH
Phtalimide	8.3			$(-CO)_2NH$
Hydrides				
H-H	35		400.4	H-H

TABLE 4.3. pK_a lists of the most common H-bond acceptors

H-BOND ACCEPTORS: 1. ORGANIC C=O, N=O, S=O, AND SE=O BASES

Organic base	pK_{BH+} (H_2O)	pK_{BH+} (DMSO)	PA(B) kcal mol^{-1}	Type
Nitro compounds				
R-NO_2	−12		180	R-N(O)=O:
Ar-NO_2	−11 to −12		191	Ar-N(O)=O:
Aldehydes				
H-CHO	−4		170.4	H_2C=O:
R-CHO	−10		184	RHC=O:
Ar-CHO	−7		200	ArHC=O:
Ketones				
R-CO-R	−7		194	R_2C=O:
Ar-CO-R	−6		206	ArRC=O:
Ar-CO-Ar			211	Ar_2C=O:
9,10-Anthraquinone	−8.27			
Esters				
R-CO-OR	−6.5		196	R(RO)C=O:
Ar-CO-OR	−7.4		203	Ar(RO)C=O:
Acids				
R-COOH	−6.5		187	R(HO)C=O:
Ar-COOH	−7 to −8		196	Ar(HO)C=O:
Ethers				
R-O-R	−4 to −5		196	R_2O:
Tetrahydrofuran, THF	−2.1		196.5	R_2O:
Ar-O-R	−6.5		201	ArRO:

TABLE 4.3. *Continued*

H-BOND ACCEPTORS: 1. ORGANIC C=O, N=O, S=O, AND Se=O BASES

Organic base	pK_{BH+} (H_2O)	pK_{BH+} (DMSO)	PA(B) kcal mol^{-1}	Type
Alcohols				
R-O-H	−2 to −5		180–192	RHO:
Water	−1.74		165	H_2O:
Thioles and Sulfides				
R-S-H			184–189	RHS:
R-S-R	−5 to −6		199	R_2S:
Tetrahydrothiophene	−4.4		202.9	R_2S:
Amides				
Ar-CO-NH_2	−1.5		213	Ar(H_2N)C=O:
R-CO-NH_2	0–1		206	R(H_2N)C=O:
Me-CO-NHPh	0.4			
2(1*H*)-Pyridinone	0.75			
Ureas, thio- and seleoureas				
H_2N-CO-NH_2, Urea	0.15			$(H_2N)_2C$=O:
H_2N-CS-NH_2, S-Urea	−1.4			$(H_2N)_2C$=S:
H_2N-CSe-NH_2, Se-Urea	0.6			$(H_2N)_2C$=Se:
Barbituric Acid	4.01			
Uric Acid	3.89			
Sulfo- and selenoxides				
R-SO-R	−2		211	R_2S=O:
R-SeO-R	3			R_2Se=O:
Phosphine and arsine oxides				
Triphenylphosphine oxide	−2.10			Ph_3P=O:
Triphenylarsine oxide	−1.01			Ph_3As=O:
***N*-Oxides**	0–5			
Pyridine-*N*-oxide	0.8		220.7	R(R=)N→O:
Me_3N→O	4.65		235.1	R_3N→O:

H-BOND ACCEPTORS: 2. ORGANIC ≡P: AND ≡N: BASES

Organic base	pK_{BH+} (H_2O)	pK_{BH+} (DMSO)	PA(B) kcal mol^{-1}	Type
Phosphines				R_3P:
Me-PH_2	2.7		203.5	
Me_2PH	6		218	
R_3/Ar_3P	9		229–235	
Nitriles				
R-C≡N	−10		186	R-C≡N:
Ar-C≡N	−10 to −11		194	Ar-C≡N:

TABLE 4.3. *Continued*

H-BOND ACCEPTORS: 2. ORGANIC ≡P: AND ≡N: BASES

Organic base	pK_{BH+} (H_2O)	pK_{BH+} (DMSO)	PA(B) kcal mol^{-1}	Type
Azocompounds				
Ar-N=N-Ar	−3 to 3			Ar(ArN=)N:
Anilines				ArH_2N:
2,4,6-Trinitroaniline	−9.4			
2,4-Dinitroaniline	−4.5			
Mononitroanilines	−0.3–2.5		207	
Dihalogenoanilines	2			
Monohalogenoanilines	3.9–4.6		210	
Ph-NH-Ph	0.84			Ar_2HN:
Aniline	4.60/4.6	3.6	210.2	
β-Naphtylamine	4.16			
α-Naphtylamine	3.92		216.8	
Ph-NH-Me	4.85		219.1	ArRHN:
Ph-NMe_2	5.2	2.5	224.9	ArR_2N:
Azoles				Cyclic R(R=)N:
Thiazole	2.44		216	
1*H*-Imidazole	7.00		225.3	
2-Aminothiazole	5.36		222.4	
1-Me-Imidazole	6.95		229.3	
2,4-Me_2-Imidazole	8.36			
Benzimidazole	5.53		228.0	
2-Phenylbenzimidazole	11.91			
Azines				Cyclic R(R=)N:
Pyridine	5.22		222	
3-Cl-Pyridine	2.84		215.9	
2,5-H_2N-Pyridine	6.48			
4-HO-Pyridine	3.2			
4,4′ -Bipyridyl	4.71			
	2.7			
2-HOOC-Pyridine, Picolinic acid	5.25			
3-HOOC-Pyridine, Nicotinic acid	4.85			
4-HOOC-Pyridine, Isonicotinic acid	4.96			
2,4,6-Me_3-Pyridine	7.43			
Pyridazine, 1,2-Diazine	2.24		216.8	
Pyrazine, 1,4-Diazine	0.65		209.6	

TABLE 4.3. *Continued*

H-BOND ACCEPTORS: 2. ORGANIC ≡P: AND ≡N: BASES

Organic base	pK_{BH+} (H_2O)	pK_{BH+} (DMSO)	PA(B) kcal mol^{-1}	Type
2-Me-Pyrazine	1.45			
2-MeHN-Pyrazine	3.39			
Pyrimidine, 1,3-Diazine			211.7	
2-H_2N-Pyrimidine	3.45			
Quinoline, 1-Aza-naphthalene	4.9		227.8	
Isoquinoline, 2-Aza-	5.42		227.5	
Cinnoline, 1,2-Diaza-	2.37		223.8	
Quinazoline, 1,3-Diaza-	3.43			
Quinoxaline, 1,4-Diaza-	0.56		216.0	
1,10-Phenanthroline	4.84			
Amines				
R_3N, *ter*-Amines	10–11	9–10	226–235	R_3N:
HC≡$(CH_2$-$CH_2)_3$≡N, Quinuclidine	11.15	9.8	235.0	
N≡$(CH_2$-$CH_2)_3$≡N, DABCO	8.82			
	2.97			
N-Me-Morpholine	7.38			
R_2NH, *sec*-Amines	10–11	11	222–228	R_2HN:
$(CH_2)_4$=NH, Pyrrolidine	11.27		226.6	
HN=$(CH_2$-$CH_2)_2$=NH,	9.83		225.5	
Piperazine	5.56			
O=$(CH_2$-$CH_2)_2$=NH, Morpholine	8.36		220.9	
RNH_2, *prim*-Amines	10–11	11	215–218	RH_2N:
NH_3, Ammonia	9.24	10.5	204.0	H_3N:
Other amines				
NH_2-NH_2, Hydrazine	8.0		203.9	
HO-NH_2, Hydroxylamine	5.96			
Me-N=NH, Me-Diazene	10.0			
Amidines				H_2N(R)C=HN:
HN=C(Me)NH_2	12.4		232.0	
HN=C$(NH_2)_2$, Guanidine	13.6		235.7	
HN=C$(NMe_2)_2$	13.6		246.6	
Proton sponges	12–16			≡N: :N≡
1,8-Me_2N-naphthalene	12.1	7.5	245.8	
1,8-Et_2N-2,7-MeO-naphthalene	≈16.3			

5

THE EMPIRICAL LAWS GOVERNING THE H-BOND: A SUMMARY

After the long excursion on the many different species of H-bonds carried out in Chapters 3 and 4 it seems timely to try to summarize the results obtained to specify the common features that make the H-bond unique and distinguishable from all other types of chemical bond. The discussion will be centered on the three main tenets of this book: (*i*) opportunity of dividing the full set of H-bonds into six classes, called the six chemical leitmotifs (CLs); (*ii*) applicability of the classical VB theory to the interpretation of H-bond geometries and formulation of the electrostatic-covalent H-bond model (ECHBM); and (*iii*) predictability of H-bond strengths by the use of acid–base indicators and assessment of the PA/pK_a equalization principle.

5.1 Summary of chemical leitmotifs (CLs): The three main classes of H-bonds

The main feature of the present treatment is that the H-bond is not considered to constitute an undifferentiated group but is assumed to have a rich internal structure represented by the six CLs displayed, in their final version, in Fig. 5.1. and whose properties are collected in Table 5.1. The need for this division can hardly be questioned in pure objective terms, being the outcome of the unequivocal ordering of large numbers of crystal structures and gas-phase dissociation enthalpies, but, obviously, it does not help to simplify the study of the H-bond as a global phenomenon. The problem can be partially overcome by sorting the six CLs according to their properties (Table 5.1) in a smaller number of more homogeneous classes. The six CLs are thus reduced to *three H-bond classes* that come to represent the three basic ways by which the H-bond is formed, that is:

Class 1 *or proton-transfer H-bonds* This class includes all acid-base H-bonds associated with the three OHB, PAHB and (±)CAHB CLs. With reference to the general classification of the H-bonds formed by the main-group elements (Section 2.3; Chart 2.3), bonds of Class 1 include *conventional H-bonds* (Group 1.1) together with the three groups of weak bonds, that is *weak H-bond donors* (Group 1.2), *weak H-bond acceptors* (Group 1.3), and *weak H-bond π-acceptors* (Group 1.4). In other words, Class 1 comes to embrace all H-bonds formed by main-group elements that are *associations of an acid (the H-bond donor)*

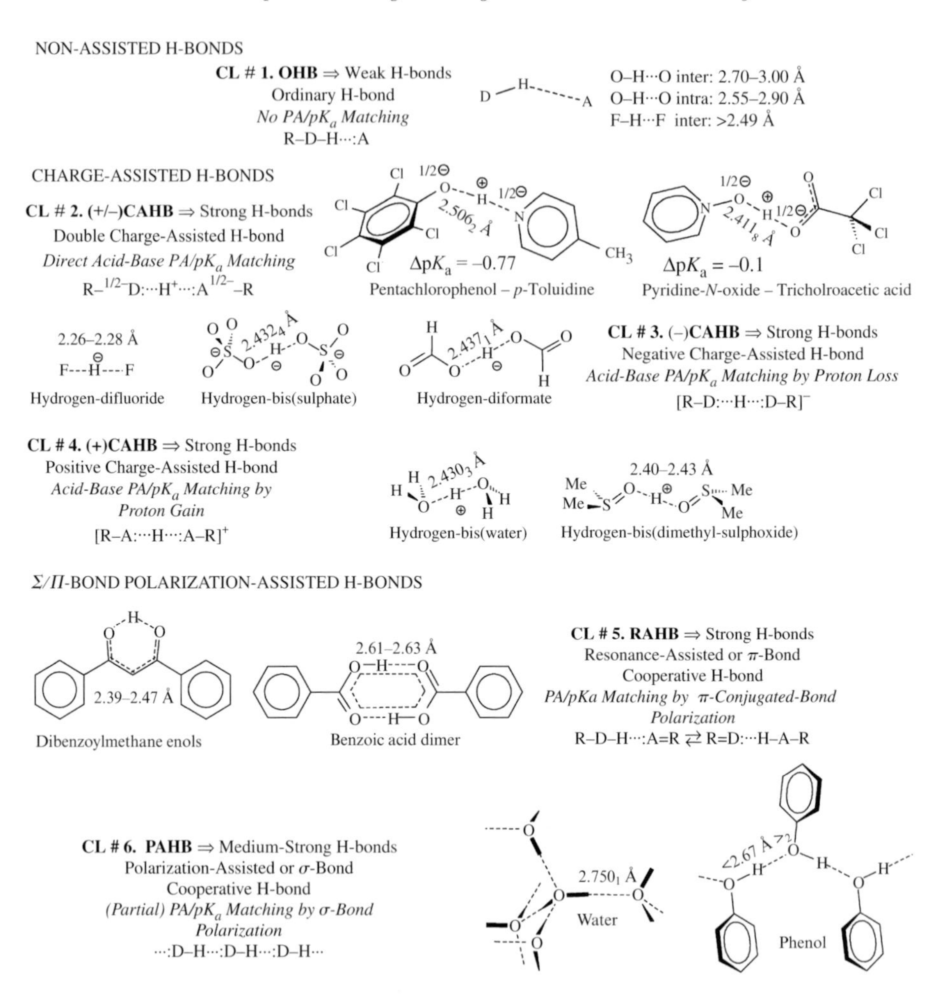

FIG. 5.1. A comprehensive tabulation of the six chemical leitmotifs (CLs).

with a base (the H-bond acceptor), thus constituting a vast superclass that contains the greatest part of the H-bonds occurring in nature (and substantially all those traditionally called H-bonds). According to the pK_a equalization principle (Chapter 4) and neglecting accidental steric effects, all the H-bonds of this immense class could, in principle, be monotonically ordered from the weakest to the strongest by arranging them in decreasing order of $\Delta pK_a = |pK_a(\text{acid}) - pK_a(\text{base})| = |pK_a(\text{donor}) - pK_a(\text{acceptor})|$, where the use of the modulus arises from the existence of two types of weak H-bonds, i.e. neutral D–H···A and ionized $^-$D···H–A$^+$ bonds.

Class 2 or proton-sharing H-bonds These are charged acid–acid or base–base H-bonds associated with the two (–)CAHB and (+)CAHB CLs. Bonds of this

TABLE 5.1. Comprehensive table of chemical leitmotif (CL) properties and definition of the three H-bond classes

CL Acronym	1. OHB	2. (±)CAHB	3. (−)CAHB	4. (+)CAHB	5. PAHB	6. RAHB
CL Name	Ordinary	Double charge-assisted	Negative charge-assisted	Positive charge-assisted	Polarization-assisted or σ-bond cooperative	Resonance-assisted or π-bond cooperative
Maximum H-bond strength	Weak	Very strong	Very strong	Very strong	Moderate	Strong
Mechanism of action. I	None	Charge assistance	Charge assistance	Charge assistance	Σ-bond cooperativity	Π-bond cooperativity
Mechanism of action. II	None	Direct PA/pK_a matching	PA/pK_a matching by proton loss	PA/pK_a matching by proton gain	PA/pK_a matching by σ-bond polarization	PA/pK_a matching by π-bond polarization
D···A proton exchange	Proton-transfer	Proton-transfer	Proton-sharing	Proton-sharing	Proton-transfer	Proton-sharing
D···A acid–base association	Acid–base	Acid–base	Acid–acid	Base–base	Acid–base	(Acid–base/ base–acid)
Thermodynamic indicator	ΔpK_a	ΔpK_a	ΔpK_a and ΔPA	ΔpK_a and ΔPA	ΔpK_a	None
H-bond class	Class 1	Class 1	Class 2	Class 2	Class 1	Class 3

class tend to be dominated by the molecular symmetry because the $\Delta pK_a = pK_a(\text{acid}_1) - pK_a(\text{acid}_2)$ or $\Delta pK_a = pK_a(\text{base}_1) - pK_a(\text{base}_2)$ is identically equal to zero whenever $\text{acid}_1 = \text{acid}_2$ or $\text{base}_1 = \text{base}_2$, so that all homomolecular associations inevitably give rise to very strong H-bonds. In principle, weaker homonuclear bonds of this sort are possible when the two interacting acids or bases are different and then their $\Delta pK_a \neq 0$. The observation of these mixed bonds in crystals is limited by the fact that, when mixing two different components (say, A and B), the two homomolecular complexes (AA and BB) are necessarily more stable than the heteromolecular one (AB) and then tend to crystallize preferentially. Gas-phase chemical equilibria allow, instead, simultaneous detection of the three AA, BB, and AB complexes and represent the method of election for their study (see Figures 4.1 and 3.35(c) as examples). Conversely, a number of heteronuclear (−) and (+)CAHBs are routinely encountered in crystals, particularly formed by O–H and N–H donors with acid anions as acceptors or by $N–H^+$ protonated amines with neutral acceptors. They are, however, normally weak because of the considerable pK_a mismatching.

Class 3 or resonance-assisted H-bonds (RAHBs) RAHBs constitute a class by themselves for a number of reasons that make them different from any other H-bond. The most important is that, though RAHBs can well be considered proton-sharing interactions (eqn 4.6c), it is not clear which partners share the proton because the π-delocalization of the resonant fragment transforms the acid (e.g. −O−H) into the base (e.g =O), and vice versa. Due to this transformation, RAHB strengths cannot be predicted from the ΔpK_a of the two originally interacting groups but can only be assessed as the *residual* ΔpK_a, that is the ΔpK_a remaining after π-delocalization of the conjugated chain, a quantity clearly inaccessible to experiments. Nevertheless, the proton-sharing nature of these bonds make them extremely sensitive to molecular symmetry and, in fact, really strong RAHBs are normally found to be associated with homonuclear bonds in symmetrically substituted molecules (Section 3.2.3.2; Fig. 3.20).

5.2 Summary of VB methods: The electrostatic-covalent H-bond model (ECHBM)

The twofold nature (electrostatic and covalent) of the O−H···O bond has been exhaustively discussed in Section 3.3.1.2 and the conclusions summarized in the so-called electrostatic-covalent H-bond model (ECHBM; Gilli *et al.*, 1994a; Gilli and Gilli, 2000). This model can now be generalized to include all other types of H-bonds studied, so leading to the following modified formulation:

a) The transition of any D−H···:A bond from weak (long) to strong (short) is associated with a parallel transition from dissymmetric (proton-outcentered and bent) to symmetric (proton-centered and linear);
b) This process can be interpreted as a progressive transformation from a mostly electrostatic bond (not requiring strict symmetry conditions) to an

essentially covalent 3c–4e bond where the two D–H and H–A bonds have identical bond number of ½ (which must be proton-centered and linear);

c) The degree of covalency is determined by the degree of mixing of the two $\Psi_{COV1,NCT}$(D–H···:A) and $\Psi_{COV2,CT}$(D:$^-$···H–$^+$A) VB wavefunctions that, in turn, depends on the energy difference $\Delta E = E(\Psi_{COV2,CT}) - E(\Psi_{COV1,NCT})$ and reaches the maximum of 1:1 when $\Delta E = 0$;

d) The four strong CLs [(±)CAHB, (−)CAHB, (+)CAHB and RAHB] are essentially molecular devices able to annihilate the normally very large ΔE associated with weak and electrostatic H-bonds [OHBs]. This annihilation can preferentially occur according to two different mechanisms:

   d′) in both homo- and heteronuclear (±)CAHBs, which are acid–base proton-transfer bonds, by accidental mixing of H-bond donors and acceptors having identical, or very similar affinities for the proton (i.e. similar PA/pK_a values);

   d″) in homonuclear (−)CAHBs, (+)CAHBs and RAHBs, which are acid–acid, base–base, or acid–base proton-sharing bonds, by mixing two H-bond donors or acceptors that are *symmetrical* as far as the distribution of chemical groups on the two sides of the H-bond is concerned; this guarantees that the two VB wavefunctions are *isoenergetic*, being identical by symmetry.

The model is then based on a simple partition of the H-bond energy in two terms only, electrostatic and covalent, and attempts to give a formal VB justification of the latter leaving the former undiscussed. Hence, it is a model for strong (mostly covalent) and not weak (mostly electrostatic) bonds and it seems interesting to investigate if it can give us, beyond the pure formal aspects, a physically grounded picture of this strong and covalent part of the H-bond.

An important feature of the *covalent H-bond* is its *dual nature*. This bond is not really *a bond* donated by the donor to the acceptor but consists of (at least) *two bonds* formed by the same proton with two acceptors that both carry an electron pair, a concept introduced since the very beginning of H-bond studies ('the hydrogen nucleus held between 2 octets constitutes a *weak bond*'; Latimer and Rodebush, 1920). In this sense, the H-bond energy, E_{HB}, is not properly an association energy of the donor with the acceptor, D–H···:A, but just the smaller of the two bonding energies, E(D–H) and E(H–A), by which the donor and acceptor atoms are competitively bound to the same central proton. If the first is stronger the second is weaker, the overall H-bond will be weak, so that strong bonds will only occur when the proton can bind, with very similar strengths, to both donor and acceptor atoms, giving rise to the condition called by Jeffrey (1997; p. 40) the *hesitating proton*.

The dual nature of the H-bond is the obvious reason why H-bond geometries can be so easily interpreted by traditional VB methods: dual bonds can always be modelled, in formal if not necessarily substantial terms, by mixing two resonant D–H···:A ↔ D:$^-$···H–$^+$A forms that share a single bond among three centers according to the bond-number conservation rule (eqns 3.4–7) and whose strongest possible bond remains defined as a 3c–4e covalent interaction with two equal bond numbers of ½ and very similar D–H and H–A bond lengths (Pauling, 1960; p. 484). If this is true, the maximum covalent contribution to the H-bond energy, $E_{HB,COV}$, should be directly related to the dissociation energy, D_0(X–H), of the two bonds involved. This can be easily proved by correlating the experimental energies of the strongest homonuclear $[X\cdots H\cdots X]^-$ and $[X\cdots H\cdots X]^+$ bonds with $\Delta E = 0$, $E_{HB,EXP} = E_{HB}$(CAHB,min) (data from Table 3.9 with the exclusion of $[F\cdots H\cdots F]^+$), with the corresponding D_0(X–H) values (Huheey *et al.*, 1993), so obtaining the regression equation

$$E_{HB,COV} = E_{HB,EXP} = -\ 31.3 + 0.55\ D_0(\text{X–H})\ \text{kcal mol}^{-1}\ (r = 0.899,\ n = 8), \quad (5.1)$$

which can be used to evaluate $E_{HB,COV}$ and, apart from the rather mysterious meaning of the negative intercept, seems to indicate that the H-bond energies are actually proportional to *one half of the X–H bond energies*, in agreement with the naive concept that strong H-bonds are covalent interactions that split a single bond into two half-bonds.

Equation 5.1 can help to understand how the nature of the atoms involved (and in particular their electronegativity) affects the maximum strength of the H-bond formed. The relevant relationship is

$$E_{HB,COV}(X\cdots H\cdots X) = f\,[D_0(X{-}H),\ \Delta E = E(\Psi_{COV2,CT}) - E(\Psi_{COV1,NCT})], \quad (5.2a)$$

which tells us that the binding energy of any X—H···X bond is a function of two independent variables, the energy of the covalent X–H bond and the energy difference between the proton affinities of the donor and acceptor atoms. At the limit of $\Delta E = 0$, the binding energy becomes a maximum and the expression reduces to

$$E_{HB,COV} = f\,[D_0(X{-}H)]. \quad (5.2b)$$

Therefore, the two independent variables play different roles: ΔE is the driving variable whose decrease causes a continuous strengthening of the H-bond, while D_0(X–H) is the limiting factor that determines the strongest possible bond in condition of $\Delta E = 0$.

A comparison between $E_{HB,COV}$ values computed from eqn 5.1 for the main-group elements and the corresponding experimental $E_{HB,EXP} = E_{HB}$ (CAHB,min) values measured so far in condition of $\Delta E = 0$ for the homonuclear $[X\cdots H\cdots X]^-$ bond is displayed in Table 5.2. The maximum predicted H-bond energies range from 43 to 5 kcal mol^{-1} going from F to Se and are seen to be quite large also for elements that are not known to form strong H-bonds, such as the C atom, whose predicted maximum is 22.7 kcal mol^{-1}. This surprising behavior is not a violation of the H-bond binding rules but is to be imputed to purely chemical factors. In fact, it is easy to produce, say, $[O\cdots H\cdots O]^-$ and $[O\cdots H\cdots O]^+$ bonds by simply changing the pH of the solution but (almost?) impossible to synthesize the corresponding $[C\cdots H\cdots C]^-$ and $[C\cdots H\cdots C]^+$ bonds. We are, however, confident that, whenever produced, they will display H-bond energies not far from the value predicted in Table 5.2.

TABLE 5.2. Maximum $[X\cdots H\cdots X]^-$ bond energies (kcal mol^{-1}) as calculated for the main-group elements from eqn 5.1, $E_{HB,COV}$, or measured from gas-phase dissociation experiments, $E_{HB,EXP}$ = E_{HB}(CAHB,min), of Table 3.9. Bond dissociation energies, D_0(X–H), from Huheey *et al.*, 1993. χ_P = Atomic Pauling electronegativity.

Atom (χ_P)	F (*3.98*)	O (*3.44*)	Cl (*3.16*)	C (*2.55*)	N (*3.04*)	S (*2.58*)
D_0(X–H)	135	109.6	102.3	98.3	92	87
$E_{HB,COV}$	42.9	29.0	25.0	22.7	19.3	16.6
$E_{HB,EXP}$	42(3)	27(1)	24(3)	–	12	13(1)
Atom (χ_P)	**Br (*2.96*)**	**P (*2.19*)**	**Si (*1.90*)**	**I (*2.66*)**	**Se (*2.55*)**	**H (*2.20*)**
D_0(X–H)	86.6	77	76	70.4	66?	
$E_{HB,COV}$	16.3	11.0	10.5	7.4	5.0	
$E_{HB,EXP}$	19(3)	–	–	–	–	

5.3 Summary of the PA/pK_a equalization principle

Because of the dual logic of the H-bond, any physical quantity able to quantify the competition of the H-bond donor and acceptor for the central proton will be suited to interpret and rationalize the H-bond itself. This is true for VB methods where such a competition is expressed by the relative energies of the two Coulson $\Psi_{COV1,NCT}$(D—H···:A) and $\Psi_{COV2,CT}$(D:$^-$···H—$^+$A) VB wavefunctions whose mixing in different ratios determines the strength of the H-bonds formed. It must be even truer for thermodynamic acid–base indicators, such as proton affinities (PAs) in the gas phase and related acid–base dissociation constants (pK_as) in solution, which are the most obvious indicators of molecular affinity for protons. Therefore, the PA/pK_a equalization principle that was derived empirically in Section 3.3.1.3 could have been formulated as an hypothesis (that is, a theory) naturally stemming from the definition of dual H-bond itself. Nevertheless, all authors who have so far taken an interest in this topic have limited themselves to formulate it as an empirical correlation between H-bond-strength descriptors (mostly IR stretching frequencies and, later on, gas-phase dissociation enthalpies) and acid–base indicators (PA, pK_a, ΔPA, and ΔpK_a) without trying to develop a method general enough to treat all types of H-bond. A full thermodynamic analysis has been performed only recently (Gilli *et al.*, 2007; Section 4.2) coming to assess two important limitations: (*i*) ΔPAs are computable only for (−) and (+)CAHBs, at variance with ΔpK_as that can be obtained for all H-bonds except RAHBs; and (*ii*) H-bond strengths can be assessed for all H-bonds from crystal geometries, but only for charged (−) and (+)CAHBs from gas-phase dissociation enthalpies.

In conclusion, *correlations between crystal geometries and ΔpK_a are the method of choice for rationalizing the H-bond strengths of all H-bonds but RAHBs.* This, of course, is not to say that RAHB strengths are not ruled by PA/pK_a equalization but that, for a number of unfortunate circumstances, neither ΔPA nor ΔpK_a values can be measured for this type of bond.

A more formal definition of the PA/pK_a equalization principle can start from the observation that both ΔPA and ΔpK_a have the meaning of energy difference, ΔE_r, between the two D–H···:A $\rightleftharpoons$ D:···H–A tautomers and then play the same role as the $\Delta E = E[\Psi_{COV2,CT}(D:^{-}\cdots H-^{+}A)] - E[\Psi_{COV1,NCT}(D-H\cdots :A)]$ difference in the ECHBM treated above. This leads to the rules:

a) The transition of any D–H···:A bond from weak, long, asymmetric, and electrostatic to strong, short, symmetric and covalent is associated with the parallel equalization of the acid–base indicator values of the H-bond donor (D–H) and acceptor (:A);
b) The equalization is quantified by the differences $\Delta PA = PA(D^-) - PA(:A)$ or $\Delta pK_a = pK_{AH}(D\text{–}H) - pK_{BH+}(A\text{–}H^+)$, which represent the thermodynamic energy difference, ΔE_r, between the two minima of the tautomeric D–H···:A $\rightleftharpoons$ D:···H–A equilibrium; very strong H-bonds can only occur when both ΔPA and ΔpK_a values are approaching zero;
c) To conclude, the energy difference between the two potentially tautomeric forms D–H···:A $\rightleftharpoons$ D:···H–A is the *thermodynamic driving force* controlling the strength of the H-bond.

Since ΔPA, ΔpK_a, and ΔE_r are just different ways for measuring the same donor–acceptor proton-affinity difference, previous considerations on ECHBM equally hold here with only small changes. So, the relevant relationship (eqn 5.2a) becomes now

$$E_{HB}(X\text{–}H\cdots X) = f\,[D_0(X\text{–}H),\ \Delta PA = PA(D^-) - PA(:A)] \text{ or} \quad (5.3a)$$
$$E_{HB}(X\text{–}H\cdots X) = f\,[D_0(X\text{–}H),\ \Delta pK_a = pK_{AH}(D\text{–}H) - pK_{BH+}(A\text{–}H^+)] \quad (5.3b)$$

that still tell us that the homonuclear X–H···X bond energy is a function of two independent variables, the energy of the covalent X–H bond and the donor–acceptor proton-affinity difference. Only at the limit $\Delta PA/\Delta pK_a = 0$, can the H-bond energy become a maximum, while the expression reduces to

$$E_{HB,COV} = f\,[D_0(X\text{–}H)], \quad (5.3c)$$

which is always identical to eqn 5.2b. Even here the two independent variables play different roles, $\Delta PA/\Delta pK_a$ being the driving variable that, by decreasing, causes the strengthening of the H-bond and D_0(X–H) the limiting factor that determines the strongest possible bond when $\Delta PA/\Delta pK_a = 0$.

5.4 On the chemical nature of the H-bond

The fact that the H-bond is controlled by the acid–base properties of the interacting molecules suggests a number of interesting considerations. Dispersive and multipolar forces, the most widespread molecular interactions, can actually be appreciated, at any level of organization of matter, by the application of relatively simple physical laws to a small number of physical variables. These variables, which are the molecular anisotropic polarizability tensor and ionization potential for the former and the molecular multipolar moments for the latter, can, in turn, be evaluated by quantum-mechanical methods starting from the individual properties of atoms. Therefore, there appears to be an uninterrupted reductionist chain leading from atomic properties to the evaluation of these two important types of molecular interactions. This could also be true for the H-bond, as far as it is weak enough to be driven by the same dispersive and multipolar forces.

Things look considerably different for strong H-bonds. Also in this case there are physical quantities that drive the interaction process, namely ΔPA or ΔpK_a, but these are now equilibrium constants of true chemical reactions in the gas phase or in water solutions where molecules not only associate but also undergo significant geometrical changes (relaxations). These constants are not directly accessible from atomic properties but can only be evaluated, at least in principle, for large molecular ensembles by the methods of statistical thermodynamics.

It may be supposed, therefore, that strong H-bonds are, in some way, different from dispersive and multipolar forces, the latter being physical properties of atomic ensembles and the former chemical properties of molecular ones. What seems to have changed is not properly physical but something of a more elusive character that could be described as an increased *level of complexity* of the forces acting in the course of the association process. Since one of the main tenets of complexity theories is that, at a higher level of complexity, new laws will emerge that will not be *reducible* to the lower level (Anderson, 1972; Schweber, 1993), we might suppose that the greater complexity of acid–base indicators will naturally set any complete H-bond theory (i.e. including both weak and strong bonds) at a higher level (the chemical or thermodynamic level) than theories accounting for dispersive and multipolar interactions (the physical level). This seems to imply that H-bond formation requires molecules that already exist, opening the intriguing perspective that, during the chemical evolution of life, molecules have been designed in view of the H-bonds they were entrusted to form.

The main fact in favor of the hypothesis that H-bond is a chemical (and not physical) phenomenon is that, after some 87 years, no H-bond theory based on atomic quantities has been really successful in explaining the full range of H-bond strengths and geometries. The fact that this physical solution has been pursued with such a constancy for so many years can only be ascribed to the persistence of the old belief that very strong H-bonds could be disregarded as curious exceptions in view of their scarce occurrence, a position clashing with the more recent discovery of thousands of such curious bonds. Inclusion of strong bonds in a comprehensive H-bond model naturally shifts the problem from physics to chemistry because strong bonds are true chemical reactions that occur with remarkable geometry relaxation and are affected by the environment, such as the presence of water and the pH of the solution.

Present considerations seem to rule out any possibility of formulating a comprehensive physical theory encompassing van der Waals, multipolar, and H-bonded interactions for the impossibility of rendering in a single equation the chemical complexity of strong H-bonds. This is not completely true, however, and models of this sort have been quite successful when limited to moderately strong OHBs that, not relaxing sensibly, can be formally reduced to a sum of van der Waals and electrostatic forces and, as such, have been included in the most popular force fields, such as the comprehensive UNI force field (Filippini and Gavezzotti, 1993; Gavezzotti and Filippini, 1994). There is little doubt, however, that the method of choice for studying strong H-bonds in complex molecular systems is quantum molecular-dynamics simulation, as briefly illustrated by the examples of Sections 8.5 and 8.6.

6

OUTLINE OF A NOVEL TRANSITION-STATE H-BOND THEORY (TSHBT)

6.1 Empirical laws, models and scientific theories: An introduction

In common practice we are used to distinguishing between true scientific theories, endowed with a mathematical apparatus suited to interpret, predict, and quantify experimental facts, from empirical laws (natural laws) that still interpret and predict experiments but on a base of empirical correlations. We may think that the many rules developed so far to rationalize the H-bond look more like empirical rules but do not know if and how they can be combined together and with some other known theoretical tool into a new formulation that can, by right, be defined as a true H-bond theory. It is obvious that, to go on with this discussion, we need a clearer definition of what scientific theories and natural laws may be, and the following simplified treatment (Gilli *et al.*, 1996b) may perhaps be sufficient to keep the discussion within reasonably correct terms.

Empirical observation of nature leads to the definition of invariant regularities that are called *empirical* or *natural laws.* Examples could be: (*i*) $pV = nRT$, the law of ideal gases; (*ii*) $\mathrm{d}U = \mathrm{d}q - \mathrm{d}w$, the first law of thermodynamics; (*iii*) 'All men are mortal', for sure one of the most disturbing laws for everybody. Natural laws are widely independent of us in the sense that we may discover them but not modify them and their consequences. In fact, they are constraints that the natural environment puts on our wishes and fantasy and, in this sense, are the more important the more they actually forbid (Popper, 1959).

Our wish to understand why and how natural laws occur leads to formulation of *scientific theories.* For our limited aims, any complete scientific theory can be thought of as consisting of two essentially independent parts: the *model* and its *physicomathematical development.* When the latter is missing or incomplete, the theory is downgraded to a simple model.

The *model* is the original idea of how things may work, a kind of logical nucleus intended to define the physical quantities that drive the phenomenon studied (the *independent driving variables*). In physical sciences models start from the intimate constituents of matter, elementary particles in nuclear physics and nuclei and electrons (or, at least, atoms and molecules) in solid-state physics, chemistry, molecular biology, and so on. For example: (*i*) the kinetic theory interprets the gas laws by the famous ideal-gas model (non-interacting molecules in incessant motion undergoing elastic collisions among themselves and with the vessel walls); (*ii*) thermodynamic laws are interpreted by a model consisting of a reservoir of particles that accumulates energy in different modes (traslational, vibrational, rotational, electronic,...); and finally, (*iii*) a model for explaining the natural law 'all men are mortal' can be 'accumulation of cell duplication errors due to thermal perturbations', though such a model has never been developed in a concluded theory and will then remain a simple model.

The *physicomathematical apparatus* of a theory is the encoding of the model in terms of the laws of a more fundamental theory or discipline (most frequently, classical or quantum mechanics, electrodynamics, classical and statistical thermodynamics). The fact that the model can be successfully developed into a coherent theory is considered to validate the model itself, even because it shows that the new theory, though elaborated to account for new facts, agrees with our previous network of scientific theories.

These simple considerations may help us to assess the form that a true *H-bond theory* should have. The classification in CLs, the CLs themselves, and the correlations among H-bond geometries, gas-phase dissociation enthalpies, and donor–acceptor acid–base parameters are the product of systematic data correlations and have, therefore, the obvious state of *empirical laws*. As shown in Chapter 5, the *general model of the H-bond* arises from *its dual nature*, for which donor and acceptor compete for the same central proton, being driven by the variables measuring the energy difference between the two competing moieties. Following these considerations, the general form of any *H-bond theory* can be logically expressed in terms of the equation

$$\text{H-bond properties} = \mathscr{F}\{\text{H-bond driving variables}\}, \tag{6.1}$$

where $\mathscr{F}$ is the physicomathematical operator (the theory at a more fundamental level) quantitatively transforming the independent *H-bond driving variables* into the *H-bond properties* that are, practically, any physical properties characterizing the H-bond itself (geometries, H-bond energies, PT barriers, dipole moments, IR stretching frequencies, NMR chemical shifts, etc.).

The first attempts at a theory of the H-bond were carried out in the framework of the VB theory, firstly by Coulson and Danielsson (1954a,b) and then by Lippincott and Schroeder (Lippincott and Schroeder, 1955; Schroeder and Lippincott, 1957) who produced the much more detailed LS-HB model (Section 3.1.3.3). According to our criteria, this last model can be considered a scientific theory because it can be expressed in terms of eqn 6.1 as

$$\text{H-bond properties} = \mathscr{VB\,Theory}\,\{(V_1, V_2),\ V_3,\ V_4\}, \tag{6.2}$$

where (V_1, V_2), V_3, and V_4 (eqns 3.8–12) are the independent variables chosen to represent the covalent, electrostatic and exchange contributions to the H-bond, respectively. It is, however, an imperfect theory because it provides analytical solutions only at the cost of an extensive semiempirical parametrization, though it cannot be denied that the results obtained for some well-parametrized systems (particularly the O–H···O bond) are remarkably good (see Section 3.1.3.3 for a discussion).

The problem of finding new and more comprehensive theories is strictly associated with that of finding a new good set of independent driving variables. The long discussion summarized in Chapter 5 has come to the conclusion that such a driving force is always an energy difference between the two states of the dual H-bond that, when expressed as a difference between VB resonance forms, ΔE, gives rise to the ECHBM, and, when expressed by the thermodynamic parameters ΔPA and ΔpK_a, to the PA/pK_a equalization principle. For the latter, in particular, the driving variables are known (Section 5.3) and the corresponding theory, for any generic D–H···:A bond, could be formulated as

$$\text{H-bond properties} = \mathscr{Therm}\ \{[D_0(\text{D–H}),\ D_0(\text{A–H})],\ \Delta\text{PA}/\Delta pK_a\}, \quad (6.3)$$

where $\mathscr{Therm}$ is the thermodynamic operator necessary to transform the independent variables into properties. This solution is not practicable, however, because classical thermodynamics '...is entirely independent on the molecular structure of matter...' and then cannot give rise to atomistic scientific theories (Fermi, 1956; Primas, 1983). Whatever thermodynamics may be, we must change our way and a new attempt can be made by remembering that both ΔPA and ΔpK_a have the meaning of ΔE_r, the thermodynamic energy difference between the two potentially tautomeric forms D–H···:A $\rightleftharpoons$ D:···H–A along the proton-transfer (PT) reaction pathway, a new driving variable for which we have now to decide the proper $\mathscr{T}$ operator. Given the nature of ΔE_r, it seems clear that we are entering the field of chemical kinetics and that the $\mathscr{T}$ operator has to be identified in the well-known *transition-state (or activated-complex) theory* (Eyring and Polanyi, 1931; Glasstone *et al.*, 1941). These considerations can be summarized in the new equation

$$\text{H-bond properties} = \mathscr{TS\,Teory}\ \{[D_0(\text{D–H}),\ D_0(\text{A–H})],\ \Delta E_r\}, \quad (6.4a)$$

which makes reference to all the possible H-bonds simultaneously and that, when considering them one at a time, simplifies to the more manageable form

$$\text{H-bond properties} = \mathscr{TS\,Teory}\ \{\Delta E_r\}. \quad (6.4b)$$

Equation 6.4 defines, therefore, a new application of the transition-state theory that has been called *transition-state H-bond theory (TSHBT)* (Gilli *et al.*, 2005a, 2006). This new theory needs to be applied and verified in practice and, in doing so, we are committed to use the tenets of the main transition-state theory (Glasstone *et al.*, 1941; Leffler and Grunwald, 1963) and entitled to use all its paraphernalia, such as the Leffler–Hammond postulate (Leffler, 1953; Hammond, 1955), the Marcus rate-equilibrium theory (Marcus, 1960, 1968; Grunwald, 1985), the empirical VB method (Åqvist and Warshel, 1993; Braun-Sand *et al.*, 2004), and the VB state correlation diagrams (Shaik *et al.*, 1992; Shaik and Shurki, 1999), as well as the linear free-energy relationships (Chapman and Shorter, 1978).

6.2 A new way of looking at the H-bond: The TSHBT

6.2.1 *Introduction*

In the previous section the opportunity of a TSHBT has been inferred from considerations of a rather philosophical nature. It is not impossible, however, to infer it by criteria of sheer common sense. In fact, if the very driving force of H-bond strengthening arises from PA/pK_a equalization, the study of single spare R_1–D–H⋯:A–R_2 bonds does not make much sense but it seems more reasonable, and probably more conclusive, to investigate sets of bonds formed within the same donor–acceptor system and modified by the electron-attracting or donating properties of the R_1 and R_2 substituents in such a way as to generate a continuous series of ΔPA/ΔpK_a differences. In other words, H-bonds are better studied as points of minimum along the pathway of a PT chemical reaction, minima that move from D–H⋯A to D⋯H–A while the actual values of ΔPA/ΔpK_a are going to change. This is clearly the point of view normally adopted in the framework of the kinetic transition-state theory to treat any generic A + BC $\rightleftharpoons$ AB + C reaction perturbed by the changing electronic properties of the substituents, a reaction illustrated in Fig. 6.1 which is bimolecular in both directions and proceeds via the D⋯H⋯A transition-state (TS) along a reaction pathway (the PT pathway) having two activation energies $\Delta^{\ddagger}E$ (the PT barriers in the opposite directions) and a reaction energy ΔE_r (corresponding to the ΔPA or ΔpK_a values previously used).

6.2.2 *Criteria for the choice of a suitable PT reaction*

Verification of the theory requires the choice of a *well-behaving* PT reaction, that is one whose profile has the correct shape for verifying the theory itself. Figure 6.2 summarizes the most common types of PT profiles associated with H-bonds of different strengths. In the following the nomenclature of Schemes **2.3.Ia–d** (Section 2.6) is used with some minor changes, that is: (*i*) all very strong H-bonds with symmetric single-well potentials (sSW; **2.3.Id**) are shortly indicated as SWHBs; (*ii*) H-bonds with symmetric double-well potentials (sDW; **2.3.Ic**) are also called LBHBs = low-barrier H-bonds (Cleland, 1992; Cleland and Kreevoy, 1994; Frey *et al.*, 1994; Harris and Mildvan, 1999); and finally, (*iii*) perfectly *symmetric* DW potentials are called *intrinsic* according to the Marcus nomenclature (see below).

Very strong SWHBs, curves **1** and **2** of Fig. 6.2(a), are not suited for our aims because they lack the DW feature. The correct form is that of LBHBs that, for $\Delta E_r = 0$, display the symmetric (or intrinsic) DW pathways **3** or **4**. In this fortunate case, successive modifications of ΔE_r induced by PA/pK_a dissymmetrization can originate the complete set of PT profiles **4–8** of Fig. 6.2(b). In synthesis a well-behaving R_1–D–H⋯A–R_2 system suited for our study must have

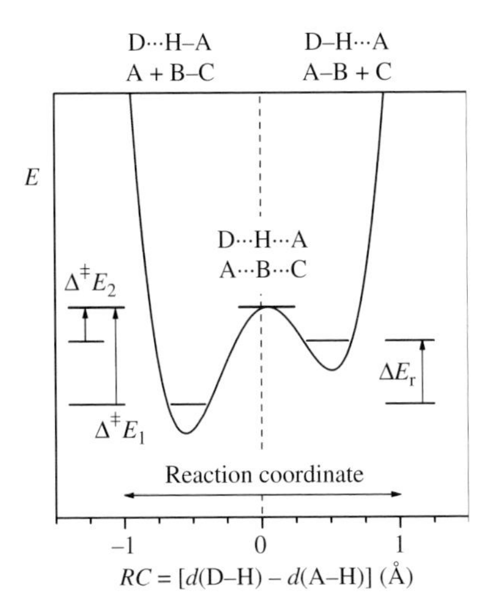

FIG. 6.1. Any D–H···A bond can be considered as a stationary point of a bimolecular reaction pathway leading from D···H–A to D–H···A through the D···H···A transition state. $\Delta^{\ddagger}E_1$ and $\Delta^{\ddagger}E_2$ are the *PT barriers* in the two directions and ΔE_r the *reaction energy* of the PT reaction, which also represents the PA (or pK_a) difference between the H-bond donor and acceptor groups. (Reproduced by permission from Gilli *et al.*, 2006)

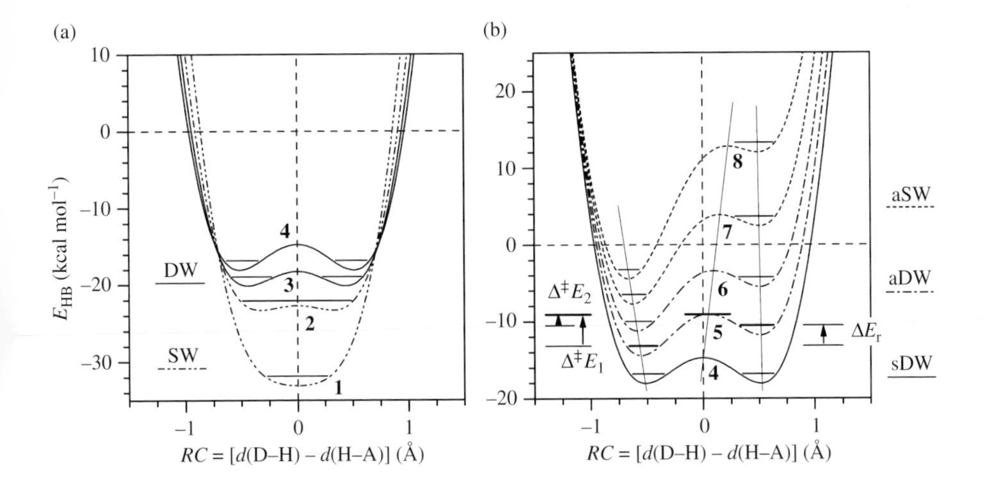

FIG. 6.2. (a) Possible shapes that may be adopted by a symmetric (intrinsic) PT reaction pathway (SW = single-well; DW = double-well); and (b) dissymmetrization of the symmetric (intrinsic) PT pathway **4** of (a) induced by the increasing value of the reaction energy, ΔE_r (sDW = symmetric DW; aDW = asymmetric DW; aSW = asymmetric SW). (Reproduced by permission from Gilli *et al.*, 2006)

the following properties: (*i*) its intrinsic PT profile is double-well; (*ii*) changes of the electronic properties of one selected substituent (say R_1) must be able to encompass the full range of H-bond properties, i.e. from D–H···A to D···H–A; (*iii*) the substituent must be in such a position not to produce steric perturbations; (*iv*) all compounds investigated must belong to the *same coherent series* (Leffler and Grunwald, 1963; Grunwald, 1985), that is a series of compounds having a common intrinsic barrier and a formal reaction zone far removed from the substitution zone.

6.3 A practical verification of the TSHBT

6.3.1 *A suitable reaction: The ketohydrazone $\rightleftharpoons$ azoenol system*

One of the H-bonded systems that can be more easily modulated for the effect of substituents is that associated with the N–H···O/N···H–O competition in ketohydrazone–azoenol derivatives forming intramolecular RAHBs (Olivieri *et al.*, 1989; Gilli *et al.*, 2002, 2005a). Its PT profiles and VB interpretation have been already discussed in Section 3.2.3.4 (Scheme **3.2.XXV** and Figs. 3.27–30). The compounds used here (Scheme **6.1.I**) are a series of 1-arylazo-2-naphthols where the electron-attracting/donating properties of the phenyl substituent are changed with continuity to induce the transition from N–H···O to N···H–O through the N–H···O $\rightleftharpoons$ N···H–O tautomeric bond. This system has been investigated (Gilli *et al.*, 2002, 2005a, 2006) by three different methods: (*i*) variable-temperature X-ray crystallography (VTXRC); (*ii*) DFT emulation; and (*iii*) application of the Marcus rate-equilibrium theory to the emulated data.

6.3.2 *Methods of study*

The VTXRC method. Variable-temperature X-ray crystallography (VTXRC) is a particularly useful tool in the study of DW tautomeric H-bonds. It implies accurate crystal-structure determinations at different temperatures (most often in the 100–300 K range) including refinement of the populations (occupancies) of the tautomeric proton, p and $1-p$, which are directly related to the equilibrium constant $K = p/(1-p)$ of the PT reaction. Further application of the van't

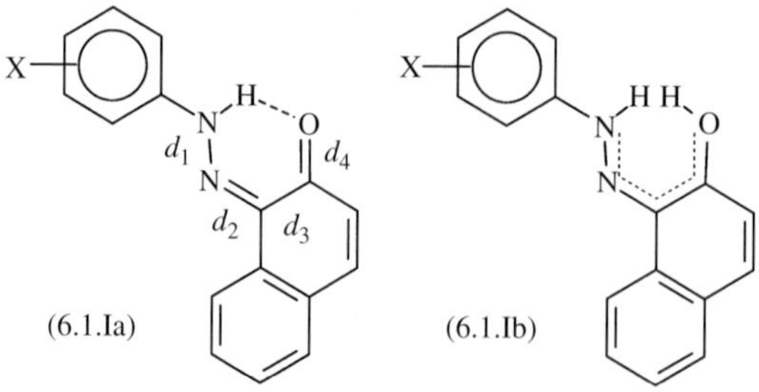

(6.1.Ia) Ketohydrazone (hydrazonaphthone)

(6.1.Ib) Azoenol (azonaphthol)

pNO2	X = *p*-NO_2
mOM	X = *m*-OCH_3
pH	X = *p*-H
pCl	X = *p*-Cl
pF	X = *p*-F
pNM2	X = *p*-$N(CH_3)_2$
pO⁻	X = *p*-O^-

Hoff equation, $\ln K = \Delta S°/R - \Delta H°/R\ (1/T)$, can so provide standard enthalpies ($\Delta H°$) and entropies ($\Delta S°$) of the tautomeric equilibrium and then the energy difference between the two minima of the DW potential. Moreover, the method allows static and dynamic disorder of the proton in tautomeric crystals to be distinguished in the sense that only fast dynamic exchange at all temperatures guarantees the rapid population equilibration that makes the van't Hoff plot possible, at variance with static disorder that, causing nearly invariant or randomly distributed populations, cannot follow any thermodynamic rule.

Computationals. DFT calculations were performed at the B3LYP/6−31+G(d,p) level of theory (Gaussian 98: Frisch *et al.*, 2002) with full geometry optimization in the C_S point group of N−H···O and N···H−O ground-states and N···H···O transition state followed by vibrational analysis for zero-point correction (ZPC) and planarity check at the three stationary points. Full PT profiles were computed by the QSTn method (Ayala and Schlegel, 1997).

Data interpretation: The Marcus rate-equilibrium theory. The energetic and geometrical quantities evaluated by DFT methods are as follows. ΔE and ΔE_{ZPC} are the non-corrected and ZP-corrected energies of the stationary points relative to the TS chosen as zero. In terms of the complete PT reaction pathway of Fig. 6.3, they correspond to the negative of the *energy barriers*, $\Delta^{\ddagger}E$ or $\Delta^{\ddagger}E_{ZPC}$, of the PT process in the two directions, while the energy differences between the two minima, ΔE_r or $\Delta E_{r,\,ZPC}$, assume the meaning of non-corrected and ZP-corrected

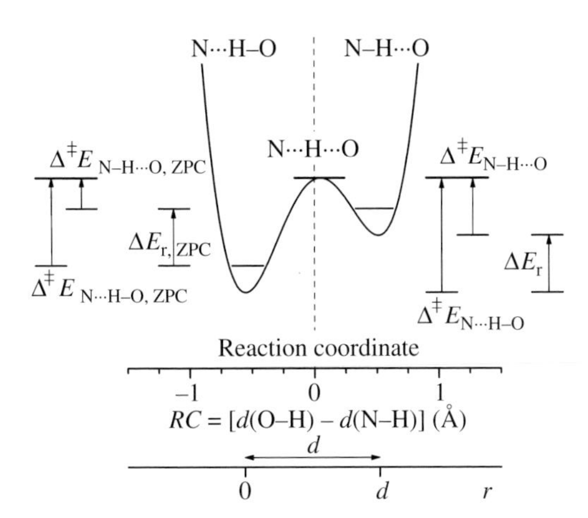

Fig. 6.3. Symbols for energetic and geometric quantities computed by DFT methods and used in the subsequent Marcus analysis. $\Delta^{\ddagger}E$ and $\Delta^{\ddagger}E_{ZPC}$ are the non-corrected and ZP-corrected PT barriers in the two directions, ΔE_r and $\Delta E_{r,\,ZPC}$ the corresponding reaction energies, RC the reaction coordinate, defined as d(O−H) − d(N−H), and d the total length of the PT pathway. (Reproduced by permission from Gilli *et al.*, 2006)

reaction energies (the two are fairly similar because ZPC has a nearly constant value of 2.52 ($\pm$0.11) kcal mol^{-1} for all compounds studied). $RC = [d(\text{O–H}) - d(\text{N–H})]$ is the *reaction coordinate* for the PT process, while $d = RC_{\text{N–H}\cdots\text{O}} - RC_{\text{N}\cdots\text{H–O}}$ is the *total length of the PT pathway.* It is advantageous to rescale RC to the *relative reaction coordinate*, $r = RC - RC_{\text{N}\cdots\text{H–O}}$ ($0 \leq r \leq d$), or to the *fractional reaction coordinate*, r/d ($0 \leq r/d \leq 1$).

Results can be interpreted in terms of the Marcus rate-equilibrium theory (Marcus, 1960, 1968; Grunwald, 1985; Shaik *et al.*, 1992) where the PT pathway is represented as the sum of two harmonic oscillators that cross at the TS (where $r = r^{\ddagger}$) and have parabolic equations $E_{\text{N}\cdots\text{H–O}} = ½\, k\, r^2$ and $E_{\text{N–H}\cdots\text{O}} = \Delta E_{\text{r}} + ½\, k\,(r - d)^2$. The theory is based on the well-known Marcus equations that connect energy barriers, $\Delta^{\ddagger}E$, to reaction energies, ΔE_{r}, in the form

$$\Delta^{\ddagger}E = \Delta^{\ddagger}E_{\text{o}} + \Delta E_{\text{r}}/2 + (\Delta E_{\text{r}})^2/(16\, \Delta^{\ddagger}E_{\text{o}}), \quad (6.5)$$

$$r^{\ddagger}/d = \Delta E_{\text{r}}/(k\, d^{\,2}) + 1/2, \quad (6.6)$$

$$k = 8\, \Delta^{\ddagger}E_{\text{o}}/d^{\,2}, \quad (6.7)$$

and where $\Delta^{\ddagger}E_{\text{o}}$ is the *intrinsic or symmetric barrier* occurring when $\Delta E_{\text{r}} = 0$.

6.3.3 *Analysis of crystallographic results*

Table 6.1 summarizes VTXRC results obtained at 100 K for ***mOM***, ***pCl***, ***pF***, and ***pNM2*** (Gilli *et al.*, 2002, 2005a), 298 K for ***pNO2*** (Whitaker, 1980) and 213 K for ***pH*** (Olivieri *et al.*, 1989) (the crystal structure of ***pO***$^-$ is unknown and data are taken from the DFT emulation). Compounds are ordered from left to right according to the decreasing value of the mesomeric constant, $\sigma^{\text{o}}_{\text{R}}$, of the *p*-phenyl substituent X, i.e. accordingly to its increasing π-donating properties (Chapman and Shorter, 1978). It is evident that π-electron donation by the substituent represents the driving force of all the process that consists in a progressive transformation of the N–H···O into the N···H–O bond through intermediate establishing of the tautomeric and dynamically disordered N–H···O $\rightleftharpoons$ N···H–O equilibrium. With reference to the PT-profile nomenclature of Fig. 6.2(b), X-substituents of increasing π-donating properties are seen to produce a first transition from the ordered N–H···O bond with the aSW profiles **7** or **8** (compounds ***pNO2***, ***mOM*** and ***pH***) to the dynamically disordered N–H···O $\rightleftharpoons$ N···H–O bond having comparable populations ($\approx 0.6 : 0.4$) and described by the nearly sDW profile **5** (compounds ***pCl*** and ***pF***). Finally, an additional increase of π-donation in compound ***pNM2*** induces a further shift toward the N···H–O form, where proton populations now reverse to 21 : 79 and the decreased symmetry of the PT pathway causes an increase of the PT barrier that switches the disorder from dynamic to static, a situation that could be well represented by the mirror image of profile **6**. We can only verify by DFT emulation that such a trend will continue with ***pO***$^-$, eventually producing the pure and ordered N···H–O bond having the aSW profile that is the mirror image of **7** or **8**.

The use of VTXRC methods to discriminate between static and dynamic disorder is exemplified in Fig. 6.4, where: (*a*) the dynamic N–H···O $\rightleftharpoons$ N···H–O disorder in crystals of 1-(*p*-fluorophenylazo)-2-naphthol (***pF***) is substantiated by the regular variations of the *p*(NH):*p*(OH) ratios with the temperature, the linearity of the van't Hoff plot, and the continuous electron-density changes in the difference-Fourier maps at 100, 200 and 295 K; and (*b*) the static N–H···O/N···H–O proton disorder in crystals of 1-(*p*-*N*,*N*-dimethylaminophenylazo)-2-naphthol (***pNM2***) is illustrated by the constancy of the *p*(NH) : *p*(OH) ratios and of the difference-Fourier peaks for a change of temperature from 100 to 295 K.

6.3.4 *DFT emulation*

DFT results are summarized in Table 6.2. Agreement with crystal geometries is fairly good with average and maximum discrepancies on bond distances of 0.007 and 0.023 Å. Differences on N···O distances are 0.004–0.009 Å with the only exception of ***pCl*** (0.030 Å). The compounds are listed in order of increasing ΔE_{r} from −1.59 for ***pNO2*** to 1.34 kcal mol^{-1} for ***pO***$^{-}$. With increasing ΔE_{r}, the stable H-bond form is seen to move from pure N–H···O (***pNO2***, ***mOM*** and ***pH***) to two coexisting N–H···O ⇌ N···H–O tautomers (***pCl***, ***pF*** and ***pNM2***) up to pure N···H–O (***pO***$^{-}$). The resulting PT profiles are in perfect agreement with X-ray findings of Table 6.1. The TS always corresponds to the shortest N···O distance along the pathway (2.38–2.40 Å), while the N···O distances in the two N–H···O and N···H–O minima, besides being longer (2.53–2.56 Å), differ by an amount, Δ, which changes regularly with ΔE_{r} from 0.026 Å for ***pNO2*** to nearly zero for ***pF*** and ***pNM2***, down to −0.012 for ***pO***$^{-}$, in agreement with the Leffler–Hammond postulate (Leffler, 1953; Hammond, 1955) that the closer a minimum is to the TS, the more it participates in its geometrical structure. Since the H-bond at the TS is always the shortest, in the N–H···O ⇌ N···H–O tautomeric couple the H-bond closer to the TS (the less stable one) is always shorter than that which is farther away (the more stable one), a difference that will fade when the PT-barrier becomes symmetrical ($\Delta E_{\mathrm{r}} = 0$). This symmetry can also be appreciated by the fractional reaction coordinate at TS, $r^{\ddagger}/d$, which is ≈0.5 for ***pF*** and ***pNM2*** but becomes outcentered for the other compounds with extremes of 0.438 and 0.555 for ***pNO2*** and ***pO***$^{-}$, respectively.

The H-bond energies, E_{HB}, at the three stationary points are calculated by making reference to the N···H–O form whose E_{HB} values are more easily evaluated by direct comparison of the total energies of the closed (H-bonded) and open forms (obtained by simple rotation of the proton by 180° around the C–OH bond of the molecule (see Scheme **3.2.XVIII**, Section 3.2.3.4A)). Values are not ZP-corrected and the corrected ones would be smaller by some 0.3–0.7 kcal mol^{-1}. The E_{HB} of the most stable tautomer remains nearly constant (some 15.4 kcal mol^{-1}) from ***pNO2*** to ***pNM2***, where the contribution of the N–H···O form is prevalent or at least significant, but steeply increases to 19.4 kcal mol^{-1} for the pure N···H–O bond in ***pO***$^{-}$ for which some shortening of the total PT-pathway length is also observed (d = 1.147 Å against the average of 1.283 Å), suggesting that the N···H–O bond is intrinsically stronger than the N–H···O one, at

TABLE 6.1. Summary of X-ray crystallography results for 1-(phenylazo)-2-naphthols X-substituted at the phenyl ring. The crystal structure of ***pO***$^-$ is missing and data in parentheses are taken from the DFT emulation. SW, DW = single-, double-well; a, sa, ≈s = asymmetric, slightly asymmetric, nearly symmetric; ΔH°_{vtH} is the energy difference between the two DW minima measured from the van't Hoff plot in kcal mol^{-1}.

Code	***pNO2***	***mOM***	***pH***	***pCl***	***pF***	***pNM2***	***pO***$^-$
X	*p*-NO_2	*m*-OCH_3	*p*-H	*p*-Cl	*p*-F	*p*-$N(CH_3)_2$	*p*-O^-
σ°_R	0.17	–	0	−0.29	−0.40	−0.53	−0.60
T (K)	298	100	213	100	100	100	–
H-Bond	N–H···O	N–H···O	N–H···O	N–H···O ⇌ N···H–O	N–H···O ⇌ N···H–O	N–H···O ⇌ N···H–O	N···H–O
d(N···O)	2.56(1)	2.548(2)	2.553(1)	2.516(2)	2.535(2)	2.534(2)	(2.539)
p(NH)%	100	100	100	69	64	21	0
p(OH)%	0	0	0	31	36	79	100
PT Profile	aSW	aSW	aSW	saDW	≈sDW	saDW	(aSW)
Order/ disorder	Ordered	Ordered	Ordered	Dynamically disordered	Dynamically disordered	Statically disordered	(Ordered)
ΔH°_{vtH}	–	–	–	−0.120(15)	−0.146(25)	–	–
PT barrier	High	High	High	Low	Low	Medium-high	(High)

least in this class of compounds. Since H-bond energies are known to decrease while the bonds become increasingly bent, the greater stability of the N···H–O bond can be perhaps ascribed to the different equilibrium values of the N–N–H and C–O–H angles (some 116 and 106 degrees, respectively) which make the N–H–O angle much less bent in the N···H–O case (152.3° in ***pO***$^-$) than in the N–H···O one (on average, 137° from ***pNO2*** to ***pF***).

6.3.5 *Marcus analysis of DFT data*

Recursive application of eqns 6.5 and 6.6 allows calculation of values of the intrinsic or symmetric barrier, $\Delta^{\ddagger}E_0$, and of the vibrational force constant, k, for each line of Table 6.2. Averaged results are $\Delta^{\ddagger}E_0 = 3.08$ kcal mol^{-1} and $k = 15.0$ kcal mol^{-1} Å^{-2}, two values that can be shown to be mutually consistent through eqn 6.7. The total length of the PT pathway is evaluated to be $d = 1.283$ Å. The agreement between Marcus- and DFT-derived values is generally fairly good

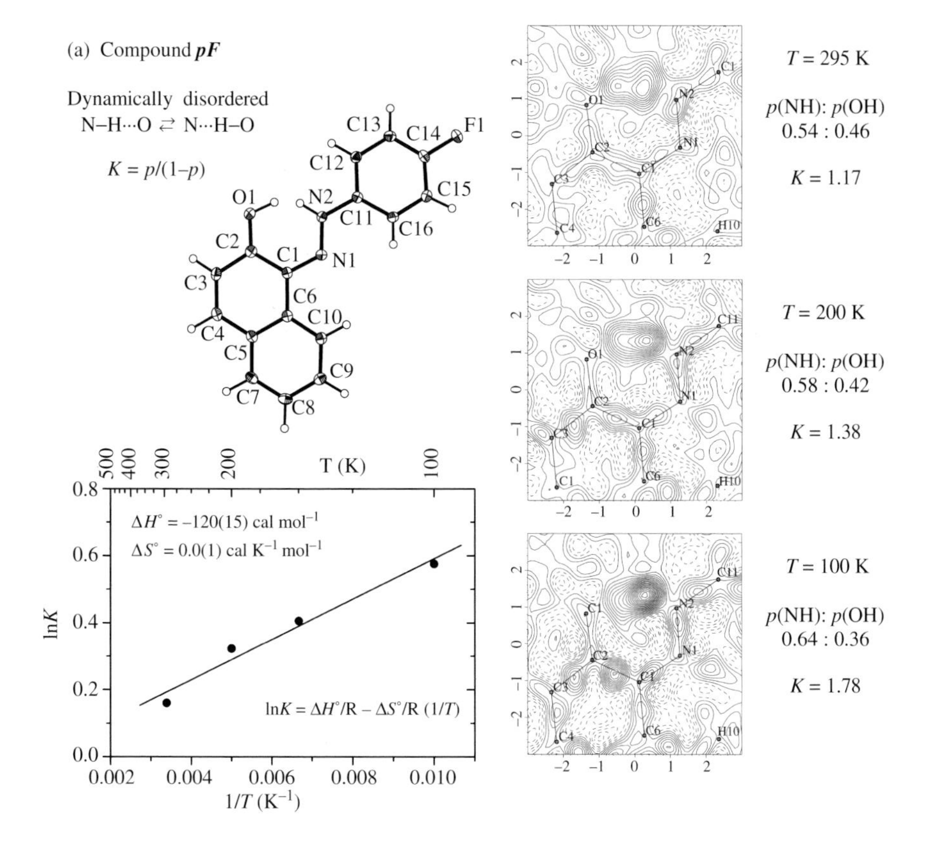

Fig. 6.4 Continues

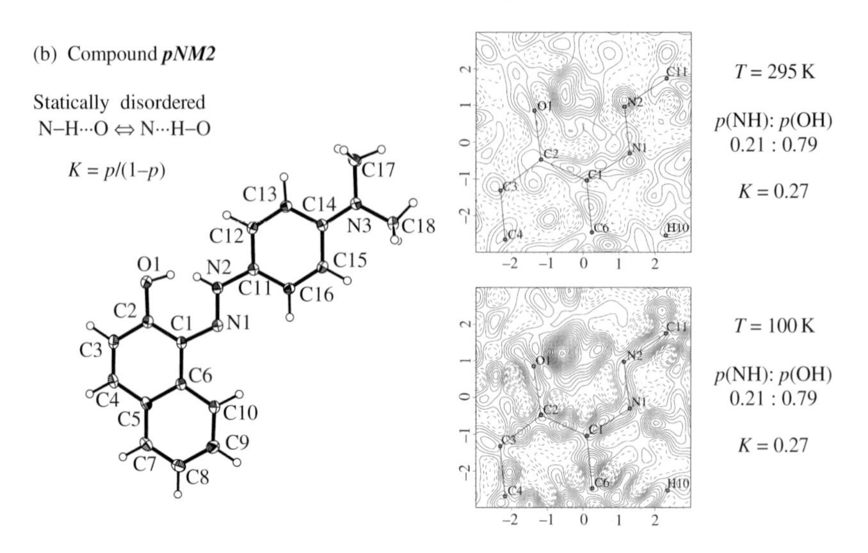

FIG. 6.4. Discrimination between static and dynamic disorder by VTXRC methods. (*a*) Dynamic N–H···O ⇌ N···H–O proton disorder in crystals of 1-(*p*-fluorophenylazo)-2-naphthol (***pF***) as shown by the continuous proton population changes with the temperature and by the linearity of the van't Hoff plot. Difference Fourier maps at the temperatures of 100, 200 and 295 K. Negative (dashed) and positive (continuous) contours at 0.04 e Å^{-3} intervals. (*b*) Static N–H···O/N···H–O proton disorder in crystals of 1-(*p*-*N*,*N*-dimethylaminophenylazo)-2-naphthol (***pNM2***) as shown by the independence of proton populations from the temperature and by the difference Fourier maps at the temperatures of 100 and 295 K. Negative (dashed) and positive (continuous) contours at 0.04 e Å^{-3} intervals. (Reproduced by permission from Gilli *et al.*, 2006)

except for the $\Delta^{\ddagger}E_0$ of $\boldsymbol{pO^-}$, which is 1.0 kcal mol^{-1} too high. Finally, as the average ZP correction is some 2.52 kcal mol^{-1} (see above), the intrinsic barrier corrected for proton vibration, $\Delta^{\ddagger}E_{\text{o,ZPC}}$, can be estimated to be $3.08 - 2.52 = 0.56$ kcal mol^{-1}.

These results are graphically summarized in Fig. 6.5(a). The two parabolas having the same height ($\Delta E_{\text{r}} = 0$) and shifted by d (continuous lines) represent the ideal *intrinsic profile* and cross at an *intrinsic barrier* $\Delta^{\ddagger}E_{\text{o}} =$ 3.08 kcal mol^{-1} at $r^{\ddagger}/d = 0.5$. The other curves (dashed and dotted lines) represent the actual compounds studied and are shifted upward or downward according to their respective ΔE_{r} values. Their parabolas cross at $r^{\ddagger}/d$ smaller or larger than 0.5 and $\Delta^{\ddagger}E$ lower or higher than $\Delta^{\ddagger}E_{\text{o}}$ according to whether

TABLE 6.2. DFT stationary-point energies (kcal mol^{-1}) and geometries (Å and degrees) along the PT pathway for the intramolecular N–H···O/N···H–O bond in 1-(phenylazo)-2-naphthols variously substituted at the phenyl ring. ΔE and ΔE_{ZPC} = non-corrected and zero-point-corrected (ZPC) total energies relative to the TS chosen as zero; Δ = difference between the N···O distances of the N–H···O and N···H–O bonds; $RC = [d(\text{O–H}) - d(\text{N–H})]$ = reaction coordinate; $d = RC_{\text{N–H···O}} - RC_{\text{N···H–O}}$ = total reaction-pathway length; $r = RC - RC_{\text{N···H–O}}$ = relative RC; $r^{\ddagger}/d$ = TS fractional RC; $<\lambda>$ = π-delocalization index of the N–N–C–C–O π-conjugated fragment; E_{HB} = H-bond energies computed with reference to the N···H–O form; PT-profile = DFT-emulated PT profile; SW, DW = single-, double-well; a, sa, ≈s = asymmetric, slightly asymmetric, nearly symmetric; σ^0_R = mesomeric constant of the p-substituent. In columns N···O and E_{HB}, the more stable form is indicated by a full point and the tautomeric pairs by two asterisks.

Code	H-bond	$\Delta E = -\Delta^{\ddagger}E$	$\Delta E_{ZPC} = -\Delta^{\ddagger}E_{ZPC}$	N···O	N–H	H–O	N–H–O	RC	r	$r^{\ddagger}/d$	$<\lambda>$	E_{HB}	PT-profile	σ^0_R
pNO2	N–H···O	−3.83	−1.44	2.555•	1.037	1.699	136.7	0.662	1.287		0.41	15.5•		
	TS	0	0	2.384	1.262	1.201	150.8	−0.061	0.564	0.438	0.60	11.7	aSW	0.17
	N···H–O	−2.23	0.12	2.529	1.633	1.008	145.3	−0.625	0		0.74	13.9		
	ΔE_r	−1.59	−1.56	Δ = 0.026				d = 1.287						
mOM	N–H···O	−3.82	−1.32	2.557•	1.037	1.700	136.9	0.663	1.293		0.44	16.3•		
	TS	0	0	2.386	1.256	1.209	150.9	−0.047	0.583	0.451	0.63	12.5	aSW	–
	N···H–O	−2.55	−0.05	2.534	1.638	1.008	145.5	−0.630	0		0.78	15.0		
	ΔE_r	−1.26	−1.27	Δ = 0.023				d = 1.293						
pH	N–H···O	−3.69	−0.99	2.555•	1.038	1.695	137.1	0.657	1.289		0.44	15.7•		
	TS	0	0	2.386	1.253	1.211	150.9	−0.042	0.590	0.458	0.63	12.0	aSW	0
	N···H–O	−2.58	−0.11	2.535	1.640	1.008	145.4	−0.632	0		0.78	14.6		
	ΔE_r	−1.11	−0.88	Δ = 0.020				d = 1.289						

TABLE 6.2. *Continued*

Code	H-bond	$\Delta E = -\Delta^{\ddagger}E$	$\Delta E_{ZPC} = -\Delta^{\ddagger}E_{ZPC}$	N···O	N–H	H–O	N–H–O	RC	r	$r^{\ddagger}/d$	$<\lambda>$	E_{HB}	PT-profile	σ^{0}_{R}
pCl	N–H···O	−3.64	−0.94	2.550*	1.039	1.688	137.2	0.649	1.285		0.45	15.2*		
	TS	0	0	2.384	1.249	1.214	150.9	−0.035	0.601	0.468	0.63	11.6	saDW	−0.29
	N···H–O	−2.72	−0.31	2.536*	1.643	1.007	145.3	−0.636	0		0.75	14.3*		
	ΔE_r	−0.91	−0.63	$\Delta = 0.014$				$d = 1.285$						
pF	N–H···O	−3.33	−0.65	2.546*	1.039	1.681	137.5	0.642	1.282		0.44	14.6*		
	TS	0	0	2.384	1.246	1.218	150.9	−0.028	0.612	0.477	0.63	11.3	≈ sDW	−0.40
	N···H–O	−2.93	−0.36	2.538*	1.646	1.006	145.3	−0.640	0		0.78	14.2*		
	ΔE_r	−0.40	−0.29	$\Delta = 0.008$				$d = 1.282$						
pNM2	N–H···O	−2.65	−0.15	2.541*	1.043	1.663	138.6	0.620	1.261		0.49	14.5*		
	TS	0	0	2.388	1.234	1.232	151.0	−0.002	0.639	0.507	0.66	11.9	saDW	−0.53
	N···H–O	−3.14	−0.70	2.541*	1.647	1.006	145.6	−0.641	0		0.81	15.0*		
	ΔE_r	0.49	0.55	$\Delta = 0.000$				$d = 1.261$						
pO^{-}	N–H···O	−1.44	0.97	2.527	1.060	1.596	143.3	0.536	1.147		0.60	18.1		
	TS	0	0	2.404	1.225	1.251	147.5	0.026	0.637	0.555	0.69	16.6	aSW	−0.60
	N···H–O	−2.78	−0.30	2.539•	1.625	1.014	152.3	−0.611	0		0.76	19.4•		
	ΔE_r	1.34	1.28	$\Delta = -0.012$				$d = 1.147$						

their specific ΔE_r value is smaller or greater than zero, so following the Leffler–Hammond rule that the position of the TS is more shifted towards the reagents (in this case N···H–O) the more the reaction is exoergonic (in the present case, the more ΔE_r is negative). Hence, the Marcus treatment seems quite able to reduce all compounds of Table 6.2 to a same coherent reaction series having a common intrinsic barrier continuously modulated by the chemical properties of the substituents outside the formal reaction zone. This seems a first important indication that the formalism arising from the TSHBT may be substantially correct.

The information arising from Marcus analysis and DFT-computed E_{HB} values is combined in Fig. 6.5(b) that is intended to represent the total energy of the system with respect to a reference state, different for each different molecule, which corresponds to the energy of its hypothetical non-bonded (open) state. The continuous curve denotes the intrinsic PT profile whose E_{HB} has been estimated as the average of ***pF*** and ***pNM2***, the two compounds approaching the condition $\Delta E_r = 0$ from opposite sides. Some compounds that do not add much to the general picture (***mOM***, ***pH*** and ***pF***) are omitted to avoid overcrowding in the N–H···O region. The figure also reports, as horizontal straight lines, the approximate vibrational levels of the proton, always assumed to have the constant value of 2.52 kcal mol^{-1}. As expected, in compounds forming pure N–H···O (***pNO2***) or N···H–O bonds (***pO***$^-$) the vibrational levels of the unstable forms are higher than the PT barrier, while in the two tautomeric compounds (***pCl*** and ***pNM2***) both levels lie slightly below it. These results are in good agreement with the diffraction experiments (Table 6.1) and seem to be an indication that the DFT-calculated barriers cannot be very far from the real ones.

It is of particular interest to examine the results so far obtained to find out whether the different quantities arising from a more or less formal application of the Marcus model really delineate an underlying physical model of the H-bond phenomenon compatible with its empirical laws summarized in Chapter 5. This analysis is carried out in the following:

(*a*) *The intrisic barrier.* The value of the intrinsic barrier found for this series of compounds amounts to $\Delta^{\ddagger}E_o = 3.08$, to be reduced to $\Delta^{\ddagger}E_{o,ZPC} = 0.56$ kcal mol^{-1} because of ZPC. This value can be considered a true estimate of the adiabatic barrier because, being obtained by MO methods, does not need the corrections normally applied in VB treatments (Åqvist and Warshel, 1993; Braun-Sand *et al.*, 2004; Gilli *et al.*, 2005a). A value as low as 0.56 kcal mol^{-1} seems reasonable because, although the bimolecular D–H···A ⇌ D···H···A ⇌ D···H–A reaction proceeds via the D···H···A transition state, its reactants and products are both *pre-bound* by the H-bond itself. Comparison with the diffraction data of Table 6.1 indicates that this barrier is low enough to allow formation of a dynamically disordered LBHB when ΔE_r is close to zero, while any increase of ΔE_r contributes to make the PT profile more and more dissymmetric

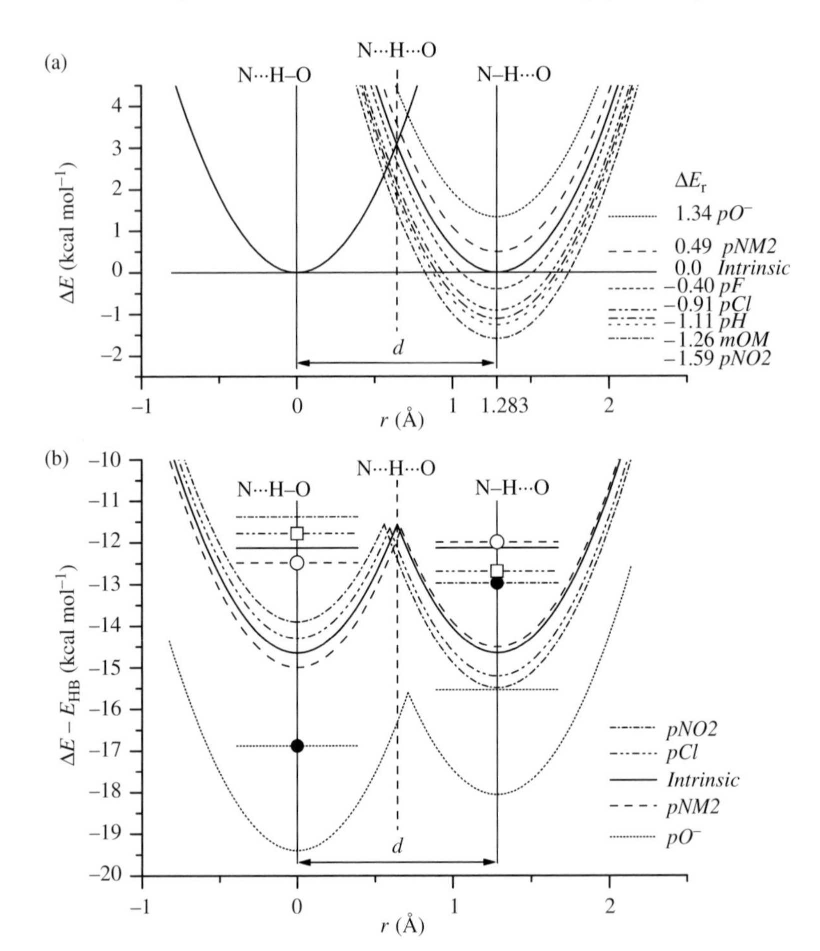

FIG. 6.5. (a) Marcus modelling of the N···H–O → N–H···O PT reaction for the compounds of Table 6.2 (without ZPC). The two symmetric parabolas (continuous) crossing at the intrinsic barrier $\Delta^{\ddagger}E_{\mathrm{o}} = 3.08$ kcal mol^{-1} for $r^{\ddagger}/d = 0.5$ represent the intrinsic PT profile. Curves for the other compounds (dashed and dotted) are shifted upwards or downwards by their respective reaction energies, ΔE_{r}, and cross at $r^{\ddagger}/d$ values larger or smaller than 0.5 with PT barriers $\Delta^{\ddagger}E$ higher or lower than $\Delta^{\ddagger}E_{\mathrm{o}}$, respectively. (b) Marcus modelling of (a) as modified by subtracting the values of the N···H–O bond energy, E_{HB}, to each individual curve. The continuous curve corresponds to the *intrinsic H-bond* with $\Delta E_{\mathrm{r}} = 0$ and $\Delta^{\ddagger}E = \Delta^{\ddagger}E_{\mathrm{o}}$. Horizontal lines mark the approximate vibrational levels of the proton; full points indicate single-well and open symbols double-well H-bonds. (Reproduced by permission from Gilli *et al.*, 2006; see also Color Plate 5)

causing the PT barrier of the more stable tautomer to increase, so inducing the shift from dynamic to static disorder and, finally, the transition to the pure D–H···A or D···H–A bond having a completely asymmetric aSW profile. Notice, however, that the intrinsic barrier in this type of compounds is small but not zero, so ruling out the possibility of forming the very short N···H···O bonds with symmetric sSW PT profile characteristic of some O···H···O RAHBs (Section 3.2.2.3).

(*b*) *The harmonic constant.* The constant of the parabolic expression $E = ½\, k\, r^2$ takes the value $k = 15.0$ kcal mol^{-1} Å^{-2} = 0.027 mdyn = 0.027 N cm^{-1} ≈ 75 cm^{-1}. Rather surprisingly, this force constant is some 250 times smaller than that associated with normal N–H or O–H stretchings (6–8 N cm^{-1}) and must therefore be associated with the NH···O or N···HO stretching vibrations of the H-bond itself, in agreement with some early IR measurements that assign frequencies of 80–250 cm^{-1} to these vibrations (Vinogradov and Linnell, 1971). Hence, all the N–H···O ⇌ N···H–O process can be described through these two shallow non-bonded vibrations that are shifted vertically by the proper value of ΔE_r and horizontally by a constant amount d = 1.283 Å (the length of the PT pathway).

(*c*) *The driving variable: The reaction energy,* ΔE_r. Equations 6.5–7 contain four parameters that fully characterize the intrinsic profile, namely $\Delta^{\ddagger}E_o$, ΔE_r, d, and k. One of these is the independent variable, that is the physicochemical quantity that drives the process and so determines the H-bond strength. The long discussion carried out in the previous chapters has concluded that such a driving variable must be the reaction energy, ΔE_r, which is here the equivalent of other similar quantities in other theories, namely ΔPA and ΔpK_a in the PA/pK_a equalization principle or $\Delta E = E(\Psi_{COV2,CT}) - E(\Psi_{COV1,NCT})$ in the VB-based ECHBM. Fixing a parametric value for the independent variable ΔE_r, the three remaining variables still remain related by the equation $k = 8\, \Delta^{\ddagger}E_o/d^2$ (eqn 6.7), so that the intrinsic PT profile ($\Delta E_r = 0$) of the H-bond studied is completely characterized by only two parameters ($\Delta^{\ddagger}E_o$ and k, $\Delta^{\ddagger}E_o$ and d, or d and k).

(*d*) *Reaction profile and invariance of the TS.* The only aspect of the reaction profile that is practically unaffected by ΔE_r changes is the geometry of the TS, which displays a nearly constant N···O distance of 2.38–2.40 Å associated with an almost complete delocalization of the interleaving resonant fragment, suggesting that the H-bond at the TS corresponds to the nearly invariant 3c–4e covalent bond in theory predicted for any H-bonded system (see also Section 3.2.2.5). Conversely, ΔE_r changes cause dramatic effects on the remaining parts of the PT pathway. While the intrinsic profile ($\Delta E_r = 0$) is associated with a tautomeric H-bond with perfectly symmetric DW potential (that can then be called the *intrinsic H-bond*), any increase (or decrease) of ΔE_r splits such a DW H-bond into two distinct bonds: (*i*) a less stable one that is shorter because it is closer to the TS; and (*ii*) a more stable one that is longer because it is farther from the TS. This leads, in agreement with the Leffler–Hammond postulate, to the conclusion that the H-bond actually observed (the more stable) is always the *longer* of the tautomeric D–H···A ⇌ D···H–A couple considered.

(*e*) *Linear free-energy relationships.* In RAHBs, the reaction energy ΔE_r can be evaluated by quantum-mechanical methods but not by using thermodynamic ΔPA and ΔpK_a parameters, and this is because of the large changes of π-delocalization occurring in the interleaving conjugated spacer (in this case the ···HN–N=C–C=O··· ⇌ ···N=N–C=C–OH··· moiety) during H-bond formation. However, it can still be appreciated through the extrathermodynamic LFER parameters of the substituents. In this study, ΔE_r values do not correlate with the usual *para* and *meta* Hammett constants (σ_p and σ_m) but rather with the mesomeric constant, σ°_R, most probably because of the resonance-assisted nature of the H-bond treated. This correlation is shown in Fig. 6.6 for all compounds except ***mOM***, for which σ°_R has no precise meaning. The correlation looks impressive and thus supports the idea that the driving force able to modify the PT profile is actually (all other factors remaining constant) the proton affinity of nitrogen as determined by the σ°_R of the phenyl *p*-substituent. The plot appears, however, slightly different from what could be expected, that is a unique straight line encompassing all compounds. What is actually observed, conversely, are two intercrossing lines with rather different slopes. Leffler and Grunwald, in their 1963 book 'Rates and equilibria of organic reactions', have considered the possibility of such double-slope plots and interpreted them as a typical feature of chemical reactions that change their mechanism beyond a certain value of σ. This indicates that the two N–H···O → N···H–O and N···H–O → N–H···O reactions have different characteristics, a fact that has been interpreted in Section 6.3.4 in terms of the greater linearity and H-bond energy of the N···H–O with respect to the N–H···O bond that is induced by the different values of the N–N–H (≈ 116°) and C–O–H (≈ 106°) angles. For example, H-bond energies and N–H–O angles are 15.5 kcal mol^{-1} and 136.7° in the N–H···O bond of ***pNO2*** but 19.4 kcal mol^{-1} and 152.3° in the N···H–O bond of ***pO***$^-$ and it can be supposed that geometrical and energetic differences of such an entity can well justify the change of mechanism occurring on the two sides of the intrinsic barrier (the horizontal line with $\Delta E_r = 0$ in Fig. 6.6).

6.3.6 *Conclusions*

The aim of this chapter was to show that the H-bond can naturally be treated as a bimolecular D–H···A ⇌ D···H···A ⇌ D···H–A reaction where D and A compete for a same central proton and that, accordingly, the process of H-bond formation can be formalized as an application of the traditional transition-state theory called here transition-state H-bond theory (TSHBT). A first verification of this thesis has been attempted on a sample system, a series of variously phenyl-substituted 1-(phenylazo)-2-naphthols forming intramolecular RAHB, obtaining results that throw both lights and shadows on the possible use of the TSHBT as a standard tool for the study of any sort of H-bond. The positive aspect is that all substituent-induced changes turn out to be quantitatively interpretable

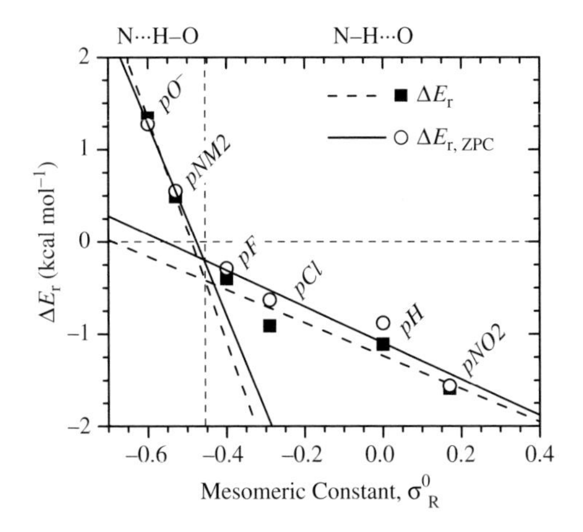

FIG. 6.6. Plot of the DFT-computed reaction energies, ΔE_r, as a function of the mesomeric constant σ°_R of the phenyl *p*-substituent. Following Leffler and Grunwald (1963), the two straight lines can be supposed to indicate that the two N–H···O → N···H–O and N···H–O → N–H···O reactions have different PT mechanisms. Full and open symbols refer to uncorrected and ZP-corrected DFT-computed data, respectively. (Reproduced by permission from Gilli *et al.*, 2005a)

by the traditional instruments of chemical kinetics, such as Leffler–Hammond postulate, extrathermodynamic Marcus rate-equilibrium theory, and the usual LFER correlations with the electronic parameters of the substituents. What is less positive is that, to obtain these results, the chemical system studied had to be carefully planned to avoid unwanted steric effects that could perturb the reaction center, i.e. the H-bond itself. In spite of that, Fig. 6.6 displays a two-slope ΔE_r versus σ^0_R correlation that seems a clear indication that the reaction center has actually changed during the PT process. This last fact seems to cast some doubts on the general applicability of the method to all types of H-bond, doubts that can be resolved only by further investigations carried out on other carefully selected H-bond systems.

7

THE STRENGTH OF THE H-BOND: DEFINITIONS AND THERMODYNAMICS

7.1 The H-bond strength in gas-phase, non-polar solvents and molecular crystals

7.1.1 *Enthalpy–entropy compensation and its influence on the H-bond strength*

Normally, H-bond strengths are defined in terms of bond energies (or, more exactly, enthalpies) and, for instance, Jeffrey (1997) defines as strong the bonds that have enthalpies from 14 to 40 kcal mol^{-1}, an interval to be reduced to some 14–30 kcal mol^{-1} when neglecting the strongest, and most controversial of them, the $[F \cdots H \cdots F]^-$ bond. However, what a H-bond may do in the chemical or biochemical system where it occurs does not depend on enthalpies but rather on how strong it is, that is on the actual value of its thermodynamic association, K_a, or dissociation, $K_d = 1/K_a$, constants. If the standard free enthalpy, $\Delta G° = \Delta H° - T\Delta S° = -RT \ln K_a = RT \ln K_d$, of the association equilibrium were determined only by the standard enthalpy, $\Delta H°$, the association constant for $\Delta H° = -30$ kcal mol^{-1} would reach the absurd value of some 10×10^{21} M^{-1} at 298 K. The reason why this does not happens is that any H-bond is an association process occurring with loss of degrees of freedom and then with a systematic decrease of the standard entropy, $\Delta S°$, which turns out to be the greater the more tightly the two interacting molecules are held together. And, in fact, two molecules combining to form a H-bonded complex lose three translational and three rotational degrees of freedom and statistical thermodynamics tells us that this loss produces a decrease of entropy, in the gas-phase or 1 M solution, of some 29–36 cal mol^{-1} K^{-1} for translations and of 10 to 30 cal mol^{-1} K^{-1} according to their increasing molecular dimension for rotations (Page and Jencks, 1971). This decrease is only partially compensated by the new vibration established across the H-bond formed whose contribution can become significant (1–4 cal mol^{-1} K^{-1}) for low-frequency vibrational modes (from 400 to 100 cm^{-1}).

Hence, H-bond enthalpies and entropies are not independent but related by a relationship of *enthalpy–entropy compensation*

$$\Delta H° = \beta \, \Delta S°, \tag{7.1}$$

with compensation temperature β, a type of relation that is known as *extrathermodynamic* because it does not span from the laws of thermodynamic but from the physical constraints occurring within the system investigated. Relationships of this sort play a substantial role in all phenomena of molecular association in gas-phase and non-polar solvents (Pimentel and McClellan, 1971; Joesten and Schaad, 1974) as well as in aqueous solutions, where they are known to be ubiquitously involved in the binding of molecules to macromolecules in biological systems (Leffler and Grunwald, 1963; Lumry and Rajender, 1970; Krug *et al.*, 1976; Tomlinson, 1983; Gilli *et al.*, 1994b; Grunwald, 1997, 2001).

These preliminary considerations indicate that the mutual relationships among $\Delta H°$, $\Delta S°$, and $\Delta G°$ may be important in understanding the H-bond itself as well as to estimate, whenever the value of the compensation temperature β is known, enthalpies from experimental equilibrium constants and vice versa. Since β is an extrathermodynamic parameter, it is not known a priori and has to be empirically estimated by intercorrelation of the experimental data. In the following, such an estimate will be separately attempted for the different enviroments where H-bonds may occur.

7.1.2 *H-bond strength in the gas phase*

Gas-phase studies are of particular importance because they include all the very strong H-bonds presently known (say, with $-\Delta H° \geq 20$ kcal mol^{-1}). They are normally performed by van't Hoff analysis of association constants measured at different temperatures and have the important limitation that, since the experimental methods used can detect only globally charged complexes, the bonds studied can only be of the (−)CAHB and (+)CAHB types. Empirical correlations of the data from the NIST database show that only the former bonds are actually linearly correlated, most probably because the latter are not associations of two but of three bodies, i.e. two molecules and a proton. The best correlation has been obtained for the set of homomolecular $[X\cdots H\cdots X]^-$ bonds with regression equations

$$\Delta S° = 1/\beta\,\Delta H° + c = 0.84(7)\,\Delta H° - 6(2) \quad (n = 14,\ r = 0.96)$$
$$\text{and } \Delta G° = 0.75(6)\,\Delta H° + 1(2) \quad (n = 14,\ r = 0.96)$$

($\Delta H°$ and $\Delta G°$ in kcal mol^{-1}, $\Delta S°$ in cal mol^{-1} K^{-1}). The value of β amounts to 1196 K, which is extremely high in respect to the values found in solution (see below). As an extreme example, a bond having a $\Delta H°$ of −30 kcal mol^{-1} will have $\Delta S° \cong -31.2$ cal mol^{-1} K^{-1}, $T\Delta S° \cong -9.2$, and $\Delta G° \cong -21.5$ kcal mol^{-1} at 298 K with a loss of binding enthalpy of nearly one third because of the enthalpy–entropy compensation. In spite of this decrease, the equilibrium constants remain very large, that is $\log K_a \cong 16.16$ or $K_d \cong 0.40$ pM. These exceptional values do not occur in solution where compensation effects are much larger (see below) and we may wonder how this difference comes about. Since translational and rotational effects are comparable in the two environments, the only explanation has to be based on the internal degrees of freedom of the H-bonded complex, in the sense that a greater number of low-frequency modes associated with conformational vibrations and rotations must be active in the gas phase where the constraints due to the environment are irrelevant. In other words, the extra stabilization of the H-bond observed in the gas phase must be an entropy-driven phenomenon of vibrational origin.

7.1.3 *H-bond strength in non-polar solvents*

The thermodynamics of H-bond formation between phenols and various bases in CCl_4 solutions was first reviewed by Pimentel and McClellan (1971) who found standard enthalpies and entropies to be related by the average equation $\Delta S°$ (cal K^{-1} mol^{-1}) = $1/\beta\, \Delta H° + c = 2.4\, \Delta H°$ (kcal mol^{-1}) + 0.12 with $\beta = 1/0.0024 =$ 417 K. Later, these results were essentially confirmed by Joesten and Schaad (1974) who, by analyzing much larger sets of H-bonds formed by phenols and organic acids in a variety of non-polar solvents, found comparable values of $1/\beta$ (and β), that is 0.00225 K^{-1} ($\beta = 444$ K) and 0.00205 K^{-1} ($\beta = 488$ K), respectively. Pimentel's correlation can then be considered with some confidence and is used here to evaluate, for each hypothetical $\Delta H°$, the corresponding values of $\Delta G°$, $\log K_a$, and K_d, which can be reasonably expected.

The results (Table 7.1) show that, because of enthalpy–entropy correlation effects, the value of $\Delta G°$ is reduced to nearly one third of that of $\Delta H°$ (more exactly, $\Delta G° = 0.284\, \Delta H°$) and, accordingly, the $\Delta G°$ associated with a $\Delta H°$ of -30 kcal mol^{-1} cannot become greater than -8.72 kcal mol^{-1} with a K_d of some 400 nM, values that come to represent, in practice, the strongest H-bond predictable with the only exception of hydrogen bifluoride. To have an idea of what this value may mean, it is useful to notice that drug–receptor and enzyme-inhibitor complexes (which are the most efficient non-bonded associations in nature with the only exception of enzyme transition-state complexes) are

TABLE 7.1. H-bond association, K_a, and dissociation, K_d, constants as a function of the relative H-bond association enthalpy, $\Delta H°$. $K_a = \exp[-\Delta G°/RT]$; $K_d = 1/K_a = \exp[\Delta G°/RT]$; $\Delta G° = \Delta H° - T\Delta S°$ where $\Delta S°$ is the loss of entropy due to H-bond formation and is calculated according to the empirical relationship $\Delta S° = 1/\beta\, \Delta H° + c = 2.4\, \Delta H° + 0.12$ derived from the thermodynamic data of Pimentel and McClellan (1971); $T = 298.15$ K.

$\Delta H°$ (kcal mol^{-1})	$\Delta S°$ (kcal mol^{-1} K^{-1})	$-T\Delta S°$ (kcal mol^{-1})	$-\Delta G°$ (kcal mol^{-1})	$\log K_a$ (M^{-1})	$K_d = 1/K_a$
−1	−0.00024	0.72	0.28	0.21	619 mM
−3	−0.0072	2.15	0.85	0.62	238 mM
−5	−0.012	3.58	1.42	1.04	91.0 mM
−10	−0.024	7.16	2.84	2.08	8.28 mM
−15	−0.036	10.73	4.27	3.13	741 μM
−20	−0.048	14.31	5.69	4.17	67.7 μM
−25	−0.060	17.89	7.11	5.21	6.14 μM
−30	−0.072	21.27	8.72	6.40	406 nM

characterized by ranges of values of $-4.7 \geq \Delta G° \geq -15.0$ kcal mol^{-1} or 100 μM $\geq$ $K_d \geq$ 10 pM (Houk *et al.*, 2003 ; Gilli *et al.*, 1994b). Hence, a single strong H-bond can give, at least in principle, a contribution to molecular association strictly comparable with the average binding constant of the most important receptorial drugs in practical use.

7.1.4 *H-bond strength in molecular crystals*

Aggregation in molecular crystals is largely controlled by weak van der Waals forces. Since H-bonds are stronger than any single van der Waals interaction, they will display a greater propensity to contribute to the lattice energy of the crystal. In this sense, it has been already remarked (Donohue, 1952; Etter, 1991) that 'all acidic H atoms available in a molecule will be used in H-bonding in its crystal structure' (of course, if there are enough acceptors). The general acceptance of C–H···O interactions as true H-bonds does not contradict this rule but suggests to rewrite it in a different way: 'All the H-bond acceptors available in a molecule will be engaged in H-bonding as far as there are available donors. These acceptors will be saturated, as a rule, in order of decreasing strength of the H-bonds formed' (Bertolasi *et al.*, 2001). This second formulation stresses the pre-eminent role of strong H-bonds in crystal packing and suggests that weak C–H···O and C–H···N bonds will complete the H-bond pattern of the crystal by the maximum number of bonds only at the end of the packing process.

Molecular crystals contain H-bonds of any known class whose enthalpies can be reasonably evaluated from the crystal geometries. Appreciation of free enthalpies can start from the observation that homomolecular crystals are nothing more than solid solutions of molecules dissolved in themselves whose behavior cannot be so different from that of the same molecules dissolved in a non-polar solvent of comparable molecular weight, even because solid and solution have similar densities and then comparable intermolecular interactions. This suggests that, at least to a first approximation, the relationship $\Delta G° = 0.284\ \Delta H°$ found above for non-polar solvents can be adopted also for molecular crystals.

7.2 The H-bond strength in aqueous solutions

7.2.1 *Introduction: Drug–receptor binding as a sample system*

H-bonds between molecules and macromolecules dissolved in aqueous solutions are well known to be of the utmost importance in many phenomena of molecular recognition occurring in chemical and biochemical systems. For this reason, they have been the subject of numerous studies that, however, did not lead to a full solution of the many complex problems connected with the process of molecular association in water.

To start with, water is both a good H-bond donor and acceptor with respective pK_as of 15.74 and –1.74, which are perfectly suited for interacting with a quite large number of common functional groups (see the pK_a slide rule of Fig. 4.2). This implies that, when a donor–acceptor couple is dissolved in water, the H-bonds formed will be a complex equilibrium of many H-bonded species, including the donor–acceptor itself together with the donor–water and water–acceptor complexes formed with an unspecified number of solvent molecules. Moreover, water is a solvent of high dielectric constant that tends to stabilize the charge-transfer form of the bond, $R_1{-}^{-}D{:}{\cdots}H{-}A^{+}{-}R_2$ (Huyskens and Zeegers-Huyskens, 1964; Sobczyk, 1998), and then to dissociate it by solvation of the two charged moieties, a phenomenon that is, by the way, the logical basis of the famous Eigen mechanism of proton transfer in aqueous solutions (Eigen, 1964). These properties of water as a solvent are shared, though to a lesser extent, by other hydroxylated solvents and, in particular, by alcohols.

Because of the overwhelming role played by the solvent, H-bond equilibrium constants cannot be practically measured in aqueous solutions in spite of the fact that there is no serious reason for believing that the specific donor–acceptor H-bond interaction in water as a pure solvent (i.e. not participating in the multiple equilibrium) has to be substantially different from that occurring in a non-aqueous medium. These difficulties can be circumvented in the framework of a *host–guest logic* for which a larger host is endowed with an accepting (donating) pocket where the H-bonds formed with the donor (acceptor) are more or less shielded from the direct contact of the surrounding waters. Two main approaches to the problem are known. In the *molecular recognition* approach, hosts are chemically synthesized macrocycles having large cavities (crown ethers, lariats, cryptands,...) suitably designed to recognize cations, anions as well as small H-bonded organic molecules (Cram and Cram, 1994; Atwood *et al.*, 1984, 1991), while in the *molecular–macromolecular* approach the host is a natural biomacromolecule (enzyme, antibody, membrane or cytoplasmic receptors) that contain *specific binding sites* deputed to recognize and bind a variety of endogenous (substrates, coenzymes, hormones,...) or exogenous (enzyme inhibitors, allergens, receptorial drugs,...) ligands. A case of small-molecule association to crown ethers and cryptands in alcohols will be separately considered in Section 7.2.8, while, in the following, the discussion will focus on *drug–receptor binding to membrane and cytoplasmic receptors in water*, a subject that is reasonably known to most experts of molecular interactions for its relevance in molecular pharmacology and biology and to which, by the way, the authors have dedicated constant attention during the last 15 years. Since the mechanisms of H-bond formation in water are reportedly similar in most systems studied, it can be presumed that this choice will not hamper the generality of the conclusions drawn.

7.2.2 *Hydrophilic and hydrophobic contributions to drug-receptor binding*

An ever-growing number of crystal structures of molecule–macromolecule (with a prevalence of inhibitor–enzyme) complexes are being published that show that binding specificity is often achieved through a net of well-localized H-bonds connecting the guest to the host binding site. Unfortunately, the precision achievable even at the highest resolution level is barely sufficient to establish how strong (short) the H-bonds formed may be. On the other side, there are a lot of modern experimental methods that allow determination of accurate values of the binding constants in water solutions but, still unfortunately, the specific contribution of H-bonding seems to be hidden behind a veil of fascinating but certainly not simple thermodynamics. No surprise that the full understanding of the H-bonds formed by biological systems is still considered one of the most difficult and open problems in H-bond theory.

The reason why it is so difficult to assess the role played by the H-bond (and in general by any *polar* or *hydrophilic* force) in the molecular association in water is that, in this environment, also *apolar* or *hydrophobic* molecules (or parts of molecules) become associated because of the so-called *hydrophobic interactions*, another manifestation of the H-bond occurring among solvent instead of solute molecules. The origin of the phenomenon is that liquid water is made of strongly polar molecules interconnected by a tridimensional net of H-bonds (only 20% of which are broken with respect to ice at room temperature) and then can dissolve only strongly polar molecules, being good H-bond donors or acceptors. Conversely, apolar molecules dissolve in water becoming surrounded by *cages of water molecules* characterized, thermodynamically, by having: (*i*) *large heat capacities*, imputable to the lack of H-bonding between the enclosed molecule and the surrounding cage; and (*ii*) *low entropies*, imputable to the tightening of the water structure at the cage borderline as a compensation for the other interrupted bonds (Edsall and Gutfreund, 1983; Sturtevant, 1977). These water cages are unstable structures that tend to reduce their surface by associating two or more hydrophobic molecules in a unique cage, a process that is known to cause a large number of practical effects, such as separation of oil from water, protein folding, and binding of non-polar ligands to enzyme or receptor binding sites, and is well characterized from a thermodynamic point of view, when the molecules involved are adequately hydrophobic, by a decrease of free energy ($\Delta G < 0$) associated with a decrease of heat capacity ($\Delta C_p < 0$) and an increase of entropy ($\Delta S > 0$).

The role played by these cages, however, largely depends on the fraction of molecular surface that can be classified as hydrophobic (such as hydrocarbons) or hydrophilic (such as hydroxylic, carboxylic and aminic functions). Variables such as SASA, SAHSA, SAPSA and %SAHSA = SAHSA/SASA $\cdot$ 100 are suited to quantify this fraction. They are easily obtained from most programs of molecular graphics and have the respective meanings of total, *hydrophobic*, and *polar* (or hydrophilic) *solvent-accessible surface areas* and of per cent hydrophobicity. Xenobiotics (drugs) and endogenous ligands (hormones, neurotransmitters and autacoids) display various %SAHSA values that can range, in the two examples illustrated below, from 76–88% for steroid hormones binding to cytoplasmic receptors to 55–65% for most ligands of the adenosine A_1 membrane receptor (Gilli *et al.*, 2005b). Not surprisingly, the binding thermodynamics of these two receptors show important differences that tend to concentrate in the standard heat capacities of the binding equilibrium, ΔC_p°, and then in the related shape of the van't Hoff plots, $\ln K_a = f(1/T)$. In particular: (*i*) steroid–nuclear receptor binding, a mostly hydrophobic process, is characterized by ΔC_p° values large and negative and then by parabolic van't Hoff plots; while (*ii*) drug–membrane receptor binding, a case of intermediate hydrophobic–hydrophilic process, is characterized by ΔC_p° values not far from zero and, accordingly, by essentially linear van't Hoff plots.

The following paragraphs are devoted to the analysis of the binding thermodynamics of these two classes of receptors as a preliminary step in trying to unravel the specific role played by the H-bond in the drug–receptor binding process. It will be shown that, in spite of the large differences caused by ΔC_p° effects, something remains invariably unchanged and this is the enthalpy–entropy relationship linking all data with a same compensation temperature, β, not far from the experimental one (Gilli *et al.*, 1994b). Enthalpy–entropy compensation is therefore the only possible key, if any, to the understanding of the real nature of the binding process in water. This is quite an intriguing problem for which many different explanatory models have been suggested. The following discussion is centered on the *Grunwald and Steel model* (Grunwald and Steel, 1995; Grunwald, 2001) which has been chosen because, for its remarkable simplicity, it appears to be more directly based on pure thermodynamic arguments and because one of its proposers has been one of the leading figures in the field of enthalpy–entropy compensation and extrathermodynamic relationships since their first formulation almost 50 years ago (Leffler and Grunwald, 1963). Of course, this particular choice cannot be taken as a criticism for any of the many other theories existing on the subject and, rather, we believe that a comparative critical analysis will be necessary and extremely welcome.

7.2.3 *Hydrophobic binding: Thermodynamics of the steroid–nuclear receptor system*

The binding of the steroid hormones β-estradiol (**7.1.Ia**), ORG2058 (**7.1.Ib**), and R1881 (**7.1.Ic**) to their respective estrogen, progesterone, and androgen cytoplasmic nuclear receptors (Gilli *et al.*, 2005b) can be taken as a good example

(7.1.Ia) 17β-estradiol (Estrogen R.)

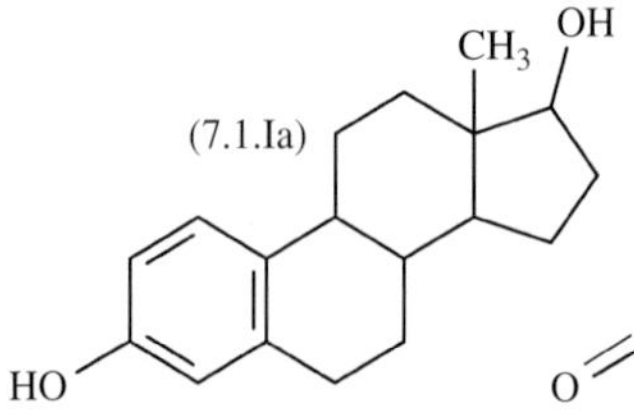

(7.1.Ib) ORG2058 (Progesterone R.)

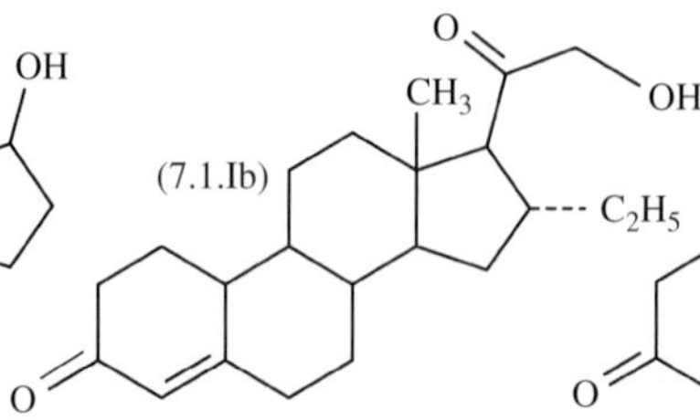

(7.1.Ic) R1881 (Androgen R.)

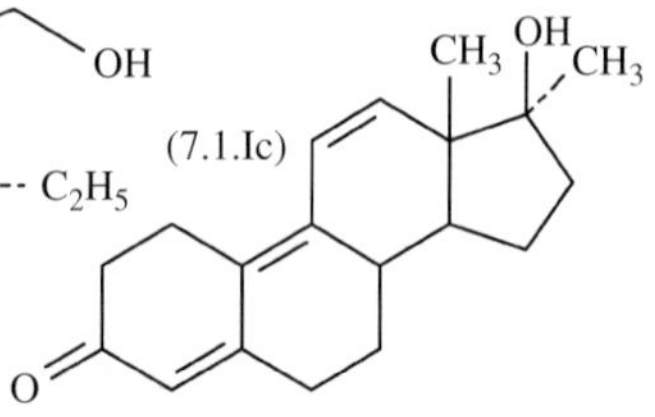

of thermodynamics controlled by hydrophobic forces. It is, moreover, one of the few cases of receptor binding for which a detailed pattern of the drug–receptor interactions can be obtained from the many crystal structures available. This is exemplified in Fig. 7.1 showing schematically the interaction scheme between 17β-estradiol and the estrogen–receptor binding site formed by a common structural pattern of mutually supporting hydrophobic and H-bonded interactions involving highly conserved residues (Brzozowski *et al.*, 1997; Tanenbaum *et al.*, 1998) that ensure, respectively, shape complementarity for the binding cavity and specific recognition of the polar functional groups.

The thermodynamic parameters of the binding equilibrium are summarized in Table 7.2, while Fig. 7.2 illustrates the $\ln K_a$ versus T and ΔG, ΔH, and $T\Delta S$ versus T plots for the binding of 17β-estradiol to the estrogen receptor. Plots for the other two receptors (not shown) are similar in shape and differ only by

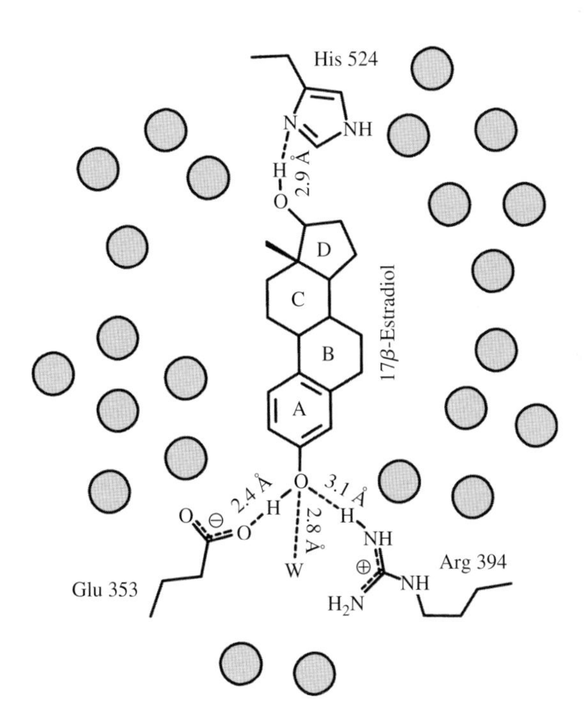

FIG. 7.1. A schematic representation of the interactions between 17β-estradiol and the estrogen-receptor ligand binding site. The approximate positions of the receptor hydrophobic residues are shown by small gray circles. The ligand makes also direct H-bonds with residues Glu 353, Arg 394, and His 524 of the receptor and a water molecule (W). (Data from Brzozowsky *et al.*, 1997; reproduced by permission from Gilli *et al.*, 2005b)

TABLE 7.2. Standard thermodynamic parameters at $T = 298.15$ K for the binding of three specific steroids to the estrogen, progesterone, and androgen cytoplasmic nuclear receptors (Gilli *et al.*, 2005b). %SAHSA is the per cent of hydrophobic over total solvent-accessible surface areas of the ligand and T_H and T_S the two characteristic temperatures at which ΔH and $T\Delta S$ cross the zero-energy line.

Receptor	Ligand	%SAHSA	$\Delta G°$ (kJ mol^{-1})	$\Delta H°$ (kJ mol^{-1})	$\Delta S°$ (kJ mol^{-1} K^{-1})	$\Delta Cp°$ (kJ mol^{-1} K^{-1})	T_H (K)	T_S (K)
Estrogen	17β-estradiol	88.2	−53.04(4)	−19.9(9)	0.111(3)	−2.4(1)	290	312
Progesterone	ORG2058	82.2	−48.62(4)	−25.3(9)	0.078(3)	−2.3(2)	286	310
Androgen	R1881	76.6	−51.56(5)	−34.5(15)	0.057(5)	−1.7(2)	278	309

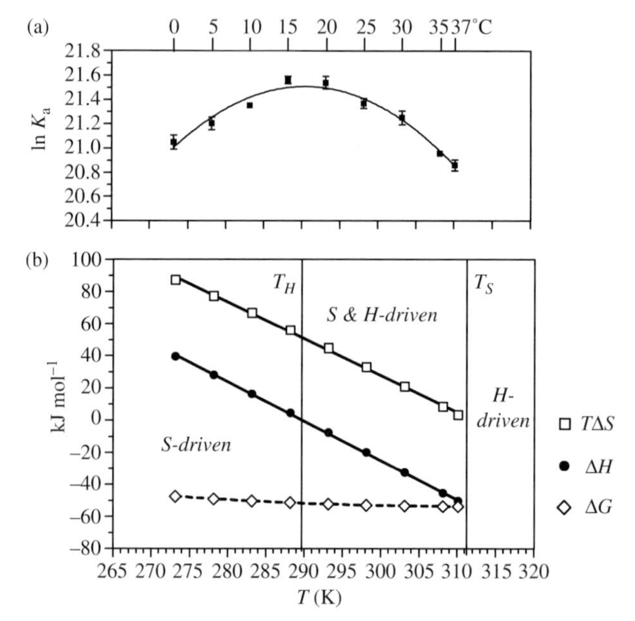

FIG. 7.2. Thermodynamics of the binding of 17β-estradiol to the estrogen receptor: (a) plot of lnK_a (M^{-1}) versus T (K); and (b) plot of ΔG, ΔH, and $T\Delta S$ (kJ mol^{-1}) versus T (K). (Reproduced by permission from Gilli *et al.*, 2005b)

being more shifted to the left because of the smaller value of T_H. The results obtained suggest the following considerations:

(*i*) the standard heat capacity of the equilibrium, ΔC_p°, is systematically large, negative and essentially temperature independent; this fact determines the parabolic convex form of the lnK_a versus T curve of Fig. 7.2(a);

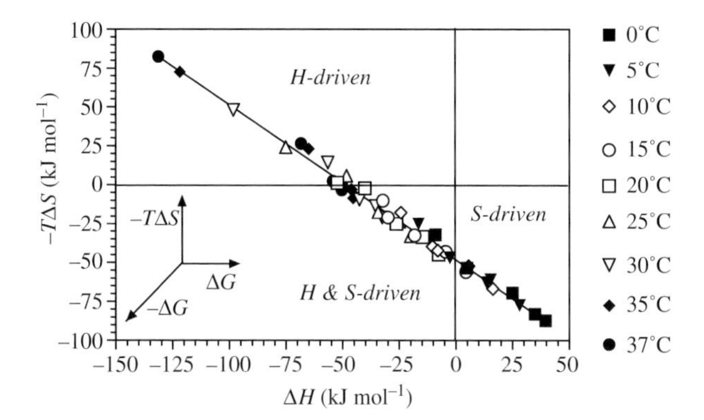

FIG. 7.3. Scatter plot of $-T\Delta S$ versus ΔH (kJ mol^{-1}) at the various temperatures investigated for the binding of three steroids to the estrogen, progesterone, and androgen receptors (data from Table 7.2) and of cortisol and dexamethasone to the glucocorticoid receptor (data from Eliard and Rousseau, 1984). The linear regression of the points [$T\Delta S = 48.2(7) + 1.00(2)\Delta H$ ($n = 45$, $r = 0.995$)] evidences a typical enthalpy–entropy compensation phenomenon for all compounds at any temperature. (Reproduced by permission from Gilli *et al.*, 2005b)

(*ii*) since $\Delta C_p = (\delta \Delta H/\delta T)_p = (\delta T\Delta S/\delta T)_p$ and $\Delta C_p < 0$, both ΔH and $T\Delta S$ decrease linearly with the increasing T, so that the binding equilibrium, which is entropy driven at low temperature, becomes increasingly enthalpy driven at higher one;

(*iii*) this decrease defines two characteristic temperatures, T_H and T_S, where ΔH and $T\Delta S$, respectively, cross the zero-energy line and out of which the former corresponds to the point where $\ln K_a = \exp[-\Delta G/RT]$ reaches its maximum;

(*iv*) because ΔH and $-T\Delta S$ are negatively intercorrelated, the variations of $\Delta G = \Delta H - T\Delta S$ with the temperature are small with respect to those of ΔH and $T\Delta S$, thus giving origin to an almost perfect enthalpy–entropy compensation relationship, $\Delta H = \beta \Delta S$, with a compensation temperature β not far from the average experimental one; this relationship is displayed in Fig. 7.3 and shows that the data for the four different nuclear receptors lie on a same correlation line.

This particular thermodynamic behaviour is in agreement with that observed for many other cases of mostly hydrophobic binding, such as enzyme inhibition, protein folding and, in general, separation of nonpolar substances from water, and can be interpreted in terms of a same model by assuming that the water cages are unstable structures of high heat capacity which tend to reduce their surface by associating the hormone and the receptor binding site in a unique cage, so causing the decrease of heat capacity, ΔC_p, actually observed. The interpretation of hydrophobic effects has been one of the outstanding topics of modern thermodynamics in the last 60 years and, for its complexity, cannot certainly be discussed here in detail (for more information the reader is addressed to the recent review by Southall *et al.* (2002)). In short, the basic idea of all models proposed is that hydrophobic behaviour is to be explained in terms of changes that the solute causes within water itself. The water-ordering or iceberg model (Frank and Evans, 1945) was the first to suggest that the nonpolar solute creates a *weakly clathrate-like cage* of first-shell waters around itself, an idea derived from the observation that nonpolar solutes are often surrounded by clathrate water cages in crystalline hydrates. A more advanced iceberg-like model was developed by Muller (1990) based on an earlier *two-state model* by Gill (Gill *et al.*, 1985) and successively generalized by Lee and Graziano (1996). It invokes two water states (in the bulk and on the cage surface) and four states of water-water H-bonding (made and broken in the bulk and made and broken in the first solvation shell of the solute). The only model that does not directly address the H-bond properties is the *small-size model* (Lucas, 1976; Lee, 1985, 1991), which rather focuses on the difficulty of generating cavities large enough to allocate the solute in consequence of the small size of water.

Going back to the thermodynamic properties of our system (Table 7.2 and Figure 7.2), their most salient feature is that the steroid-receptor binding changes deeply with the temperature, being fully entropy-driven at lower and essentially enthalpy-driven at higher one, in agreement with the negative value of the ΔC_p measured. In molecular terms, however, these findings can only be explained making use of a two-state model predicting that, at low temperature, the water molecules of the cage prefer to adopt only a few orientations (low entropy) to favor the formation of the maximum number of H-bonds (low enthalpy) while, at higher temperature, more conformations become accessible (high entropy) but at the cost of breaking a part of the H-bonds (high enthalpy). Accordingly, the binding process can only be driven by the increase of entropy or by the decrease of enthalpy, respectively, at lower or higher temperatures.

7.2.4 *Hydrophilic–hydrophobic binding: Thermodynamics of the adenosine A_1 membrane receptor*

Nearly a dozen membrane receptorial systems have been so far studied in considerable detail from a thermodynamic point of view. They include six G-protein coupled receptors (GPCRs): adenosine A_1 (Murphy and Snyder, 1982; Borea *et al.*, 1992), adenosine A_{2A} (Borea *et al.*, 1995, 1996b), adenosine A_3 (Merighi *et al.*, 2002), β-adrenergic (Weiland *et al.*, 1979; Contreras *et al.*, 1986), dopamine D_2 (Kilpatrick *et al.*, 1986; Duarte *et al.*, 1988), and serotonine 5-HT_{1A} (Dalpiaz *et al.*, 1995, 1996); four ligand-gated ion channel receptors (LGICRs): glycine (Ruiz-Gomez *et al.*, 1989), $GABA_A$ (Maksay, 1994), serotonin 5-HT_3 (Borea *et al.*, 1996a; Maksay, 1996), and nicotinic (Banerjee and Ganguly, 1995, 1996; Borea *et al.*, 1998);

Cmp.	R
PEA	$-CH_2-CH_2-Ph$
CHA	–cyclohexyl
R–PIA	$-CH(CH_3)-CH_2-Ph$

Cmp.	R_1	R_2	R_3	R_4
Caffeine	$-CH_3$	$-CH_3$	$-CH_3$	–H
DPSPX	$-nC_3H_7$	$-nC_3H_7$	–H	–S–Ph
IBMX	$-CH_3$	$-iC_4H_9$	–H	–H

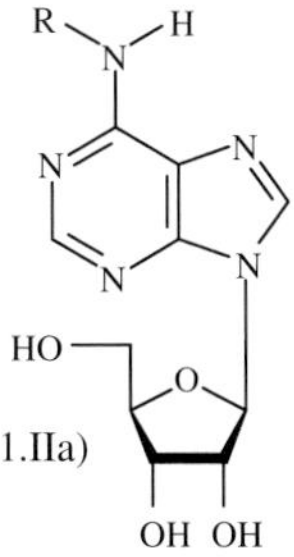

(7.1.IIa)

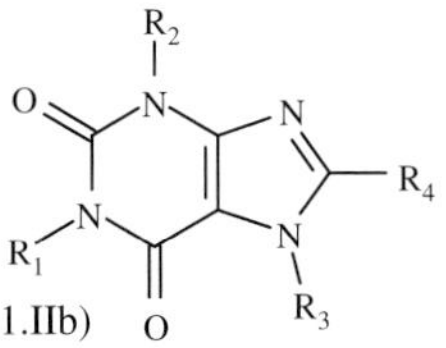

(7.1.IIb)

TABLE 7.3. Standard thermodynamic parameters at T = 298.15 K for the binding of three representative agonists and antagonists to the adenosine A_1 membrane receptor (Borea *et al.*, 1992, 1996b). ΔC_p° values not reported because they are statistically indistinguishable from zero; %SAHSA = per cent of hydrophobic over total solvent-accessible surface area of the ligand; ago = agonist and ant = antagonist.

Ligand	Type	%SAHSA	ΔG° (kJ mol^{-1})	ΔH° (kJ mol^{-1})	ΔS° (kJ mol^{-1} K^{-1})
PEA	ago	60.9	−45.3(1)	23(4)	0.230(12)
CHA	ago	60.8	−51.1(2)	39(5)	0.306(17)
R-PIA	ago	65.0	−52.0(3)	30(8)	0.276(28)
Caffeine	ant	55.5	−24.7(1)	−22(4)	0.011(13)
DPSPX	ant	81.3	−38.0(1)	−20(2)	0.059(6)
IBMX	ant	65.4	−30.1(3)	−36(9)	−0.022(13)

and one modulator receptor for benzodiazepines (Kochman and Hirsch, 1982; Maguire *et al.*, 1992). A detailed survey of drug–receptor binding thermodynamics is also available (Raffa, 2001).

The thermodynamic behavior of all known membrane receptors is considerably different from that just described for the cytoplasmic ones. The main difference is that the ΔC_p° of the binding equilibrium, instead of being large and negative, is normally zero or, at least, smaller than its experimental standard deviation. Accordingly, both ΔH° and ΔS° values are temperature independent. This behavior is exemplified by the van't Hoff plots of Fig. 7.4 for the binding to adenosine A_1 receptor of the three agonists (**7.1.IIa**) and three antagonists (**7.1.IIb**) listed in Table 7.3. The linearity of the plots within the zone of stability of the complex indicates that ΔC_p° is zero and then ΔH° and $T\Delta S^\circ$ are independent of temperature. The relative positions and slopes of the straight

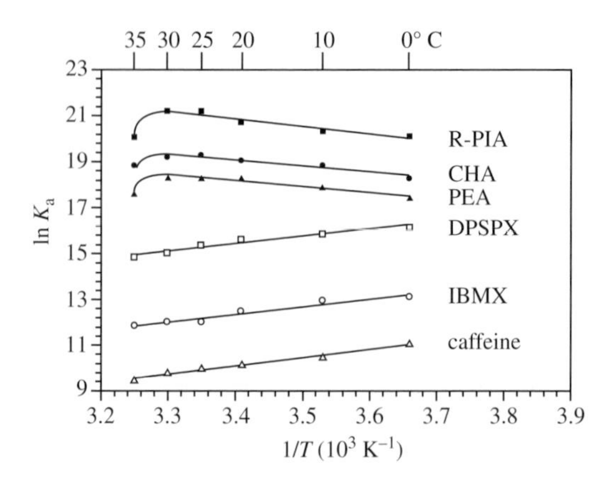

FIG. 7.4. Van't Hoff plots for the binding of three agonists (R-PIA, CHA, PEA) and three xanthinic antagonists (caffeine, DPSPX, IBMX) to the adenosine A_1 receptor. The essential linearity of the plots within the zone of stability of the complex indicates that the standard heat capacity of the equilibrium, ΔC_p°, is essentially zero (that is, ΔH° and $T\Delta S^\circ$ are constant). (Reproduced by permission from Borea *et al.*, 2001)

lines show that the binding of agonists is stronger (greater $\ln K_a$) and entropy driven (negative slope means positive ΔH°) while that of antagonists is weaker (smaller $\ln K_a$) and enthalpy driven (positive slope means negative ΔH°).

A possible explanation may be sought in the effect produced by the increasing number of hydrophilic (H-bonding) groups of the ligand on the heat capacity of water cages. The cages of fully hydrophobic compounds are associated with large heat capacities imputable to unrestrained rotational and vibrational motion of the solute within the cage. This motion (and associated heat capacity) can be severely reduced by ligands that, because of their more hydrophilic nature, can more easily form H-bonds with the cage itself. In consequence, the binding to membrane receptors can occur without the large C_p decrease typical of hydrophobic receptors.

In spite of these differences, the extrathermodynamic enthalpy–entropy relationship is conserved. The scatter plot $-T\Delta S^\circ$ versus ΔH° (kJ mol^{-1}) for the binding to the adenosine A_1 receptor of all agonists and antagonists so far studied (Borea *et al.*, 1992, 2001) is shown in Fig. 7.5. All data lie on the same diagonal with equation $T\Delta S^\circ = 39(1) + 1.18(4)\Delta H^\circ$ ($n = 39$; $r = 0.976$) that defines a classical case of enthalpy–entropy compensation. Two points deserve to be considered:

i) Though the enthalpy–entropy correlations for nuclear cytoplasmic (Fig. 7.3) and membrane receptors (Fig. 7.5) look quite similar, they have different meanings in the two cases. In the former the correlation is thermodynamic because the two variables are related by the equations $\Delta C_p = (\delta\Delta H/\delta T)_p = (\delta T\Delta S/\delta T)_p$ with $\Delta C_p < 0$, while in the latter there is no thermodynamic

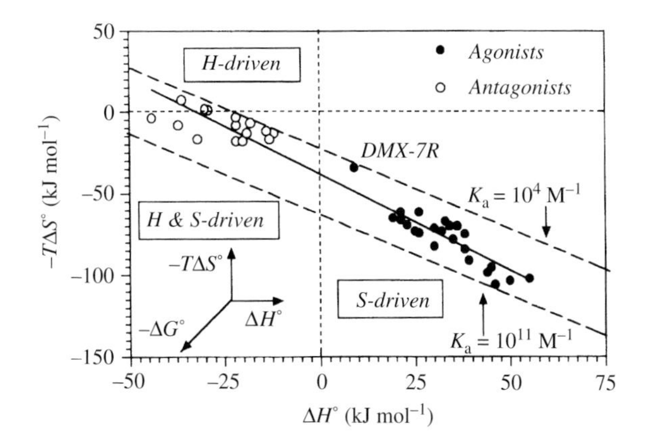

FIG. 7.5. Scatter plot of $-T\Delta S^\circ$ versus ΔH° (kJ mol^{-1}) for the binding of a number of agonists and antagonists to the adenosine A_1 receptor. The two dashed lines indicate the combinations of ΔH° and $T\Delta S^\circ$ values that give rise to the two association constants indicated ($K_a = 10^4$ and 10^{11} M^{-1}). (Reproduced by permission from Borea *et al.*, 2001)

link between $T\Delta S^\circ$ and ΔH°, which are temperature-independent constants characteristic of the process studied. In this second case, therefore, the correlation is completely extrathermodynamic, that is it does not arise from the laws of thermodynamics but from the physical constraints of the binding process. This important point will be further investigated in the next paragraph.

ii) An interesting aspect of Fig. 7.5 is that agonists and antagonists cluster in different regions of the plot, the former binding being *entropy driven* and the latter *enthalpy/entropy driven*. This phenomenon of *thermodynamic discrimination* (Weiland *et al.*, 1979; Borea *et al.*, 1992) has been observed for other membrane receptors and seems another clear indication of the extrathermodynamic nature of the enthalpy–entropy correlation observed. It will be considered again in Section 7.2.7.

7.2.5 *Enthalpy–entropy compensation: A universal property of drug–receptor binding*

The two examples above suggest that enthalpy–entropy compensation is typical of drug–receptor binding irrespective of the specific nature of receptors and ligands considered. Full verification of this hypothesis needs a much larger set of experimental data or, better, a comprehensive database of thermodynamic parameters (ΔG°, ΔH°, ΔS°, and ΔC°_p) of the drug–receptor binding equilibrium. To this aim we are using our own database of binding thermodynamic data, which was established in 1993 and has been regularly updated until now. The methods for collecting and analyzing these data require, however, a few preliminary considerations.

Binding measurements are usually performed *in vitro* on whole cells or tissue homogenates for membrane receptors and on purified cytosol fractions for enzymes and cytoplasmic receptors. The techniques for measuring association, K_a, and dissociation, K_d, constants and related standard free enthalpies ($\Delta G° = -RT \ln K_a = RT \ln K_d$) have been refined in the last decades, becoming one of the most efficient tools for screening new potential drugs. Thousands of $\Delta G°$ values have been collected but, unfortunately, very few include the determination of the $\Delta G°$ components, $\Delta H°$ and $T\Delta S°$, which are considered of little practical utility in the normal pharmacological practice.

Another factor needing to be carefully considered concerns binding specificity. In fact drugs, besides binding to their specific binding sites, can also become associated with the binding sites of other receptors as well as with any surface macromolecular sites accidentally able to recognize them. To get correct values of $\Delta G°$, $\Delta H°$, and $\Delta S°$ it is necessary to distinguish between specific and aspecific binding and this, in practice, can only be obtained by performing saturation or inhibition experiments of radiolabelled (mostly ^{3}[H] tritiated) ligands of verified binding selectivity for the receptor studied.

In saturation experiments the affinity of a drug, D, for a receptorial binding site, R, is directly measured from the binding equilibrium

$$\mathrm{R} + \mathrm{D}^* \rightleftharpoons \mathrm{R-D}^*, \tag{7.2}$$

where K_a and $K_d = 1/K_a = [\mathrm{R}]\,[\mathrm{D}^*]/[\mathrm{R{-}D}^*]$ are the association and dissociation constants and D* the radiolabelled drug under investigation. When the drug is not available in labelled form, a displacement or inhibition experiment is performed, and the following equilibria are studied

$$\mathrm{R} + \mathrm{L}^* \rightleftharpoons \mathrm{R-L}^*, \tag{7.3a}$$

$$\mathrm{R{-}L}^* + \mathrm{D} \rightleftharpoons \mathrm{R-D} + \mathrm{L}^*, \tag{7.3b}$$

where L* is a suitable radioligand of high affinity and selectivity and D is the drug under examination. When IC_{50} is defined as the drug concentration able to displace the 50% of L* from the binding site, $K_d{}^* = [\mathrm{R}]\,[\mathrm{L}^*]/[\mathrm{R{-}L}^*]$ is the already known dissociation constant of R–L*, and $K_d = 1/K_a = [\mathrm{R}][\mathrm{D}]/[\mathrm{R{-}D}]$ is the unknown dissociation constant of R–D, it has been shown (Cheng and Prusoff, 1973) that, under controlled experimental conditions,

$$K_d = IC_{50} \,/\, (1 + [\mathrm{L}^*]^0/K_d{}^*), \tag{7.4}$$

where $[\mathrm{L}^*]^0$ is the concentration of labelled ligand initially added.

The problem of determining separate values of $\Delta H°$ and $\Delta S°$ is complicated by the extremely low concentrations of receptors present in biological tissues (e.g. 1–100 fM/mg of tissue for most neurotransmitter receptors) that practically make microcalorimetric determination of $\Delta H°$ impossible. Nevertheless, methods based on K_d measurements over a range of temperatures still remain applicable. Most receptors display standard equilibrium heat capacities not far from zero and, in this case, the standard enthalpy, $\Delta H°$, and entropy, $\Delta S°$, can be easily obtained from the slope and intercept of the van't Hoff plot $\ln K = \Delta S°/R - \Delta H°/R\ (1/T)$. Methods for data treatment when the heat capacity is significantly different from zero have been recently surveyed (Gilli *et al.*, 2005b; Borea *et al.*, 1998). In their simplest form, they can be based on the least-squares fitting of the observed ΔG values by the quadratic expression $\Delta G = A + BT + CT^2$, for which it can be easily shown that $\Delta H = [\delta(\Delta G/T)/\delta(1/T)]_p = A - CT^2$, $\Delta S = -(\delta\Delta G/\delta T)_p = -B - 2CT$, and $\Delta C_p = (\delta\Delta H/\delta T)_p = -2CT$.

The 1993 version of the database contained 186 independent experiments performed on 136 ligands binding to 13 receptorial systems and was used to produce the first general enthalpy–entropy compensation study in drug–receptor binding (Gilli *et al.*, 1994b). The present updated version includes 447 experiments performed on 307 different ligands binding to 26 receptors. The latter involve 17 membrane and 4 cytoplasmic receptors together with 3 inhibited enzymes and some cases of binding to DNA oligomers.

The general $-T\Delta S°$ versus $\Delta H°$ scatter plot for the full set of data is displayed in Fig. 7.6. Data are linked by the regression equation

$$-T\Delta S° = -42.1(4) - 1.02(1)\ \Delta H° \quad (n = 447,\ r = -0.979,\ P < 0.0001), \tag{7.5}$$

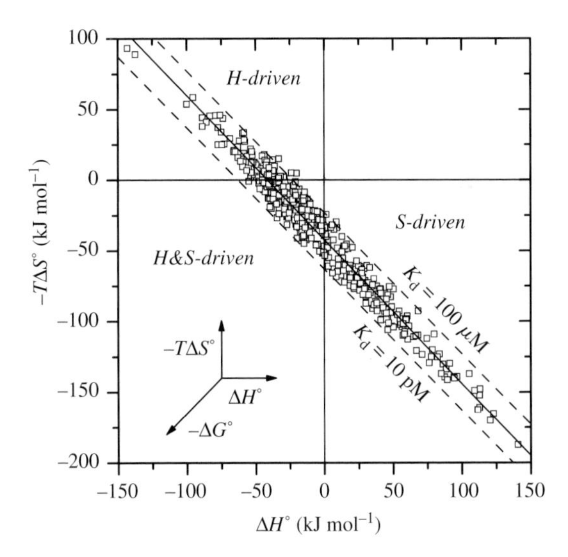

FIG. 7.6. Scatter plot of $-T\Delta S°$ versus $\Delta H°$ (kJ mol^{-1}) for 447 binding equilibrium experiments performed on 307 different agonists and antagonists bound to 26 different receptors. These latter involve 17 membrane and 4 cytoplasmic receptors together with 3 inhibited enzymes and a few cases of DNA binding. The two dashed lines indicate the *loci* of points where the combinations of $\Delta H°$ and $T\Delta S°$ values give rise to the two equilibrium constants indicated ($K_d = 100\ \mu$M and 10 pM).

which represents a remarkable example of enthalpy–entropy compensation, $\Delta H = \beta\,\Delta S$, with the compensation temperature $\beta = 292.3$ K essentially identical to the average temperature of the data treated, i.e 298(5) K. The nature of the correlation can be better understood by writing eqn 7.5 in the approximate form

$$\Delta H^\circ - 298.15\,\Delta S^\circ = -42.1 \text{ kJ mol}^{-1} = \Delta G^\circ, \qquad (7.6)$$

which immediately shows that the regression line is the *locus* of the points for which $\Delta G^\circ = -42.1$ kJ mol^{-1} or K_d (298) = 42.1 nM. Let us consider that high-affinity drugs seldom display K_d values smaller than some 10 pM ($\Delta G^\circ = -62.8$ kJ mol^{-1}) while those having $K_d \geq 100$ μM ($\Delta G^\circ = -22.8$ kJ mol^{-1}) can barely be considered drugs because they are unable to produce their effect at a reasonably low concentration. If we now draw in Fig. 7.6 the *loci* of the points having these two extreme K_d values, they will appear as two lines (dashed lines) that encompass the correlation line and delineate a narrow band where practically all experimental data remain confined.

The final result is that, while binding enthalpies and entropies are distributed over relatively wide intervals ($-143.1 \leq \Delta H^\circ \leq 140.6$ and $-180.0 \leq -T\Delta S^\circ \leq 93.3$ kJ mol^{-1}), free enthalpies remain confined within the much tighter range of $-61.1 \leq \Delta G^\circ \leq -15.1$ kJ mol^{-1}, which is nearly 1/6 of the two previous ones. This constraint is the evident phenomenologic origin of the effect: any decrease of the binding enthalpy must be compensated by a parallel decrease of binding entropy, or vice versa, to keep free enthalpies, $\Delta G^\circ = \Delta H^\circ - T\Delta S^\circ$, within their range of physical existence. The very origin of the enthalpy–entropy compensation is therefore that ΔG°, at variance with ΔH° and $T\Delta S^\circ$, is *physically bound* within a restricted range that no phenomenon of drug–receptor binding can actually overcome. The plausible origin of this constraint is the subject of the next section.

7.2.6 *Solvent reorganization and enthalpy–entropy compensation in drug–receptor binding: The Grunwald and Steel model*

The expression *solvent reorganization* is normally used to denote the transfer of solvent molecules between the bulk of the solvent and the solvation shells of solutes. It is generally agreed that solvent reorganization accompanying chemical reactions or spectral transitions may have important thermodynamic consequences, especially in H-bonded solvents, and one such consequence is the propensity toward enthalpy–entropy compensation, an example of which is reported in Fig. 7.6. The relationships between solvent reorganization and enthalpy–entropy compensation have been interpreted by Grunwald and Steel (1995; Grunwald, 2001) in terms of a particularly simple and efficient thermodynamic model, called here *the Grunwald and Steel model.* This model will now be applied to the interpretation of the particular association reaction we are presently interested in, that is drug–receptor binding in water solutions. This application is particularly simple and straightforward because the authors have grounded their model on two sets of experimental data, one of which is our first enthalpy–entropy correlation in drug–receptor binding (Gilli *et al.*, 1994b).

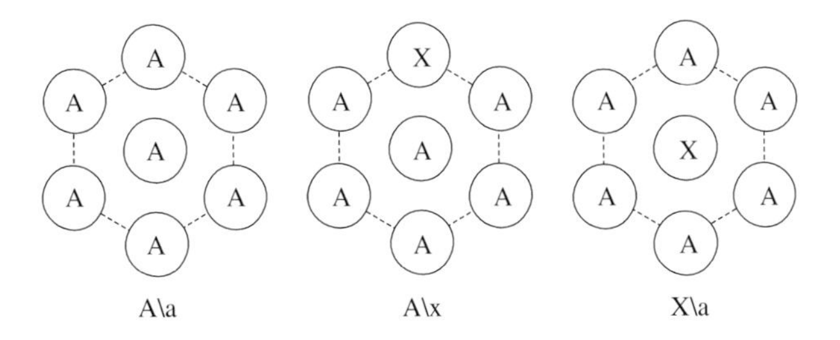

FIG. 7.7. Caged species in a diluted solution of solute X in solvent A. A\a denotes an A molecule with only A neighbors in its solvent shell, while in A\x one of the neighbors is X; similarly, X\a denotes an X molecule with only A neighbors in its solvent shell. (Reproduced by permission from Grunwald and Steel, 1995)

The basic point is that in a dilute solution of, say, the solute X in the solvent A there are (at least) three main *caged species* indicated in Fig. 7.7 by the symbols A\a, A\x, and X\a, that, respectively, denote an A molecule having only A neighbors in its solvent shell, an A molecule where one of the neighbors is X and the rest are As, and an X molecule surrounded by only A neighbors. To be chemically relevant, these species must be energetically *distinguishable*, a fact that will preferentially occur when the binding interactions among solvent molecules are large, as is certainly the case for H-bonded solvents such as water.

Taking into account now the specific role played by these solvent molecules (A), the equation for the ligand–receptor binding equilibrium

$$\mathrm{L} + \mathrm{R} \rightleftharpoons \mathrm{LR} \tag{7.7}$$

can be written in its *environmental form* as

$$\begin{aligned}\mathrm{L\backslash a} + s_{\mathrm{L}}\cdot \mathrm{A\backslash l} + \mathrm{R\backslash a} + s_{\mathrm{R}}\cdot \mathrm{A\backslash r} &\rightleftharpoons \mathrm{LR\backslash a}\\ &+ (s_{\mathrm{L}} + s_{\mathrm{R}} - s_{\mathrm{LR}})\cdot \mathrm{A\backslash a} + s_{\mathrm{LR}}\cdot \mathrm{A\backslash lr},\end{aligned} \tag{7.8}$$

which explicitly considers environmentally labeled molecular species and where the coefficients s_{L}, s_{R}, and s_{LR} denote the mean number of A molecules in the solvent cage of the L, R, and LR molecules, respectively. Let ΔG°, ΔH°, and $T\Delta S^\circ$ be the corresponding equilibrium parameters. This equation can be divided into two parts, the *nominal* equation

$$\mathrm{L\backslash a} + \mathrm{R\backslash a} \rightleftharpoons \mathrm{LR\backslash a}, \tag{7.9a}$$

with parameters $\Delta G^\circ_{\mathrm{nom}}$, $\Delta H^\circ_{\mathrm{nom}}$, and $T\Delta S^\circ_{\mathrm{nom}}$ and the *environmental* one

$$s_{\mathrm{L}}\cdot \mathrm{A\backslash l} + s_{\mathrm{R}}\mathrm{A/r} \rightleftharpoons s_{\mathrm{LR}}\cdot \mathrm{A/lr} + (s_{\mathrm{L}} + s_{\mathrm{R}} - s_{\mathrm{LR}})\cdot \mathrm{A\backslash a} \tag{7.9b}$$

with parameters $\Delta G^\circ_{\mathrm{env}}$, $\Delta H^\circ_{\mathrm{env}}$, and $T\Delta S^\circ_{\mathrm{env}}$.

Clearly, $\Delta G^\circ = \Delta G^\circ_{\mathrm{nom}} + \Delta G^\circ_{\mathrm{env}}$, $\Delta H^\circ = \Delta H^\circ_{\mathrm{nom}} + \Delta H^\circ_{\mathrm{env}}$, and $\Delta S^\circ = \Delta S^\circ_{\mathrm{nom}} + \Delta S^\circ_{\mathrm{env}}$ and the main contribution of the authors to the problem is to have shown that $\Delta G^\circ_{\mathrm{env}}$ is always equal to zero at the equilibrium, so that

$$\Delta G^{\circ}_{\text{env}} = \Delta H^{\circ}_{\text{env}} - T\Delta S^{\circ}_{\text{env}} = 0 \tag{7.10a}$$

$$\Delta G^{\circ} = \Delta G^{\circ}_{\text{nom}} = \Delta H^{\circ}_{\text{nom}} - T\Delta S^{\circ}_{\text{nom}}. \tag{7.10b}$$

These equations illustrate the important fact that, while the environmental part is fully enthalpy–entropy compensated with compensation temperature T, there is no compensation requirement in the nominal part because $\Delta G^{\circ}_{\text{nom}}$ is not generally zero. However, when (*i*) the ΔG° values observed in large series of ligands and receptors are physically confined within a relatively small $\Delta\Delta G^{\circ}$ range, (*ii*) $\Delta H^{\circ}_{\text{env}}$ and $T\Delta S^{\circ}_{\text{env}}$ values may be large because of the H-bonding properties of the solvent, and (*iii*) the flux of solvent molecules associated with the binding reaction is large, the environmental $\Delta H^{\circ}_{\text{env}} = T\Delta S^{\circ}_{\text{env}}$ enthalpy–entropy compensation (with $T=$ average experimental temperature) will dominate the set of data, producing a typical ΔH° versus $T\Delta S^{\circ}$ plot (Fig. 7.6) where the two variables are linearly related with a dispersion around the diagonal equal to $\Delta\Delta G^{\circ}$.

Enthalpy–entropy compensation may also derive from the nominal part of the reaction (eqn 7.10b) through different mechanisms, the most relevant being the smaller vibrational entropy associated with strongly enthalpic H-bonds (see the previous discussion for the H-bonds in gas phase and non-polar solvents). No matter how important this nominal compensation may be, it is not enough to overcome the environmental one and, to a first approximation, we can discuss the results in terms of solvent reorganization only.

The main conclusions that can be derived from this analysis can be summarized as follows:

i) The main consequence is that the ΔG° of the drug–receptor binding reaction does not depend on the solvent reorganization process but is an *intrinsic* or *nominal property*, $\Delta G^{\circ}_{\text{nom}}$, of the $\text{L}+\text{R}\rightleftarrows\text{LR}$ reaction itself that would be nearly the same in any reaction medium. Since solvent effects do not affect $\Delta G^{\circ}_{\text{nom}}$, also its components $\Delta H^{\circ}_{\text{nom}}$ and $\Delta S^{\circ}_{\text{nom}}$ come to represent the intrisic changes of binding enthalpy and of the related vibrational and conformational degrees of freedom consequent to the association between ligand and receptor binding site (see Section 8.5.2.1 for an application to the H-bond strength in water).

ii) The phenomenon of enthalpy–entropy compensation is observed whenever $\Delta\Delta G^{\circ} \lll \Delta\Delta H^{\circ}$, the large enthalpy variations being the product of the solvent reorganization in systems where the solvent molecules are interlinked by high-enthalpy bonds (typically H-bonded waters or alcohols). Accordingly, compensation effects in non-polar solvents are to be considered much more nominal than environmental as, on the other hand, suggested by their compensation temperature, systematically higher than the average experimental one (see Section 7.1.3).

iii) Therefore, $\Delta H^{\circ} \approx \Delta H^{\circ}_{\text{env}}$ and $\Delta S^{\circ} \approx \Delta S^{\circ}_{\text{env}}$ are environmental intercorrelated quantities that monitor the solvent rearrangement reaction and, according to the Grunwald and Steel model, bear little or no connection with the intrinsic binding process. Hence, the often reported statement that

$\Delta H°$ and $\Delta S°$ address intermolecular non-bonded interactions and solvent reorganization, respectively, is invalid. It could be true when applied to the nominal component only.

iv) $\Delta H° \approx \Delta H°_{\text{env}}$ and $\Delta S° \approx \Delta S°_{\text{env}}$ should then be thought of in different and much more physical terms as indicators of specific solvent-reorganization effects occurring, e.g. in folding/unfolding of proteins, closing/opening of channels in membrane proteins, or large conformational changes occurring in cytoplasmic receptors because of hydrophobic ligand binding. Little has been done in this sense and there is wide scope for further investigation.

v) In general, the environmental equation 7.8 does not seem to resemble any other ordinary chemical reaction because it has the unique property of obtaining an identical effect in two apparently opposite ways. In fact, Fig. 7.6 clearly shows that the same value of the binding constant (e.g. that of the regression line) can be indifferently obtained by two distinct mechanisms, either entropic (lower right) or enthalpic (upper left corner). Even more cogently, Fig. 7.2 (a typical case of a mostly hydrophobic process) indicates that this same entropy-to-enthalpy-driven shift can be achieved by a simple temperature change from 0° to 37 °C, showing again that the same process of drug–receptor binding can occur in two opposite ways, the first being entropic (and low-temperature) and the second enthalpic (and high-temperature). We are very far from a complete understanding of these facts though they seem to point to a unique model with *two extreme cage states*: (*i*) a low-temperature one where the cage is highly ordered (low entropy, low enthalpy, high heat capacity, and hydrophobic ligands), and (*ii*) a high-temperature one where the cage is vibrationally disordered at the cost of a significant cage H-bond breaking (high entropy, high enthalpy, lower heat capacity, and partially hydrophilic ligands).

7.2.7 *Thermodynamic discrimination in ligand-gated ion channels*

The Grunwald and Steel model of drug–receptor binding in water discussed above dissociates the binding process from solvent-reorganization phenomena. In particular, the most important thermodynamic parameters, i.e. equilibrium constants and free enthalpies, become intrinsic (nominal) properties of the system without any direct relation to the large $-T\Delta S°$ or $\Delta H°$ changes, which now become exclusive indicators of the fluxes of solvent molecules associated with the binding process. This complete separation between binding and solvent effects, if really true, seems to open new interesting perspectives for interpreting the fluxes of solvent that occur during the binding itself. It is not easy, however, to imagine real cases where the determination of such fluxes can be actually performed and, for this reason, we think it worthwhile to report briefly about a particular aspect of the binding thermodynamics of membrane receptors that seems hard to explain without turning to solvent effects. It concerns the so-called *thermodynamic discrimination* for which agonist binding is enthalpy driven while antagonist binding is entropy driven, or vice versa (Weiland *et al.*, 1979; Borea *et al.*, 1992), a type of behavior that is not so frequent among biological receptors where, as a rule, affinity (i.e. $\Delta G°$ and its components) is totally dissociated from efficacy (i.e. agonist, antagonist or, sometimes, inverse agonist behavior). The effect has been so far confirmed for three G-protein coupled receptors [GPCR: β-adrenergic (Weiland *et al.*, 1979; Contreras *et al.*, 1986), adenosine A_1 (Murphy and Snyder, 1982; Borea *et al.*, 1992), and adenosine A_{2A} (Borea *et al.*, 1995, 1996b) receptors] and four ligand-gated ion-channel receptors [LGICR: glycine (Ruiz-Gomez *et al.*, 1989), $GABA_A$ (Maksay, 1994), serotonin 5-HT_3 (Borea *et al.*, 1996a; Maksay, 1996), and nicotinic (Borea *et al.*, 1998) receptors].

Interpretation of this phenomenon is anything but simple, particularly because there is little structural information on membrane receptors, a class of compounds that has resisted so far any crystallization attempt. The first hypothesis was put forward for β-adrenoceptors where the enthalpic agonist binding was supposed to also drive the conformational changes determining the effect (Weiland *et al.*, 1979). Later, this idea was abandoned because most GPCRs were found to be non-discriminated and, in the few cases they are, agonist binding was found to be entropic. A new hypothesis was suggested for LGICRs, a class of receptors where ligand binding causes the opening of an ionic channel, for which it was supposed that the discrimination could directly arise from the changes of the water-exposed molecular surface consequent to channel opening (Borea *et al.*, 1998). It seems promising that all known LGICRs have been actually found to be thermodynamically discriminated, as illustrated in Fig. 7.8 by the relative $\Delta H°$ versus $\Delta S°$ scatter plots. Things are, however, considerably more complicated than expected as shown by the fact that the discrimination does not always occur in the same direction, agonist binding being definitely entropic for glycine, $GABA_A$, and serotonin 5-HT_3 receptors, but mostly enthalpic for the nACh receptor. A possible lead to the solution of the problem may come from the hydrophobic effect models (end of Section 7.2.3) predicting that water cages exist in two extreme states of low S/low H/high order S-driven binding and high S/high H/low order H-driven binding occurring, respectively, at low and high temperatures. In LGICRs there is no dependence of thermodynamic properties on temperature ($\Delta C_p^\circ = 0$), but nothing excludes that the two ordered and disordered cage states be activated by other factors, such as the hydrophobic/hydrophilic ratio of both ligand and binding site as well as the particular spatial configuration of the channel or the changes it undergoes during the process of ion transmission.

7.2.8 *Enthalpy–entropy compensation in crown ethers and cryptands*

Previous discussion on the role played by the H-bond in polar solvents has been exclusively focused on drug–receptor binding in water, a field perhaps too specialistic and somewhat far from the normal chemical experience. It seems worthwhile, therefore, also to consider a more chemical case of binding of small

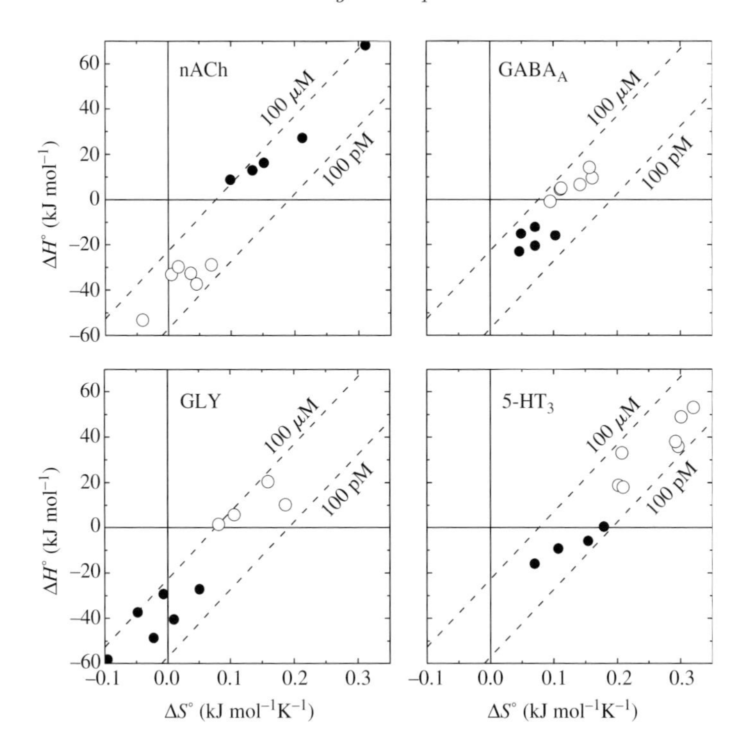

FIG. 7.8. Comparison of the ΔH° versus ΔS° scatter plots for the four ligand-gated ion-channel receptors (LGICRs) considered: nACh = nicotinic acetyl-choline; GABA$_A$ = γ-amino-butyric acid A; GLY = glycine; 5-HT$_3$ = serotonin 5-HT$_3$. The two dashed diagonal lines are the loci of the points for the limiting K_d values of 100 pM and 100 μM at 289 K. Open circles = agonists; full circles = antagonists. (Reproduced by permission from Borea *et al.*, 1998)

molecules to chemically synthesized macrocyclic hosts (Danil de Namor *et al.*, 1991) to show that enthalpy–entropy compensation is a general property of all reactions of molecular association in polar solvents not specifically dependent on the nature and dimensions of the molecules involved.

In this example, both guests and hosts are partially hydrophilic compounds consisting of a series of 18 zwitterionic amino acids, R-CH(NH_3^+)COO^-, which bind by intermolecular H-bonds to 18-crown-6 ether (**7.1.IIIa**) or cryptand-222 (**7.1.IIIb**). Thermodynamic results, which have been obtained by isoperibolic titration calorimetry in methanol or ethanol solutions, are summarized in the $-T\Delta S^\circ$ versus ΔH° plot of Fig. 7.9, not dissimilar in shape from that previously reported for drug–receptor binding (Fig. 7.6) but extended on a much narrower ΔH° range (of some 60 instead than 300 kJ mol^{-1}) shifted towards the enthalpy-driven side. As before, the key feature of the compensation phenomenon is that ΔG° is nearly invariant (total range of 9 kJ mol^{-1}) in spite of the changes of

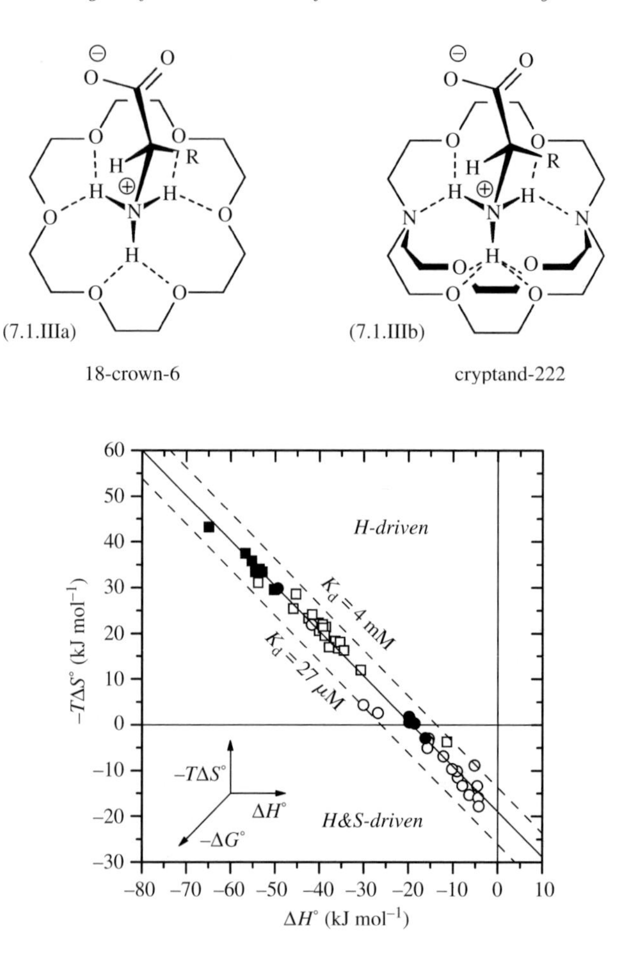

FIG. 7.9. Scatter plot of $-T\Delta S^\circ$ versus ΔH° (kJ mol^{-1}) for the binding thermodynamics of amino acids to 18-crown-ether and cryptand-222 in methanol and ethanol. The two dashed lines indicate the *loci* of the points where the combinations of ΔH° and $T\Delta S^\circ$ values give rise to the two equilibrium constants indicated (K_d = 4 mM and 27 μM). Open and full squares = 18-crown-6 in methanol and ethanol; open and full circles = cryptand-222 in methanol and ethanol. (Data from Danil de Namor *et al.*, 1991)

both reactants and solvents, while ΔH° is much more dispersed over a range of some 60 kJ mol^{-1}. Accordingly, the plot conforms to the regression equation

$$-T\Delta S^\circ = -19.0(6) - 0.99(2)\ \Delta H^\circ$$
$$(n = 48,\ r = -0.993,\ P < 0.0001), \qquad (7.11)$$

which resembles a textbook example of enthalpy–entropy compensation, $\Delta H = \beta\ \Delta S$, with compensation temperature β near 300 K.

8

THE ROLE OF STRONG H-BONDS IN NATURE: A GALLERY OF FUNCTIONAL H-BONDS

8.1 Introduction

8.1.1 *Detecting strong H-bonds*

H-bond strengths are greatly affected by the particular CL the bond belongs to. For instance, in terms of enthalpies ordinary O–H···O bonds (1–5 kcal mol^{-1}) can be strengthened up to 25–30, 15–25, and 8–9 kcal mol^{-1} in CAHBs, RAHBs, and PAHBs, respectively. This means that, when the simple inspection of a more or less complex molecular aggregate points out the possibility of, say, an O–H···O CAHB (or RAHB), we can be reasonably confident that this particular bond will be 5–6 times stronger (3–5 for RAHBs) than any other O–H···O bond of the system and that, because of its much larger energy, it will be inevitably formed unless hindered by extreme, but generally predictable, steric repulsions. This preliminary information can be enriched and complemented by other predictive methods, out of which the pK_a slide rule looks certainly the most efficient and inexpensive. This is to say that the use of very simple tools, such as CLs combined with pK_a matching, can provide us with a considerable ability to single out particularly strong H-bonds by simple inspection of the molecular association patterns, so opening the possibility of screening the most efficient ways of intermolecular interaction and recognition of complex molecular arrays 'on a paper sheet', that is from the simple graphical representation of the interacting molecular entities.

8.1.2 *The concept of 'functional H-bonds'*

There is an extremely wide literature concerning the role played by H-bonds in a great number of chemical and biochemical systems. Only in a few cases, however, do these studies make specific reference to the *real strengths* of the H-bonds involved or to particular effects induced by strong bonds. The reason is that the concept of H-bond strength is, rather paradoxically in view of the long story of this bond, a relatively modern concept and perhaps the first explicit classification in strong, moderate, and weak bonds has been reported in Jeffrey's book not before 1997 and now reproposed here in terms of CLs and pK_a equalization rules. However, there are good reasons for believing that the distinction between *strong and weak H-bonds* may be of great relevance in most H-bonded systems because of the quite different effects that these two classes of bonds can exert on two important chemical properties, i.e. molecular association and proton transfer.

Association properties are easy to quantify, at least approximately. It has been shown above (Chapter 7) that the $\Delta G°$ of the H-bond association equilibrium can be estimated to be nearly *one third* of its corresponding $\Delta H°$ because of enthalpy–entropy compensation in most environments of practical interest (non-polar solvents, molecular crystals and, most likely, in the nominal part of the binding in aqueous solutions). The normally accepted H-bond enthalpy range (say, $-3 \leq \Delta H° \leq -30$ kcal mol^{-1}) is then reduced to $-1 \leq \Delta G° \leq -10$ kcal mol^{-1} and becomes $5.41 \leq K_a = \exp(-\Delta G°/RT) \leq 2.14 \times 10^7$ in terms of association constants at room temperature. This corresponds to a total variation in equilibrium constants of seven orders of magnitude or to an increment of 158 times every 3 kcal mol^{-1}, clearly showing that strong H-bonds will inevitably dominate any H-bond-driven association occurring in solution.

More complex are the relationships between H-bond strength, shape of the PT profile, and height of the PT barrier. Generally speaking, the barrier is quite large in *weak* H-bonds and normally such to hinder the PT process but rapidly decreases until it disappears in *very strong* and proton-centered bonds (sSWHBs) often occurring in CAHBs with perfect pK_a matching. More interestingly, *strong* DWHBs (or LBHBs) are endowed with a double-well PT potential that transforms the bond in a two-state or bistable system whose barrier can be finely tuned for the effect of the substituents, a case often associated with RAHB and discussed in detail in Chapter 6.

Since it seems impossible that nature, in its four billions of years of evolution on Earth, has not taken full advantage of the many particularities of weak, strong, and very strong H-bonds, we have recently undertaken a rather widespread analysis of a number of molecular and biomolecular processes where H-bonds were known, or suspected, to play a role in the mechanism of action of important chemical (reactivity, tautomerism, recognition, crystal packing, mechanisms of functional materials, ...) or biochemical (enzymatic catalysis, transport phenomena, receptor binding, ...) phenomena. This preliminary investigation has shown that the occurrence of strong H-bonds is always very low but that, when they do occurr, they are often deeply involved in some particular aspect of the working mechanism of the system. Moreover, the range of phenomena where these strong H-bonds turn out to be involved is surprisingly wide, including, as far as we could assess, activation of the α-carbonyl in organic reactions, keto–enol and enol–enol tautomerism, bistate properties of ferroelectric crystals, excited-state proton transfer, structure and properties of α-helices and DNA base pairing (RAHBs); enzymatic catalysis in serine and aspartic proteases and other enzymes (CAHBs); proton transmission in water and in gramicidine A transmembrane channels, as well as water-without-proton transmission in aquaporin transmembrane channels (PAHBs).

Wanting to give them a name, we have tentatively called these strong H-bonds found to exert functions of control in so many different systems *functional H-bonds*, because they actually exert a functional role, though *smart H-bonds* could also be appropriate, because many of them surprisingly seem to behave in an intelligent way.

Since no systematic work has ever been carried out in this field, we cannot advance any pretence of reasonably complete coverage of cases or sufficiently complete treatment of each single case. The following exposition can only be intended as a preliminary collection of possible examples organized as a *graphic gallery of cases* with little or no discussion, a collection of themes that have already been, or deserve to be, investigated to unravel the true role played by the H-bond in natural systems. We hope that this gallery can help to convince the reader that the H-bond, though 88 years old, is always a topical subject because its everlasting relevance stems from the discovery of ever new systems where it plays a pre-eminent and irreplaceable role.

8.2 RAHB-driven prototropic tautomerism

8.2.1 *RAHB-activation of the carbon in α to a carbonyl*

Base-catalysed electrophilic alkylation of the carbon in α to a carbonyl is an important chemical and biochemical reaction. It reportedly occurs via formation of an enolate-ion intermediate and proceeds with difficulty owing to the very low acidity of the α-carbon ($pK_a \approx 19$–21) unless catalysed by strong bases (C8.1.I).

This reaction is extraordinarily facilitated even in the absence of bases in β-dicarbonyl derivatives where the α-carbon transforms in a so-called 'active methylene' having greatly enhanced acidity, its pK_a being reduced to 11–13 in β-diesters, 10–11 in β-ketoesters, and down to 4–5 in β-diketones, three classes of molecules known to form intramolecular RAHBs of increasing strength. Though enolization may also be assisted by electronic factors, it seems undoubtable that the great acidity increase has to be mainly imputed to RAHB-induced enolization (C8.1.II) stabilized by an intramolecular O–H···O bond of high energy (12–22 kcal mol^{-1} in the case of β-diketones). This, in turn, will stabilize the enolate intermediate by allowing the negative charge to be delocalized over all the H-bonded ring (C8.1.IIIb).

CHART 8.1. RAHB: α-CARBONYL ACTIVATION

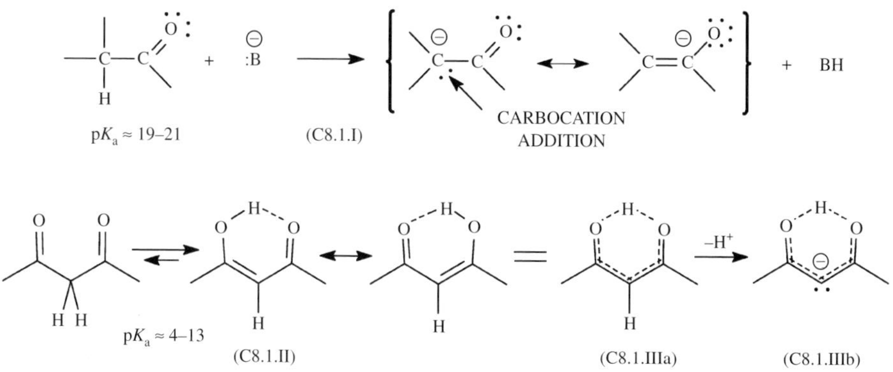

$pK_a \approx 19$–21 (C8.1.I) CARBOCATION ADDITION + BH

$pK_a \approx 4$–13 (C8.1.II) $-H^+$ (C8.1.IIIa) (C8.1.IIIb)

8.2.2 *RAHB-induced enolization in keto–enol tautomerism*

The results of the previous chart can be generalized by saying that, though the keto–enol equilibrium C8.2.I is, in itself, always strongly shifted to the left, enolization can easily occur in compounds where the enolic tautomer is stabilized by intramolecular (C8.2.II) or intermolecular RAHB (C8.2.III). The problem has been the object of extensive treatment (Gilli and Bertolasi, 1990; Rappoport, 1990) and has been considered in detail in Section 3.2.

For our aims, it is particularly important to add that enolization is actually associated with two different forms of tautomerism, the first being the usual keto ⇄ enol equilibrium and the second the keto–enol ⇆ enol–keto one illustrated in the right part of Chart 8.2. Though only the first is normally quoted in books of organic chemistry, the second is probably even more important in determining many forms of prototropic tautomerism, as will be shown in the next section.

8.2.3 *RAHB-induced tautomerism in heteroconjugated systems*

According to Finar's treatise of organic chemistry (Finar, 1973) there are ten main chemical systems suited to produce prototropic tautomerism for reasons differently imputed to a number of factors, such as acid or base catalysis,

CHART 8.2. RAHB: INDUCED KETO–ENOL TAUTOMERISM

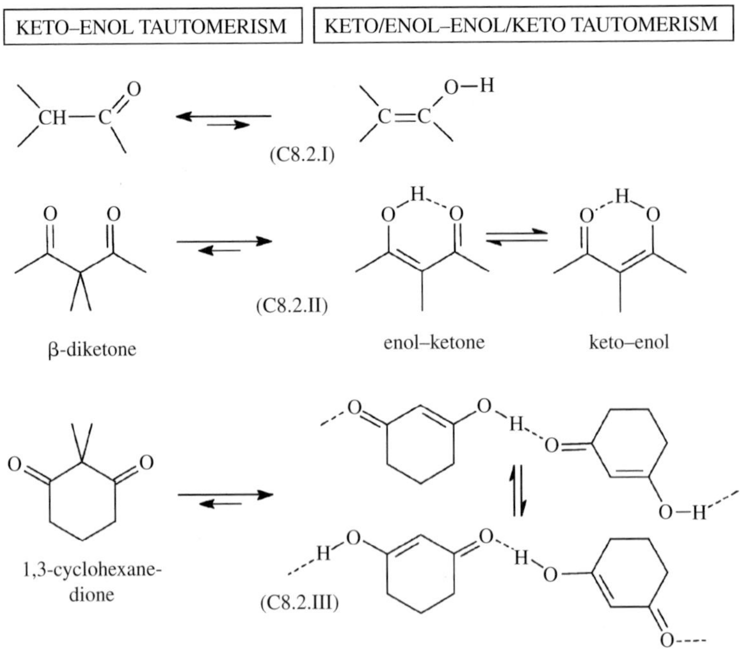

KETO–ENOL TAUTOMERISM	KETO/ENOL–ENOL/KETO TAUTOMERISM

(C8.2.I)

β-diketone (C8.2.II) enol–ketone keto–enol

1,3-cyclohexane-dione (C8.2.III)

resonance energy, intramolecular H-bonding and, inevitably, steric factors. The idea that this type of tautomerism is more easily explained in terms of stabilization of both tautomers and lowering of the PT barrier between them induced by intramolecular R_3-RAHB formation is exploited in Chart 8.3. It shows some of the most popular cases of tautomerism, some of which have already been treated in detail in the previous chapters, particularly for what concerns the role of substituents in determining the exact position of the tautomeric equilibrium. The discussion cannot be resumed here but, certainly, a comprehensive and systematic treatment of tautomerism in terms of intra- and intermolecular RAHB would certainly be fruitful.

8.2.4 *RAHB cooperativity and anticooperativity in more complex cases*

Cooperativity is an important factor in H-bond strengthening and the concepts of σ-bond and π-bond cooperativity (Jeffrey and Saenger, 1991) have already been introduced in Section 3.2.1. The former will be treated in Chart 8.10 and only the latter is considered here. As a synergism of H-bond strengthening and enhanced delocalization of the interleaving π-conjugated fragment, RAHB is a natural π-bond cooperative system that can be classified by the notation R_n-RAHB, where R_n represents a fragment of n atoms (n odd) connecting the H-bond donor and acceptor groups. So R_1-RAHB can address carboxylic acids or amides, R_3-RAHB β-diketone enols (β-enolones) or β-enaminones, and R_5- or R_7-RAHB δ- or ζ-diketone enols (Section 3.2.3.1).

CHART 8.3. RAHB: TAUTOMERISM IN HETERODIENES

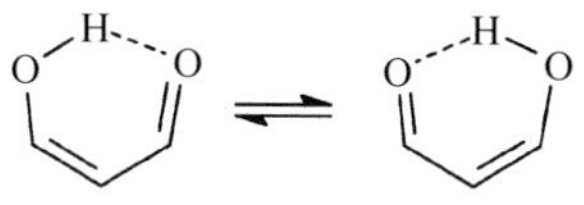

Enol–ketone ⇌ Keto–enol (C8.3.I)

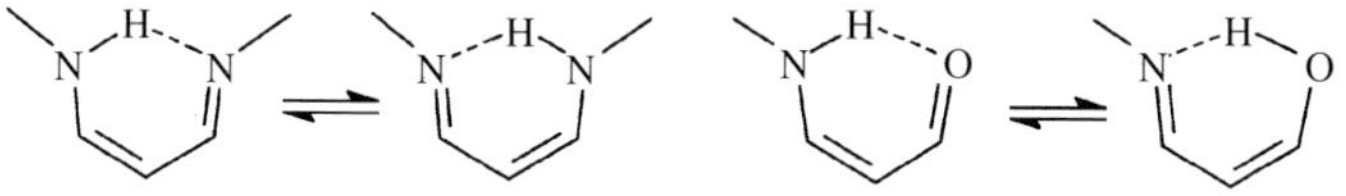

Enamino-imine ⇌ Imino-enamine (C8.3.II)

Enaminone ⇌ Imino-enol (C8.3.III)

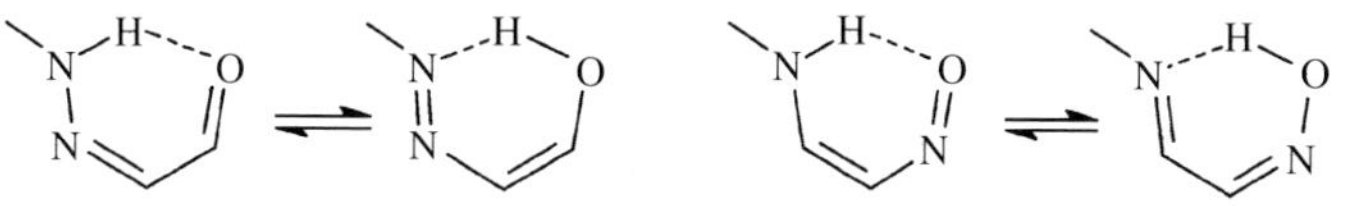

Keto-hydrazone ⇌ Azoenol (C8.3.IV)

Nitroso-enamine ⇌ Imino-oxime (C8.3.V)

CHART 8.4. RAHB: π-BOND COOPERATIVITY AND ANTICOOPERATIVITY

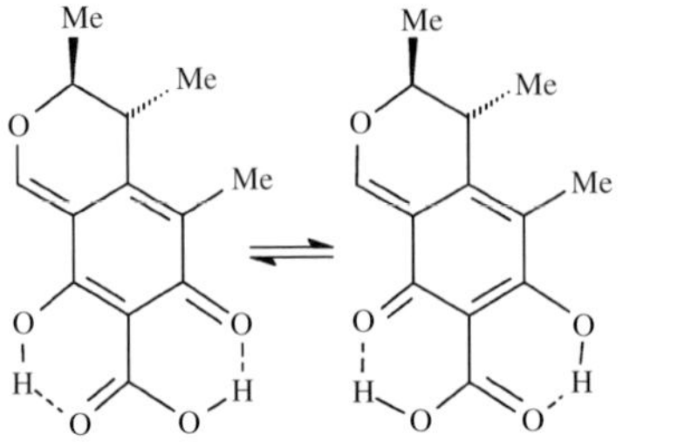

citrinin, a fungal metabolite (C8.4.I)

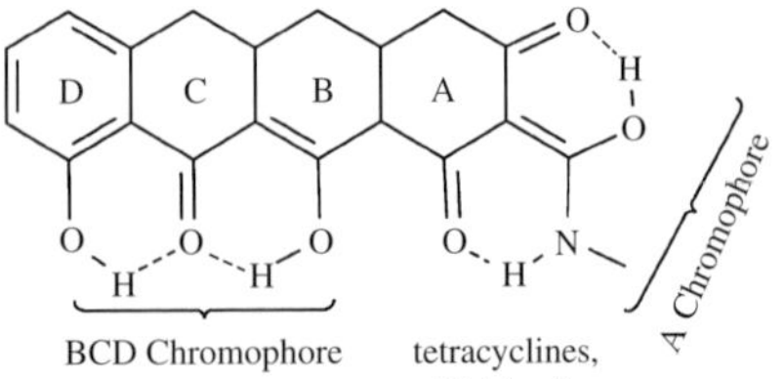

BCD Chromophore tetracyclines, antitiobics from *Streptomyces* A Chromophore (C8.4.II)

[2,2']OOOO (C8.4.Ia) (R_3/R_3)-RAHB anticooperative

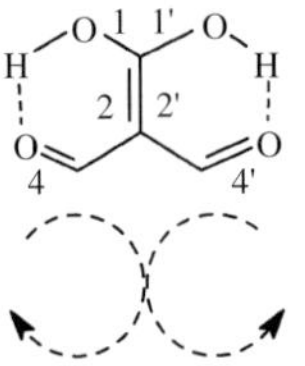

[4,4']OOO (C8.4.IIa) (R_3/R_3)-RAHB anticooperative

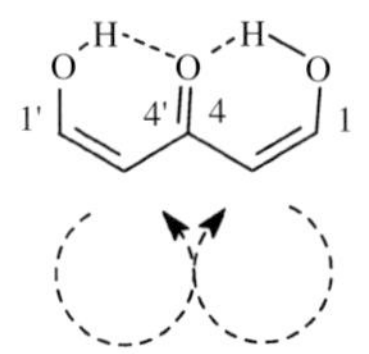

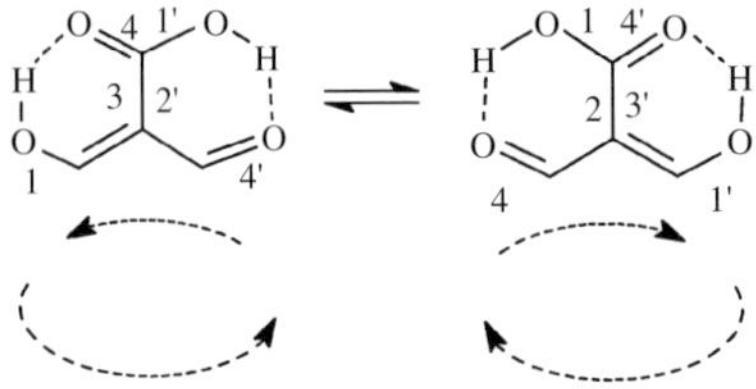

[3,2']OOOO (R_1+R_3)-RAHB cooperative (C8.4.Ib)

[2,3']OOOO (R_1+R_3)-RAHB cooperative (C8.4.Ic)

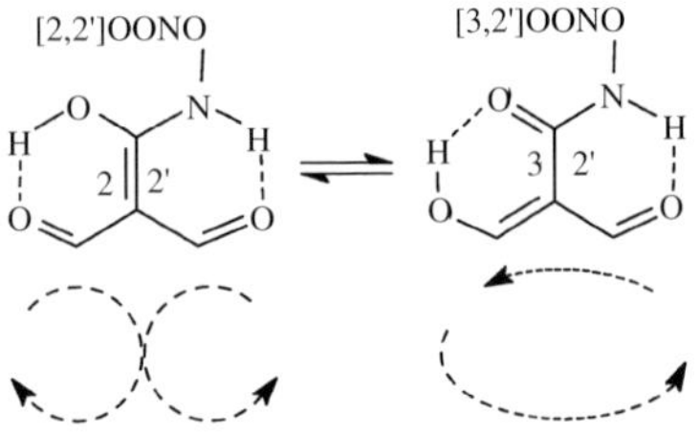

[2,2']OONO (R_3/R_3)-RAHB anticooperative (C8.4.IIb)

[3,2']OONO (R_1+R_3)-RAHB cooperative (C8.4.IIc)

This does not put an end to the possible RAHB combinations as shown by citrinine (C8.4.I) and tetracyclines (C8.4.II) that are seen to form different associations of two R_3-RAHB rings. The possible combinations of two such rings have been recently considered (Bertolasi *et al.*, 2006) showing that, by numbering as 1–4 the four covalent bonds inside the H-bonded ring, four different junctions of the two rings become possible, that is [4,4′], [2,2′], [3,2′], and [2,3′]. Moreover, the ring fusion gives rise to a new type of tautomerism never described before that could be called *homodromic (cooperative)* $\rightleftarrows$ *heterodromic (anticooperative) tautomerism* because it is due to the fact that the π-electron shifts caused by RAHB within the two H-bonded rings can have the same direction (π-cooperativity) or two opposite directions (π-anticooperativity), as indicated in Chart 8.4 by the direction of the dashed arrows marking the sense of these RAHB-induced π-electron shifts.

For citrinine three different isomeric patterns are possible, one (R_3/R_3)-RAHB anticooperative (C8.4.Ia) and two tautomeric (R_1+R_3)-RAHB cooperative (C8.4.Ib and C8.4.Ic) that represent a new type of RAHB where the delocalization runs all around the two fused rings. The two cooperative forms must be more stable because they are more easily delocalizable and, in fact, all crystals so far studied consist of a dynamical equilibrium of these two cooperative tautomers (Destro and Marsh, 1984; Destro, 1991) forming considerably shortened H-bonds with average O···O distances of 2.51 Å.

Tetracyclines contain two quite different types of ring fusions. The pattern formed by the so-called BCD chromophore has been observed in many other compounds and is particularly simple, being the anticooperative [4,4′] fusion of two β-enolone rings (C8.4.IIa) where the two hydroxyls compete for the same carbonyl according to the O–H···O···H–O scheme and where, because of this competition, both H-bonds turn out to be weakened. Chromophore A is the fusion of two different rings, β-enolone and β-enaminone, which can form both O–H···O and N–H···O bonds but not O–H···N bonds, which are unstable in these conditions (see Section 3.2.3.4). The two possible patterns are shown in Schemes C8.4.IIb and C8.4.IIc. Since one is anticooperative and the other cooperative, they behave as separate isomers that do not show tautomeric exchange in the solid state and whose occurrence depends on the chemical substituents present on the rings. It has been suggested that this isomerism may play an important role in the complex scheme of protonation and isomerization reactions typical of tetracyclines (Gilli and Bertolasi, 1990).

8.3 H-bond-controlled crystal packing

8.3.1 *The crystal packing of squaric acid and its anions*

Squaric acid is a strong biprotic acid that can crystallize as pure H_2SQ (Semmingsen *et al.*, 1977; Gilli and Bertolasi, 1990; Gilli *et al.*, 1993) as well as hydrogen squarate (HSQ^-) or squarate (SQ^{2-}) anions, the last two being particularly suited for interacting with a variety of H-bond donors. H_2SQ and its anions (C8.5.I) are flat molecules because of their full π-delocalization, and, accordingly, the four O atoms behave as planar sp^2 electron donors of one or two lone pairs. Resonance stabilizes the monoanion (C8.5.Ib) and, to a greater extent, the dianion (C8.5.Ic) with respect to the undissociated acid (C8.5.Ia) and this stabilization accounts for the fact that both H_2SQ and HSQ^- are remarkably strong acids [pK_1 = 1.4(2); pK_2 = 3.3(3)] with an unusually small difference between the two dissociation constants.

Squaric acid and its anions have been chosen here to illustrate the effect of strong H-bonding in crystal packing for two reasons, the first being that they can form two types of strong H-bonds (i.e. CAHB and RAHB) and the second that they have been frequently used as counteranions in the preparation of a number

CHART 8.5. RAHB & CAHB: SQUARIC ACID CRYSTALLIZATION

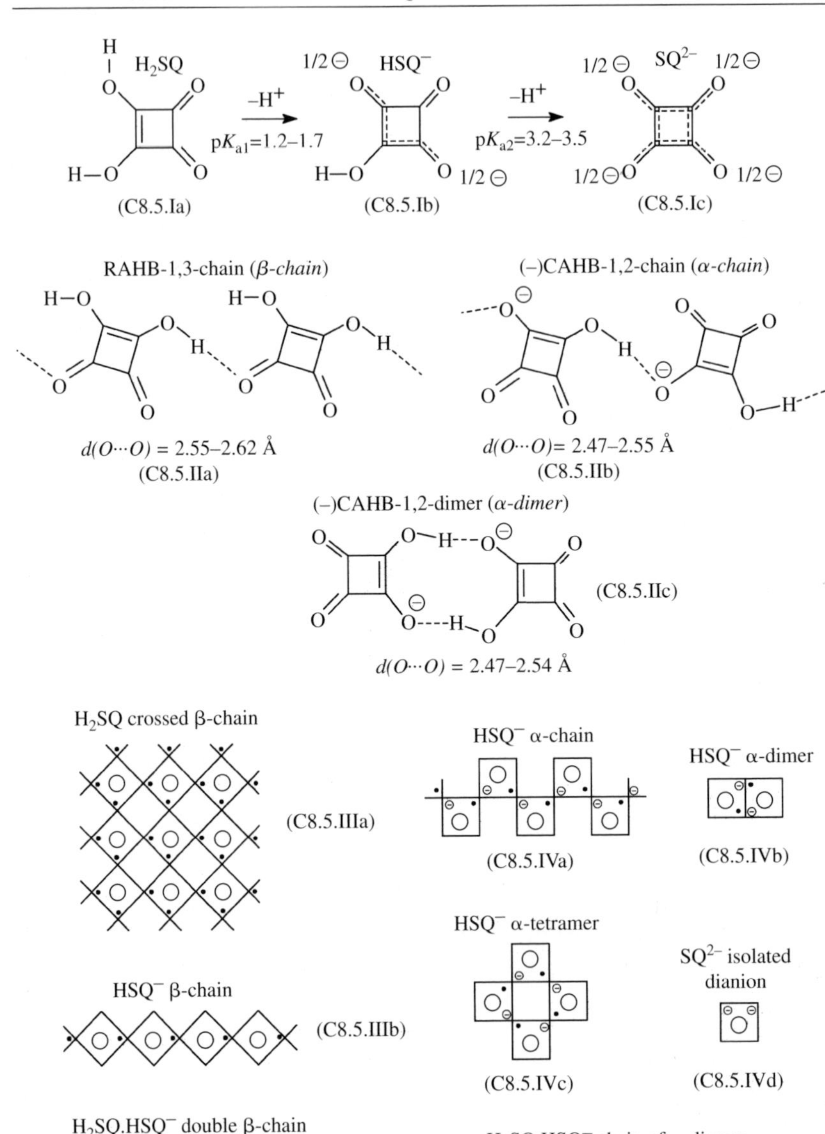

of mixed molecular crystals, so that their crystal structures are relatively abundant. Three basic association patterns are observed (Gilli *et al.*, 2001):

(*i*) *RAHB-1,3-chains (β-chains)* (C8.5.IIa): typical resonance-assisted infinite chains (Section 3.2.3.2) bound by moderately strong RAHBs with O···O distances of 2.55–2.62 Å and associated E_{HB} energies of 7–10 kcal mol^{-1}, as evaluated by the LS-HB method;

(*ii*) *(−)CAHB-1,2-chains (α-chains)* (C8.5.IIb): infinite chains connected by $[O{-}H\cdots O]^-$ bonds with O···O distances of 2.47–2.55 Å and associated energies of 10–15 kcal mol^{-1};

(*iii*) *(−)CAHB-1,2-dimers (α-dimers)* (C8.5.IIc): dimers bound by two $[O{-}H\cdots O]^-$ bonds having α-chain-like properties.

These three patterns summarize and conclude the *bonding* features of squaric acids and its anions. However, they do not conclude their crystal chemistry, which is much more complex and cannot be explained in simple terms of bonding properties. In this context, a comparison may be useful with the crystal chemistry of silicates with which squarates show not irrelevant affinities. Squaric acid, $H_2C_4O_4$, has, in fact, definite topological similarities with orthosilicic acid, H_4SiO_4, both consisting of a core (the C_4 frame or the Si atom) connected to the other units by up to four bonds (four hydrogen bonds or four covalent Si–O–Si bridges). The main differences between the two systems are that: (*i*) in space, squaric acid is flat (two-dimensional) while silicic acid is tetrahedral (three-dimensional); and (*ii*) in projection, the former is tetragonal while the latter is trigonal.

It is reasonable to presume that the crystal chemistry of squarates is a kind of two-dimensional projection of that of silicates. This is illustrated in Chart 8.5 by representing the C_4O_4 moiety by a simple square where hydroxyls are marked by dots. In analogy with silicates, the structural motifs are seen to include isolated dimeric (C8.5.IVb) and monomeric (C8.5.IVd) anions, single-chains (C8.5.IIIb and C8.5.IVa) and double-kinked chains (C8.5.IVe), ribbons of double chains (C8.5.IIIc), and tetramers (C8.5.IVc), while the structure of squaric acid (C8.5.IIIa) is the evident topological projection of the three-dimensional structure of silica into the plane.

8.4 Bistable H-bonds in functional molecular materials

8.4.1 *Generalities*

Functional materials can be defined as molecular systems that perform specific functions at a nano-technological level of scale. They are known to have important applications in many advanced fields (electronics, optoelectronics, photonics, memory storage, signal transmission and amplification, sensors, and so on) though their comprehensive classification is far from being complete. In these materials H-bonds, in particular strong H-bonds, may play two different types of roles indicated here as the structural and functional role.

The structural role is based on the well-known observation that H-bonds may be strong determinants of molecular recognition in general and of crystal packing in particular. Accordingly, they can be taken advantage of to design specific intermolecular complexes or crystal architectures (crystal engineering) that constitute the indispensable physical substrate for the functionality of the material itself. These particular aspects will not be discussed here.

The H-bond functional role is a rather more complicated topic because it requires understanding of the basic working principles on which functional materials exert their action at the molecular level. For the limited aims of this treatment, functional materials can be thought of as *assemblies of nanoscale molecular machines* (gadgets, devices,...) *based on two-state, or bistable, molecular systems* that produce their specific effects by cyclically switching between two different stability points. According to the nature and function of the molecular system considered, the driving force switching the system can arise from three main processes, energy transfer, charge (or electron) transfer, and proton transfer (PT).

Hence, H-bond-driven functional materials must be assemblies of two-state molecular systems based on H-bonds where the proton can assume two competitive positions (prototropic tautomerism). This immediately points to *strong H-bonds* characterized by a double-well PT reaction profile and reasonably small PT barrier (a class of bonds already discussed in detail with the names DWHB or LBHB), while *weak* (OHBs) or *very strong* (sSWHBs) bonds cannot be employed because of their, respectively, too high or too low PT barriers.

Of course, not all tautomeric H-bonds are suited to build a nanoscale molecular machine. To this aim, they must possess another and much less common property: the shift of the proton must be able to reverse the value of some physical quantity of the system that will thus become the steering variable for switching the system itself, from the outside, from one state to the other. Such a steering variable can be of quite a different nature, such as electric or magnetic fields, pressure, temperature, and irradiation in a proper frequency range. Two examples are illustrated below: crystals with ferroelectric properties and molecules undergoing excited-state proton transfer (ESPT). In ferroelectric crystals the quantity of interest is the electric dipole moment and, accordingly, the system can be switched, below its critical temperature, by applying an external electric field. In ESPT the tautomeric H-bond assumes different configurations (X–H···Y or X···H–Y) according to whether the molecule is in its ground or excited state. Accordingly, the system can be switched by irradiation with a suitable wavelength.

8.4.2 *RAHB and ferro/antiferroelectric behavior*

Ferroelectricity is a physical property of solid materials that exhibit a spontaneous electric dipole moment whose direction can be switched between two equivalent states by the application of an external electric field. A ferroelectric placed between two conductive plates is a ferroelectric capacitor, characterized by having a *polarization versus voltage* response that is *non-linear* but assumes

CHART 8.6. RAHB: POTENTIAL FERRO/ANTIFERRO–PARAELECTRIC SYSTEMS

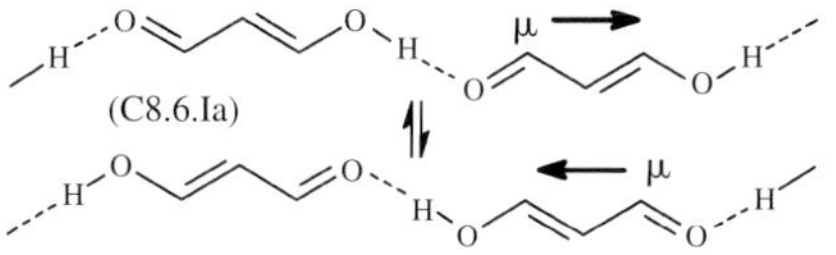

(C8.6.Ia)

Infinite chains of β-enolones (R_3-RAHBs)

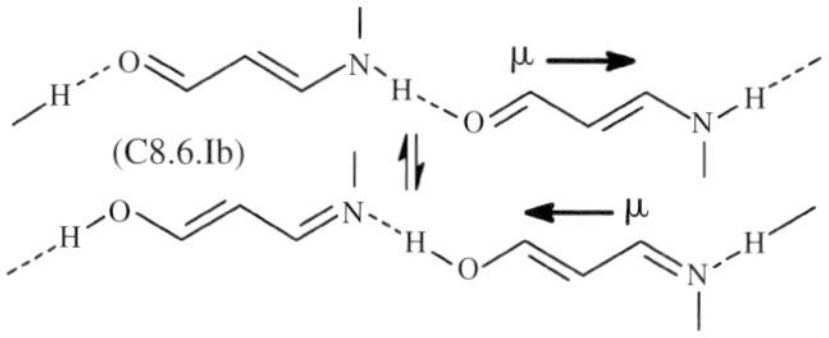

(C8.6.Ib)

Infinite chains of β-enaminones (R_3-RAHBs)

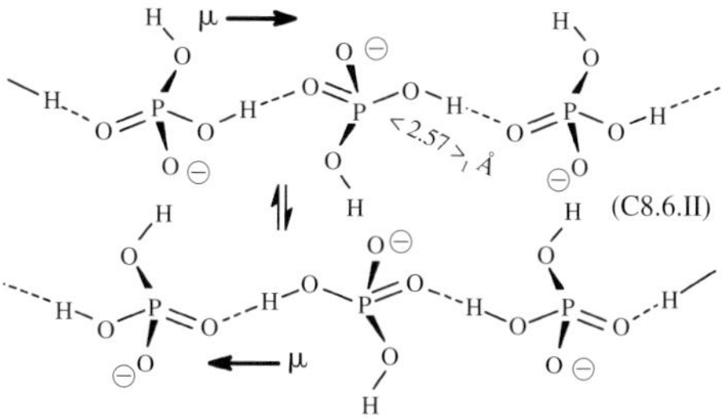

(C8.6.II)

Infinite chains of dihydrogen phosphates (R_1-RAHBs)

KH_2PO_4 (KDP)
Ferroelectric phase

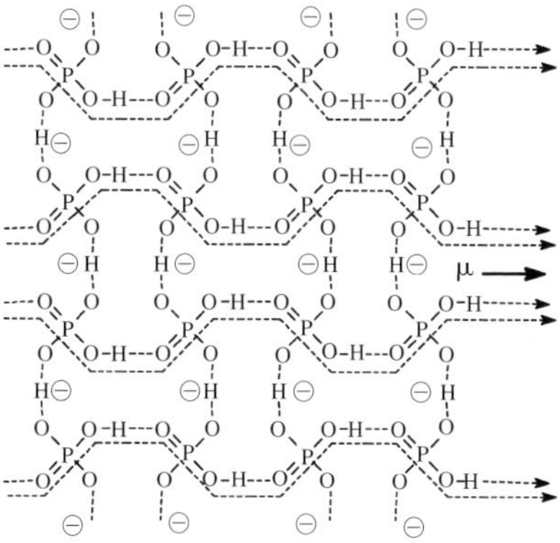

(C8.6.IIIa)

$NH_4H_2PO_4$ (ADP)
Antiferroelectric phase

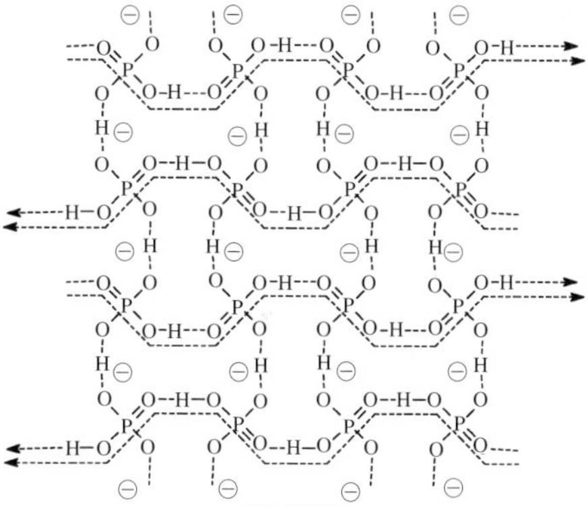

(C8.6.IIIb)

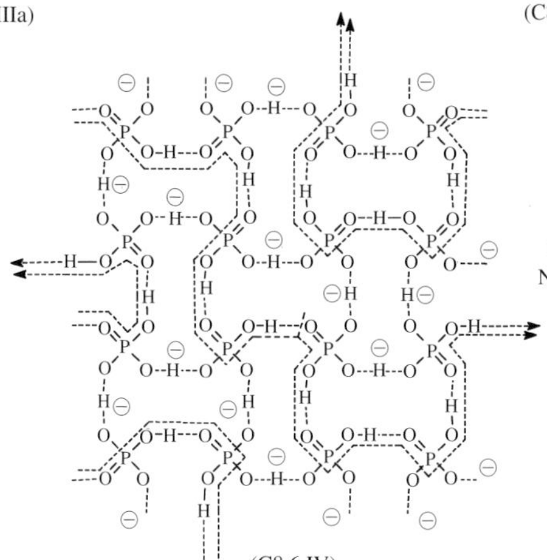

KH_2PO_4 (KDP)
$NH_4H_2PO_2$ (ADP)
Critical state

(C8.6.IV)

the form of the *hysteresis cycle* typical of ferroelectric (and ferromagnetic) materials. Practical applications of ferroelectric capacitors are countless because hysteresis means memory and thus applications in ferroelectric RAM memories and RFID cards, while non-linearity means excellent dielectric properties that make ferroelectrics key materials in microelectronics. It can be said that the combined properties of high dielectric constant, memory, piezoelectricity, and pyroelectricity make ferroelectric capacitors some of the most useful technological devices in modern society. Without them, everyday applications such as medical ultrasound machines, fire sensors, sonar, vibration sensors, compact laptop computers, cell phones and electro-optic modulators at the base of the internet would be impossible.

There are two main classes of ferroelectrics. In *displacive ferroelectrics*, such as barium titanate and lead zirconate titanate, the effect is not related to H-bonds but due to the so-called polarization catastrophe that occurs when the displacement of a particular ion from its equilibrium position forces, because of the coupling due to the local electric fields, the displacement of all equivalent ions in the crystal.

Conversely, H-bonds may have great importance in *order–disorder ferroelectrics*, which have already been introduced in Section 3.2.3.2C in connection with the antiferromagnetic behavior of squaric acid. Let us consider a crystal having one molecule with dipole moment *per* unit cell. At high temperatures the dipoles are oriented at random (*paraelectric* crystal) because the weak attractions among dipoles (quantified by the *coupling constant*) cannot overcome dynamic disorder. Upon lowering the temperature, however, this ordering can start to occur at a characteristic temperature, called the critical or Curie temperature (T_c), which marks the *disorder–order transition* from single-state to double-state properties. At still lower temperatures, the ordering can become complete with perfect orientation of all dipoles in the same direction (*ferroelectric crystals*) or in alternating directions (*antiferroelectric crystals*). In the former, application of a suitable electric potential will be able to reverse the direction of all dipoles, at a time.

The main (and perhaps the only) way for producing infinite arrays of coupled molecular dipoles in crystals is through infinite chains of H-bonded molecules (Telegren, 1975; Chemla and Zyss, 1987; Prasad and Williams, 1991; Katrusiak, 2003). Good examples are the α- and β-chains shown in Chart 8.5 for the crystal packing of squaric acid and its ions, a choice that seems particularly timely as crystals of squaric acid actually show antiferroelectric behavior. Both chains can support proton transfer but by a different mechanism of transmission, the first through isolated $O{-}H{\cdots}O^- \rightleftarrows {}^-O{\cdots}H{-}O$ proton jumps and the second by RAHB-based keto–enol $\rightleftarrows$ enol–keto tautomerism. The second may seem more likely to occur in crystals and this idea is tentatively exploited in Chart 8.6 for some of the most common cases.

Schemes C8.6.Ia and Ib report the often discussed RAHB chains of β-enolones and β-enaminones illustrating their mechanism of dipole inversion that makes these compounds natural candidates for non-linear dielectric materials. It has already been remarked, however, that simple chains of this sort do not display ferroelectric behavior and this has to be imputed (Section 3.2.3.2C) to the fact that monodimensional lattices cannot undergo order–disorder phase transitions, which are only possible for at least two-dimensional lattices (Landau and Lifshitz, 1967). This hypothesis is in agreement with the fact that squaric

CHART 8.7. RAHB: EXCITED-STATE PROTON TRANSFER

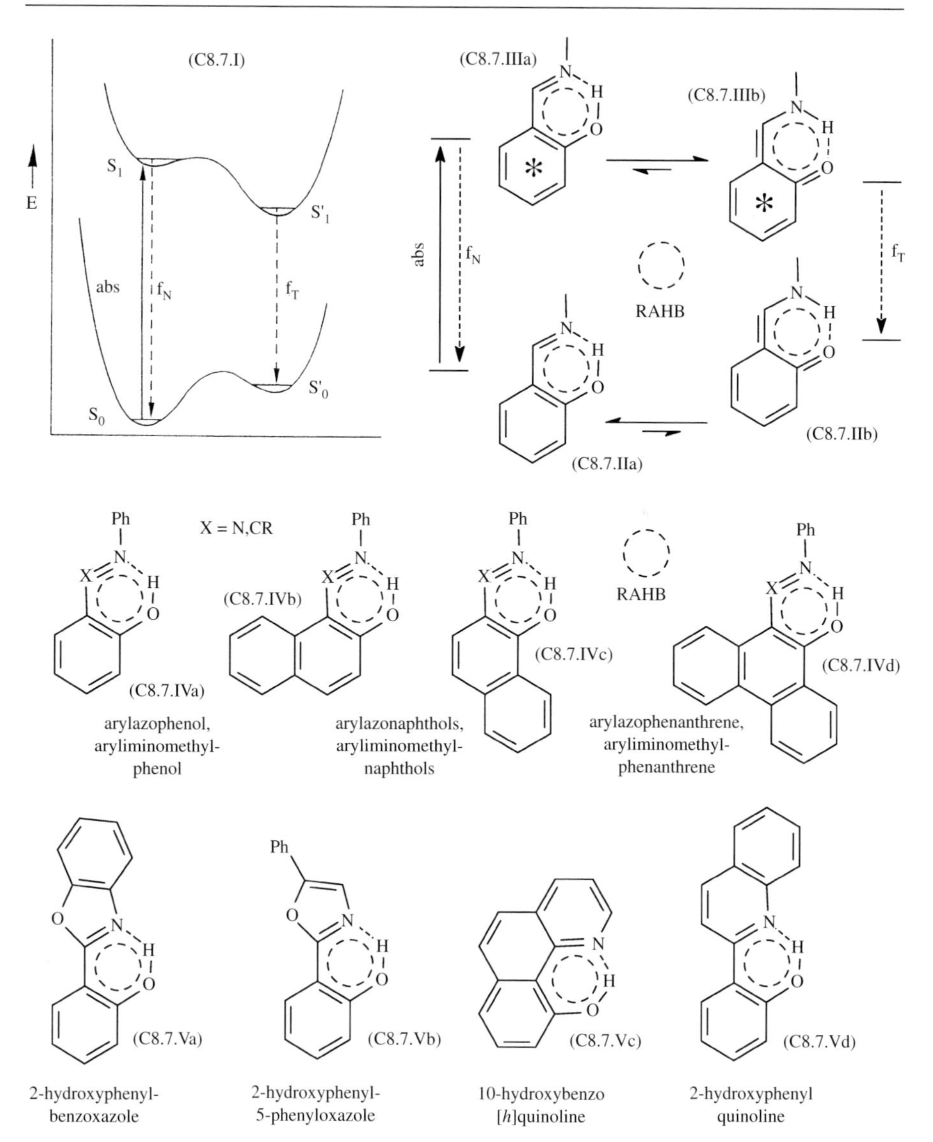

acid, endowed with the two-dimensional H-bond net C8.5.IIIa, undergoes an antiferroelectric–paraelectric transition at $T_c = 375$ K. Much more convincing evidence comes from two well-known materials of widespread practical use, ferroelectric KDP (potassium dihydrogen phosphate, $T_c = 213$ K) and ADP (ammonium dihydrogen phosphate, $T_c = 230$ K), whose crystal packing is based on the common pattern C8.6.II. It presents the infinite R_1-RAHB chain of ···O=P–O–H···O=P–O–H··· bonds whose inversion can cause the reversal of the chain dipole moment but, at the same time, can be linked by perpendicular chains (not indicated) of O–H···O$^-$ bonds in a two-dimensional array. How these two different chains can be arranged in ferroelectric KDP and antiferroelectric ADP crystals is shown is C8.6.IIIa and IIIb, respectively. Scheme C8.6.IV represents, instead, one of the infinite chain arrangements that can be associated with the critical state at the temperature T_c or with states of higher temperature. To notice that the intertwining of the two chains can occur without the need of introducing any local defect within the two-dimensional structure.

8.4.3 *RAHB and excited-state proton transfer (ESPT)*

Excited-state proton transfer (ESPT) is a phenomenon that occurs when the two minima of the PT-pathway (C8.7.I) have reversed stabilities in the molecular *ground* (lower curve) and *excited state* (upper curve). In this case, excitation of the X–H···Y ground-state configuration S_0 to its excited state S_1 may be followed by adiabatic proton transfer with formation of the excited-state X···H–Y tautomer, S'_1. Fluorescence emission (f_T) from S'_1 to the ground-state of the X···H–Y tautomer S'_0 is usually strongly *red shifted* with respect to the normal X–H···Y fluorescence f_N and, therefore, ESPT can be easily detected by comparing absorption and emission (or fluorescence) spectra in the solutions of the compound of interest. Not surprisingly, these molecules may be thermochromic, photochromic and photoluminescent (both fluorescent and phosphorescent) and are reportedly considered a good starting point for the design of a variety of molecular devices, such as dye-lasers (typically 4-level lasers), energy storage, optical-data storage, solar collectors, optical switches for triggering molecular electronic devices, luminescent probes, scintillators, and proton-transfer materials.

There are many classes of chemical compounds able to produce intramolecular and intermolecular ESPT, most of which are dyes with a low-energy $\pi^* \leftarrow \pi$ transition. ESPT molecular systems were screened in an attempt to identify a common mechanism of action at the atomic-resolution level grounded on the concepts of CL, LBHB, and functional H-bond. Though an overall classification based on these criteria was unattainable, a wide class of intramolecular ESPT molecules was identified that can be actually rationalized in terms of LBHBs formed by RAHBs (Chart 8.7). They refer to the already discussed (Section 3.2.3.4) enaminone $\rightleftharpoons$ iminoenol and ketohydrazone $\rightleftharpoons$ azoenol tautomeric equilibria illustrated here, for the enaminone case, in Schemes C8.7.II and III. In this example, the H-bonded RAHB ring is fused with a phenylene ring whose large resonance energy, stabilizing the iminoenol form C8.7.IIa in the ground state,

CHART 8.8. RAHB: SECONDARY STRUCTURE OF PROTEINS AND DNA PAIRING

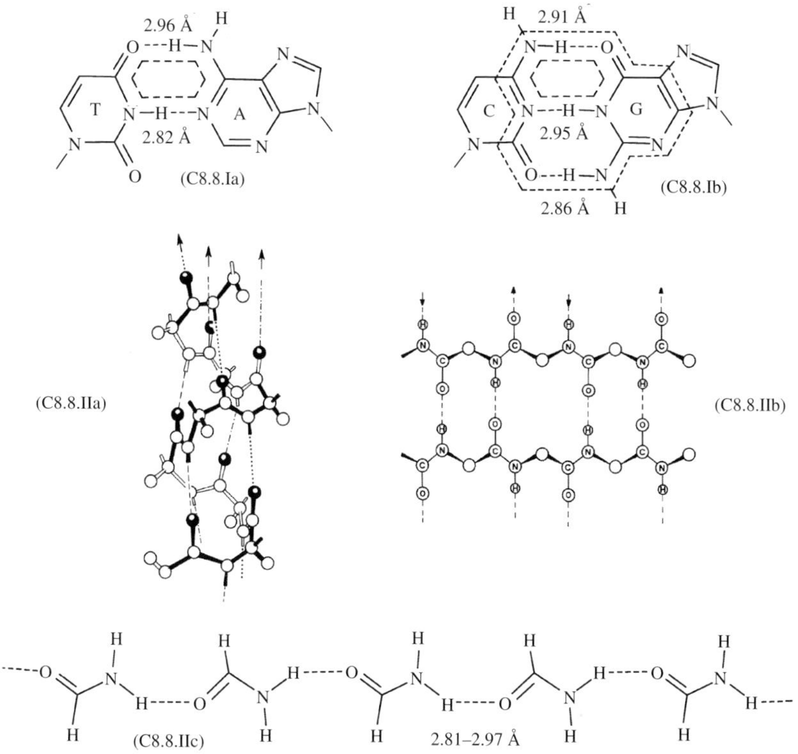
2.96 Å
2.82 Å
T
A
(C8.8.Ia)
2.91 Å
2.95 Å
2.86 Å
C
G
(C8.8.Ib)
(C8.8.IIa)
(C8.8.IIb)
(C8.8.IIc)
2.81–2.97 Å

is removed in the excited form C8.7.IIIa brought about by $\pi^* \leftarrow \pi$ excitation. Accordingly, the proton is transferred transforming C8.7.IIIa into C8.7.IIIb, that is the enaminone form normally observed in the absence of the fused phenylene. Subsequent radiative or non-radiative (not shown) decay will lead to the ground-state enaminone form C8.7.IIb where the benzene aromaticity can be re-established, bringing back the molecule to its original ground-state iminoenol form C8.7.IIa, and so closing the circle.

Hence, this particular enaminone $\rightleftharpoons$ iminoenol ESPT system looks like an almost paradigmatic application of the RAHB model developed in Section 3.2. RAHB is seen to affect and determine ESPT in two ways: (*i*) by strengthening the H-bond and then making possible the indispensable DW PT profile; and (*ii*) by creating, through the π-delocalization of the resonant fragment typical of RAHB, a direct π-bonding connection between the H-bond itself and the aromatic ring that is the site of the $\pi^* \leftarrow \pi$ excitation. It may be supposed that this behavior is a general feature of many other molecular systems whose tentative list is displayed in the chemical Schemes C8.7.IV and V.

8.5 Functional H-bonds in biological systems

8.5.1 *RAHB in the secondary structure of proteins and in DNA base pairing*

The possibility that RHAB could be functionally involved in the thymine–adenine (C8.8.Ia) and cytosine–guanine (C8.8.Ib) coupling occurring in DNA and in the formation of α-helices (C8.8.IIa) and β-pleated sheets (C8.8.IIb) present in the secondary structure of proteins has been suggested since the very first paper devoted to this bond (Gilli *et al.*, 1989). Four years later (Gilli *et al.*, 1993) it was proposed again in the rather exultant form '...nature itself may have taken advantage of the greater energy of RAHB to keep control of molecular associations whose stability is essential to life'.

All this can be taken as a tale on the risks we run into when making data extrapolations too readily. In fact, both papers were dealing with O–H···O RAHBs and their data were instead used to guess the properties of N–H···O ones. Only later were heteronuclear N–H···O RAHBs systematically studied (Gilli *et al.*, 2000, 2002; Gilli and Gilli, 2000) and found to be considerably weaker than homonuclear O–H···O ones. Moreover, as discussed in Section 3.2.3.4E, this weakness is intrinsic, being due to the different O and N proton affinities, and could be overcome only by proper substituents that, however, cannot occur in the biological amides considered.

Crystallographic H-bond geometries of secondary and primary amides are summarized in Fig. 3.33, showing that all the parameters involved (C–O and C–N bond distances and N···O contact distances) undergo only small changes because of H-bond formation, changes that are at the borderline of the standard errors of accurate molecular structures and considerably smaller than those of any macromolecular structure, even at the highest resolution level. Other difficulties arise from the scarcity of good molecular structures, frequent disorder in crystals and, for the DNA-base coupling, complex tautomerism of the bases (Saenger, 1984) which breaks the data set up. After the many attempts it must be concluded that, at best, the role of RAHB in proteins and DNA is an interesting hypothesis that cannot be excluded but neither can it be confirmed with reasonable certainty as, for instance, for the β-diketone enols of Fig. 3.9.

Nevertheless, we still believe in this hypothesis and feel somewhat comforted by the results that start to come from the quantum-mechanical simulation that, at present, seems to be the only possible way for circumventing the inaccuracy of experimental data. The DFT simulation of amide chains from 2 to 15 monomer units long (C8.8.IIc; Kobko and Dannenberg, 2003) has already been discussed at the end of Section 3.2.3.4E and summarized in Table 3.8 by comparing the properties of the amide dimer with those of a same dimer but at the center of a 15-member chain. In agreement with the crystallographic evidence, bond-length changes are quite small (0.014 for C=O and –0.016 Å for C–N) indicating only a small increase of π-bond delocalization but, notwithstanding this, H-bond energies are markedly increased from 4.5 to 13.0 kcal mol^{-1}, while the molecular dipole moment increases from 4.6 to 5.9 Debye *per* residue.

Both effects can be interpreted in terms of π-bond cooperativity typical of RAHB, so supporting the idea that RAHB itself can actually give a significant contribution to the stability of amide chains and, presumably, of α-helices and β-pleated sheets. The role played by RAHB in these particular biological systems remains, therefore, a problem open to discussion and to theoretical and experimental research.

8.5.2 *Charge-assisted H-bonds in enzymatic catalysis*

8.5.2.1 *Generalities*

The hypothesis that formation of a short, very strong, low-barrier H-bond (LBHB) in the transition state can be an important contribution to enzymatic catalysis was firstly formulated at the beginning of the 1990s (Cleland, 1992; Gerlt and Gassman, 1993; Cleland and Kreevoy, 1994; Frey *et al.*, 1994) and mainly supported by the setting up of new NMR spectroscopic methods able to detect the presence of such strong bonds inside the enzymatic binding site itself. These methods, in particular high-resolution proton chemical shifts (δ) and H/D fractionation factors (ϕ), have been recently reviewed (Harris and Mildvan, 1999) showing that, in fact, they can provide N···O or O···O H-bond lengths with accuracies $\leq$0.05 Å that are much better than those of 0.2–0.8 Å obtainable from protein X-ray crystallography.

Weak H-bonds are normally associated with proton chemical shifts, δ, not exceeding 10–12 ppm but, since in stronger bonds the proton is considerably deshielded by the lengthening of the covalent D–H bond, the signal shifts downfield approaching extreme values that can reach 21–22 ppm in general and somewhat less in enzymes. The deuterium fractionation factor, ϕ, is the equilibrium constant for the exchange of hydrogen with deuterium in D_2O solutions. It measures the fractional preference of the D–H moiety for deuterium over protium relative to solvent, a quantity that, in very weak H-bonds, is nearly one (no preference) while in strong bonds it decreases (preference for protium) reaching the estimated limit of 0.16 in an O–H···O bond with a O···O distance of 2.39 Å (Kreevoy and Liang, 1980) and experimental values of 0.3–0.4 in enzymes. The limiting range of these quantities observed in enzymes (δ = 15–19 ppm and ϕ = 0.4–0.3) consistently indicate strong LBHBs with O···O or N···O contact distances of 2.45–2.55 Å and putative H-bond enthalpies of 10–20 kcal mol^{-1} as estimated by the LS-HB method, which substantially represents the range of properties normally imputed to strong LBHBs.

These results are confirmed by molecular association studies in non-polar solvents but there has been considerable debate on the possibility that such strong bonds could really occur in aqueous solutions. The main difficulty arises from the fact that the studies of H-bonded complexes at room temperature are plagued by chemical exchange that causes the ^1H NMR lines to become broad and averaged among different competitive forms, a problem particularly severe in the presence of water that can exchange with both H-bond donor and acceptor moieties. Many of these problems can be circumvented by experiments at cryogenic temperatures where the most stable complexes can be observed as individual species. Recent ^{1}H NMR studies (Cassidy *et al.*, 1998; Frey, 2001) performed at -50°C on the strong intramolecular (−)CAHBs formed by hydrogen maleate and *cis*-cyclohexane-1,2-dicarboxylate in both aprotic solvents and partially aqueous solutions (90/10, acetone-d_6/H_2O) have measured chemical shifts of 19.0–20.2 ppm, irrespective of the solvent. Since an acetone–water 90:10 mixture contains sufficient water to allow for full solvation of the H-bonded species, these results indicate that H-bond strength is an intrinsic (nominal) property of the bond itself barely or not at all affected by what occurs in the solvent. This, by the way, looks like a substantial validation of the Grunwald and Steel model of enthalpy–entropy compensation in water already discussed in Section 7.2.6.

Strong H-bonds in enzymes are either of (±)CAHB or (−)CAHB type, both requiring a substantial pK_a matching between the H-bond donor and acceptor groups that, in polypeptides, normally means matching between amino acid substituents. For (±)CAHBs, combinations are practically limited to aspartic (pK_a = 3.90) and glutamic (pK_a = 4.30) acids binding to histidine (pK_a = 5.97 in its protonated form). This makes of these three residues, and in particular of histidine, the basic ingredients of all catalytic sites where (±)CAHBs play a functional role. All (−)CAHBs so far studied are O–H···O$^-$ bonds that, when very strong, are better described by the [O···H···O]$^-$ notation. The bond is generated by the previous deprotonation of one of the two –O–H acids and also in this case the value of ΔpK_a = pK_a(acid$_1$) − pK_a(acid$_2$) plays a quite important role, though sometimes underestimated. The strongest bonds are formed by two identical acids (Δp$K_a \equiv 0$) as in aspartic proteases (Chart 8.9C) while, when the acids are different, the bond will be the stronger the more alike the interacting groups are. For instance, in ketosteroid isomerase (Chart C8.9B) the relevant part is the phenolic –O–H group of tyrosine, which could form a strong bond with the ketosteroid itself or with an aspartic acid. Since the former will give rise to a phenol–phenolate and the latter to a phenol–carboxylate bond, it seems reasonable, in sheer terms of pK_a matching, to chose the former as the stronger bond.

A considerable number of enzymatic mechanisms have been interpreted by making recourse to the concept of LBHB that cannot be extensively surveyed here. In the following, the importance of (±)CAHB will be illustrated for chymotrypsin (Chart 8.9A), a well-known example of serine protease, and that of (−)CAHB for ketosteroid isomerase (Chart 8.9B) whose data are mostly taken from two short reviews on the subject (Frey, 2001, 2002; Mildvan *et al.*, 2002). A second example of (−)CAHB in HIV-1 protease, an aspartic protease, is briefly mentioned in view of the elegance of its crystal structure and of its popularity, being presently one of the major targets for therapy against the AIDS epidemic (Chart 8.9C).

In discussing these examples it is important to have a reasonably accurate scale of comparison between NMR chemical shifts, δ, and O···O and N···O distances, d, derived from X-ray crystallography. The recently reported correlation d(O···O) = 5.04 − 1.16 lnδ + 0.0447 δ (Harris and Mildvan, 1999), being derived from the study of 59 O–H···O bonds in small molecules, fits well O···O data but gives systematically too short distances when applied to the N···O bond. The problem has been reconsidered here making use of a previous δ_{NH} versus $d(\text{N}\cdots\text{O})$ correlation for intramolecular N–H···O bonds having average N–H–O angle of 146(3)° (Gilli *et al.*, 2000), which can be extrapolated to the present intermolecular case by taking the sum $d(\text{N–H}) + d(\text{H}\cdots\text{O})$. Final data are summarized in Table 8.1.

8.5.2.2 *The catalytic triad of serine proteases*

Serine proteases are a wide family of proteolitic enzymes (such as trypsin, chymotrypsin, elastase, subtilisin, etc.) that catalyse the hydrolysis of proteins into smaller polypeptides. Each enzyme can cut the chain in a specific position and, for instance, chymotrypsin (the example treated in Chart 8.9A) is specialized in cleaving it at aromatic aminoacyl positions.

The active site of serine proteases includes three residues that participate directly in the catalysis (the so-called catalytic triad), i.e. serine, histidine, and aspartate that are indicated as Ser195, His57 and Asp102 in the catalytic site of chymotrypsin sketched in C8.9A.Ia–c and as Ser221, His64 and Asp32 in that of *Bacillus lentus* subtilisin determined by X-ray diffraction at 100 K and 0.78 Å resolution (Fig. 8.1). The pK_as of the three aminoacids are summarized in **8.1.I** and **II** where histidine values are also compared with those of its non-substituted parent imidazole. This latter has two constants, pK_{a1} = 14.5 and pK_{a2} = 7.00, corresponding to its neutral and protonated forms, respectively, and these values are only slightly shifted for histidine to pK_{a1} = 12–13 and pK_{a2} = 6.00. Comparison with the pK_a of 3.90 of aspartic acid suggests that this acid could form only a rather weak H-bond with histidine ($\Delta pK_a \approx -8.5$) but quite a strong one, most probably of LBHB type, with protonated histidine which is nearly pK_a matched ($\Delta pK_a \approx -2.1$). This agrees with the

TABLE 8.1. Intermolecular O···O and N···O H-bond distances (Å) as a function of the NMR ^{1}H chemical shift δ, as respectively derived from Harris and Mildvan (1999) and Gilli *et al.* (2000).

δ_{OH} or δ_{NH} (ppm)	$d(\text{O}\cdots\text{O})$ (Å)	$d(\text{N–H})+d(\text{H}\cdots\text{O})$ (Å)
19	2.47	2.53
18	2.49	2.56
17	2.51	2.59
16	2.53	2.63
15	2.56	2.66
14	2.60	2.70
13	2.65	2.73
12	2.69	2.76
11	2.75	2.80
10	2.81	2.83

CHART 8.9A. (±)CAHB: THE CATALYTIC TRIAD OF SERINE PROTEASE

THE CATALYTIC TRIAD

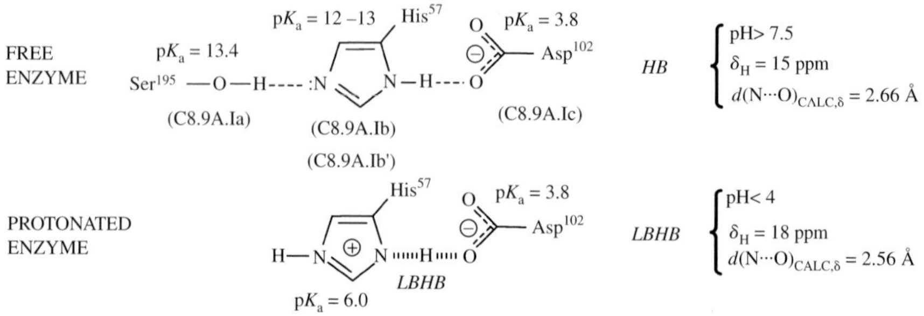

FREE ENZYME

$pK_a = 13.4$ Ser[195] —O—H---- :N ... $pK_a = 12-13$ His[57] ... N—H--- O ... O $pK_a = 3.8$ Asp[102]

(C8.9A.Ia) (C8.9A.Ib) (C8.9A.Ic)

HB { pH> 7.5; $\delta_H = 15$ ppm; $d(N\cdots O)_{CALC,\delta} = 2.66$ Å

(C8.9A.Ib')

PROTONATED ENZYME

H—N ⊕ N ııııHıııı O ... His[57] ... O $pK_a = 3.8$ Asp[102]

LBHB

$pK_a = 6.0$

LBHB { pH< 4; $\delta_H = 18$ ppm; $d(N\cdots O)_{CALC,\delta} = 2.56$ Å

THE PROPOSED MECHANISM OF ENZYME ACTION

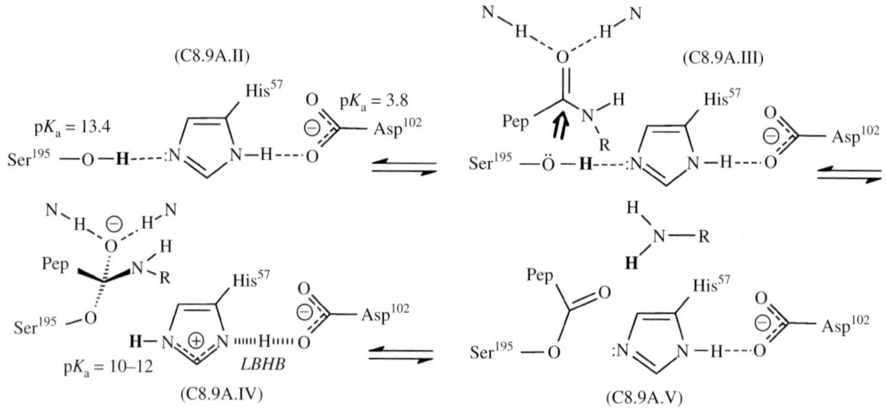

(C8.9A.II) (C8.9A.III)

(C8.9A.IV) (C8.9A.V)

STRUCTURE AND PROPERTIES OF THE TRANSITION STATE

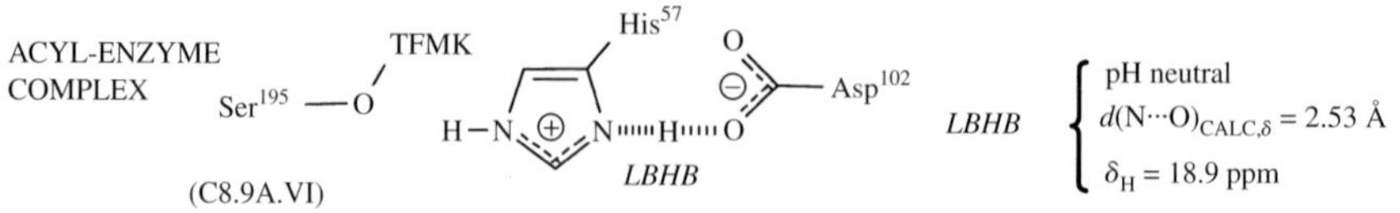

ACYL-ENZYME COMPLEX

Ser[195] —O—TFMK ... H—N ⊕ N ııııHıııı O ... His[57] ... O Asp[102]

LBHB

(C8.9A.VI)

LBHB { pH neutral; $d(N\cdots O)_{CALC,\delta} = 2.53$ Å; $\delta_H = 18.9$ ppm

first high-resolution ^{1}H NMR experiments on chymotrypsin (C8.9A.I; Robillard and Shulman, 1974) where the proton was found to resonate at 15 ppm at pH above 7.5 (C8.9A.Ib) and 18 ppm at pH below 4 (C8.9A.Ib′). The two downfield resonances indicate N—H···O bonds with respective N···O distances of 2.66 and 2.56 Å (as evaluated from Table 8.1) and the more shifted (18 ppm) was firstly classified as LBHB in 1994 (Frey *et al.*, 1994).

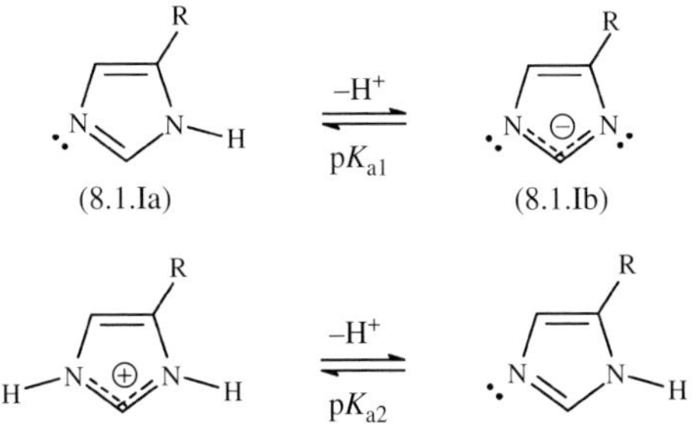

(8.1.Ia) $\underset{pK_{a1}}{\overset{-H^+}{\rightleftharpoons}}$ (8.1.Ib)

(8.1.IIa) $\underset{pK_{a2}}{\overset{-H^+}{\rightleftharpoons}}$ (8.1.IIb)

Bases:	pK_{a1}	pK_{a2}
Imidazole (R = H)	14.5	7.00
Histidine (R = R')	12–13	6.00
Acids:		
Aspartic acid (R'–COOH)	3.90	
Serine (R'–OH)	≈13.5	

R' = $-CH_2-CH(COO^-)NH_3^+$

The chemical mechanism by which chymotrypsin and other serine proteases hydrolyse acyl compounds can be considered, to a first approximation, a two-step process. In the first, the peptide bond is cleaved by reaction of serine with the acyl carbonyl group of the substrate to form a tetrahedral covalent acyl-enzyme intermediate with subsequent release of the amine product, while in the second step the acyl-enzyme undergoes hydrolysis releasing the acid product. Only the first step needs to be discussed here because it is supposed to involve a LBHB in its mechanism of action (C8.9A.II–V) with the double function of catalysing serine deprotonation and lowering the energy of the transition state (Frey, 2001, 2002). Also chymotrypsin acylation (C8.9A.III – C8.9A.IV) is a two-step process involving a metastable tetrahedral intermediate resulting from the addition of Ser195 to the carbonyl group of the substrate. The acylation proceeds via acid–base catalysis by His57, in which Asp102 is postulated to assist by stabilizing the tetrahedral addition complex through LBHB formation with His57.

The main mechanistic point is that serine, by itself, is not very reactive as a nucleophile and its action must therefore be assisted by the transfer of its O–H proton to the near nitrogen of histidine (C8.9A.IV). However, as shown in C8.9A. Ib′, the pK_a of the protonated His57 assumes a value of nearly 6, which is too low to permit the efficient transfer of the proton from serine having a pK_a of 13.4. The role of the strong Asp-His LBHB would be therefore that of modulating the pK_a of the N–H group between its extremes (14 and 7 making reference to imidazole) in such a way it can assume the correct value of, say, 10–12 which would allow histidine to be strong enough as a base to abstract a proton from Ser195 (C8.9A.IV) but not too strong to hinder the protonation of the leaving amino group in the following step of the mechanism (C8.9A.V).

Tetrahedral transition states can be stabilized by reaction with suitable inhibitors. The peptidyl trifluoromethylketones (TFMK) are good inhibitors of chymotrypsin and form tetrahedral addition complexes that structurally resemble the transition state formed by the substrate, as shown by the TFMK complex C8.9A.VI. The measured ^{1}H NMR chemical shift, $\delta = 18.9$ ppm, corresponds, according to Table 8.1, to an N···O distance of 2.53 Å, which is only slightly

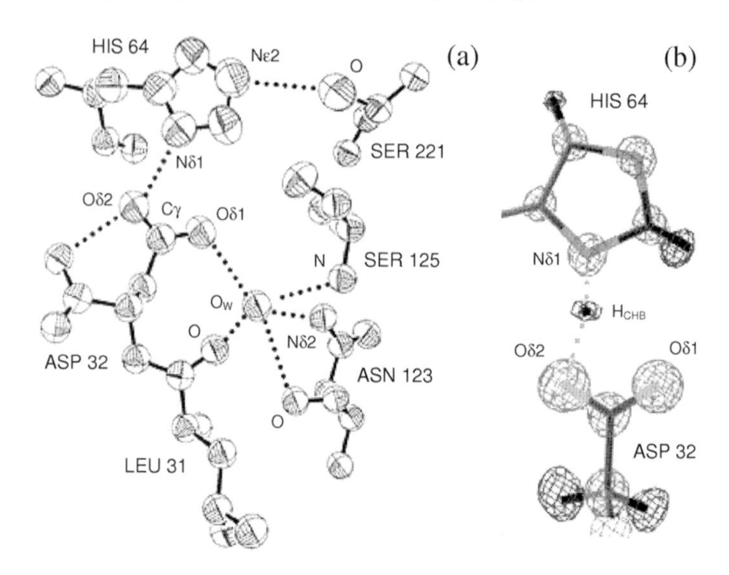

FIG. 8.1. The structure of the serine protease *Bacillus lentus* subtilisin as determined by X-ray diffraction at 100 K and 0.78 Å resolution. (a) The three Ser221, His64, and Asp32 residues building up the catalytic triad are interlinked by two H-bonds having quite different N···O distaces, 2.62 Å for His-Asp and 3.10 Å for Ser-His; (b) the difference-Fourier electron density map has identified a small peak at 1.2 Å from Nδ1 and 1.5 Å from Oδ2 which can be associated to the proton of the N–H···O bond. (Reproduced by permission from Kuhn *et al.*, 1998)

longer than the shortest N···O value of 2.506 Å obtained from a sample of 305 small-molecule structures of good accuracy (Table 3.9). All data seem, therefore, to point unequivocally to the real existence of these very strong bonds that have been called LBHBs in enzymes.

8.5.2.3 *Δ^5-3-Ketosteroid isomerase*

Ketosteroid isomerase (Chart 8.9B) catalyses the conversion of Δ^5- or Δ^4-3-ketosteroids using Asp38 as a general base and both Tyr14 and Asp99 as general acids. The mechanism of action displayed is consistent with the X-ray crystal structure at 2.5 Å resolution of the enzyme complexed with equilenin, an analogue of the dienolate intermediate (Kim *et al.*, 1997), and with the low-temperature ^{1}H NMR measurements on similar complexes with both estradiol and dihydroequilenin that revealed two deshielded resonances at 18.15 and 11.60 ppm having corresponding fractional factors, ϕ, of 0.34 and 0.97 (Zhao *et al.*, 1997). These results have been recently surveyed (Mildvan *et al.*, 2002) and relevant data are summarized in the caption of Chart 8.9B. The resonance at 11.60 ppm displays the characteristics of a normal H-bond donated by a carboxyl with an O···O distance of 2.68–2.72 Å, as evaluated from chemical shift

CHART 8.9B. (–) CAHB: THE MECHANISM OF Δ^5-3-KETOSTEROID ISOMERASE

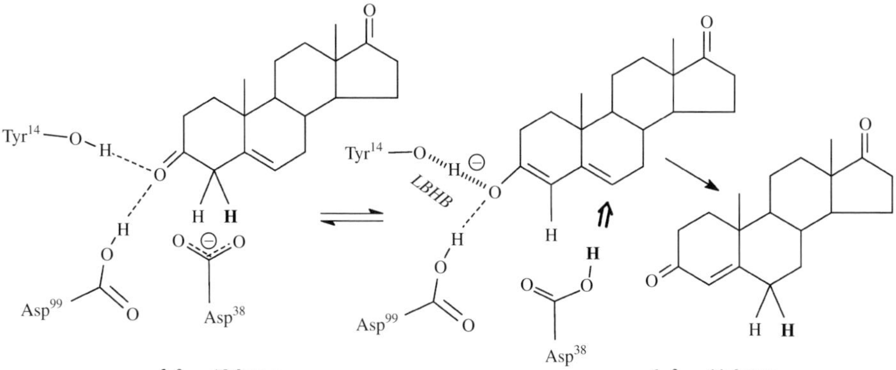

Tyr[14]···⁻O–C *LBHB*: $\delta_H = 18.2$ ppm; $\phi = 0.34$; $d(O\cdots O)_{CALC,\delta} = 2.49_2$ Å; $d(O\cdots O)_{CALC,\phi} = 2.50_1$ Å; $d(O\cdots O)_{EXP,X} = 2.58_{23}$ Å

Asp[99] ··· ⁻O–C *HB*: $\delta_H = 11.6$ ppm; $\phi = 0.97$; $d(O\cdots O)_{CALC,\delta} = 2.72_2$ Å; $d(O\cdots O)_{CALC,\phi} = 2.68_2$ Å; $d(O\cdots O)_{EXP,X} = 2.62_{24}$ Å

and H/D fractional factor. It has been assigned to the Asp99···⁻O–C interaction. The 18.2 ppm resonance corresponds to an estimated O···O distance of 2.49–2.50 Å, which is typical of (–)CAHBs endowed with a DW PT profile (that is a typical LBHB), and has therefore been imputed to the Tyr14···⁻O–C complex. This is not the place for discussing mechanistic implications. The relevant point is rather how modern NMR techniques make it possible to measure H-bond distances in enzymes with an accuracy unthinkable for macromolecular structures and not far from that obtainable from small-molecule crystallography. It seems possible that these new experimental achievements will produce considerable advances in the study of the H-bond in the macromolecular field, not differently from those produced 30 years ago by neutron diffraction in the field of small molecules.

8.5.2.4 *An aspartic protease: HIV-1 protease*

Aspartic proteases (Northrop, 2001) have the longest history in enzymology since rennet, and later chymosin, have been used for thousands of years in the making of cheese. Moreover, pepsin, usually considered the leader in this enzymatic family, was recognized as the active principle of gastric juice that dissolves meat in 1783 and was given a name in 1825. Another enzyme of the family, nepenthesin, was detected in insect-eating plants by Darwin in 1875 and Sörensen invented pH in 1909 to plot pepsin activity against hydrogen ion concentration. In spite of their long history, the mechanism of action of these enzymes has been one of the less understood aspects since the proposal made by Piana and Carloni (2000) that their active site is centered on an aspartic acid–aspartate LBHB of (–)CAHB type with an estimated O···O distance of

CHART 8.9C. (–)CAHB: HIV-1 PROTEASE

(C8.9C.I)

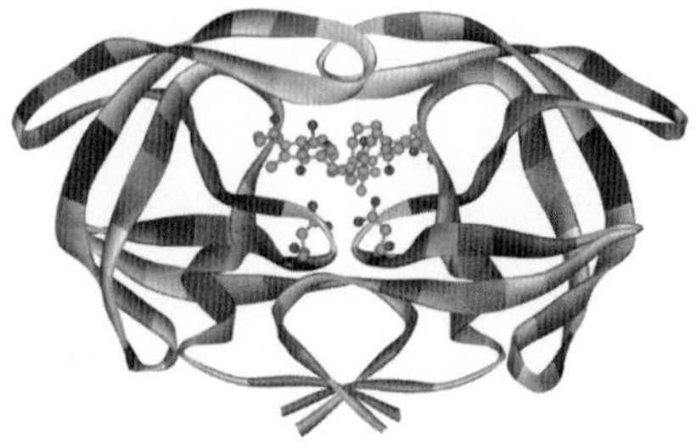

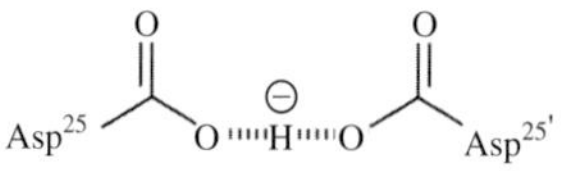

(–)CAHB
Negative charge-assisted HB
(SWHB = single-well or
LBHB = double-well)

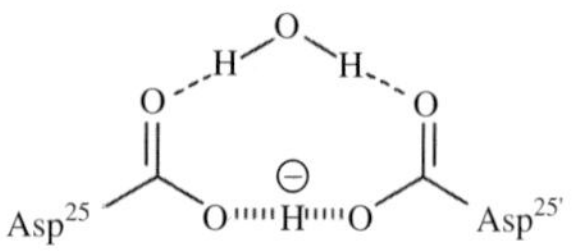

(C8.9C.IIa)

(C8.9C.IIb)

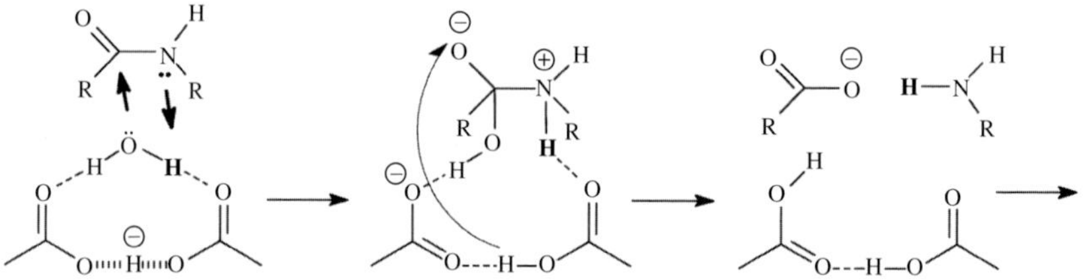

(C8.9C.IIIa)

(C8.9C.IIIb)

(C8.9C.IIIc)

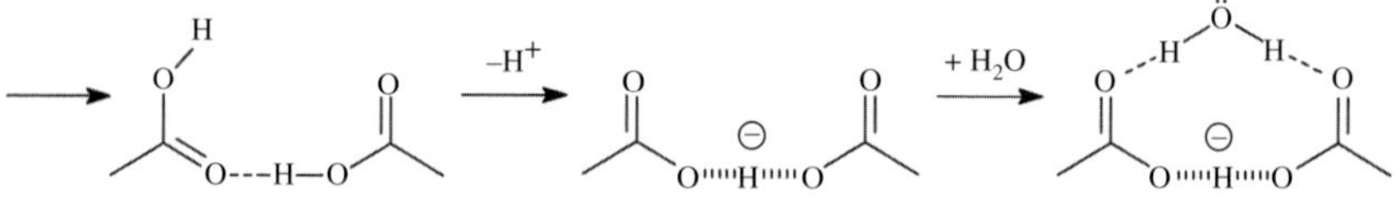

(C8.9C.IIIc′)

(C8.9C.IIId)

(C8.9C.IIIe)

2.5 ± 0.1 Å. This proposal, which does not arise from new experiments but from *ab initio* molecular dynamics simulation performed on HIV-1 protease, promises to become the new unifying hypothesis for understanding the common mechanism of action in this class of enzymes.

HIV-1 protease (human immunodeficiency virus protease) is an aspartic protease that performs an essential step in the life cycle of HIV, the virus that is the etiologic agent of the AIDS epidemic. Though unknown a few decades ago, it is now one of the most studied enzymes because its inhibitors can block virus maturation and have then become an important class of drugs in AIDS treatment. The first crystal structure dates back to 1990 (Fitzgerald *et al.*, 1990) and, a decade later, over one hundred structures were available in the PDB, including several genetic strains and mutants together with complexes with many drugs and inhibitors. One of these complexes is displayed in C8.9C.I, showing that HIV-1 protease is a small enzyme composed of two identical chains (each only 99 aminoacids long) that assemble to form a long tunnel where the enzyme wraps around the polypeptide chain to be cleaved (here substituted by the inhibitor molecule). The two short chains are connected by a strong Asp-25$\cdots$Asp-25$'$ H-bond (C8.9C.IIa) binding a water molecule used to break (hydrolyse) the substrate chain (C8.9C.IIb).

The mechanism of action of HIV-1 protease and its inhibition by pepstatins have been the object of accurate *ab initio* molecular dynamics simulations (Piana and Carloni, 2000; Cavalli *et al.*, 2006), which cannot be discussed here in detail. A simplified mechanism of action, adapted from a recent review by Northrop (2001), is shown in C8.9C.IIIa–e. The reaction center is represented by the complete H-bonded 10-member ring built up by the aspartic-acid–aspartate–water molecule complex C8.9C.IIb: during the reaction this very water breaks and adds to the substrate causing the hydrolysis of the peptidic bond. The reaction is started by the approaching of the peptide (C8.9C.IIIa) that appears to trigger the counterclockwise motion of an electron (the negative charge) from inside the $[O\cdots H\cdots O]^-$ CAHB to the $C{-}O^-$ of the tetrahedral intermediate (C8.9C.IIIb), and finally to the leaving carboxylate group (C8.9C.IIIc). After dissociation of the reaction products the aspartic acid dimer C8.9C.IIIc is deprotonated, leading to the formation of the aspartic acid–aspartate LBHB (C8.9C.IIId) and, with the final addition of a molecule of water, to the rebuilding of the original reaction center (C8.9C.IIIe).

Energetics and proton tunnelling of this mechanism are far from being completely understood and the different hypotheses certainly cannot be analyzed here. An important fact deserves, however, to be briefly discussed. It has been sometimes stated that LBHBs lower the activation barrier of enzymatic reactions by lowering, because of their strength, the energy of the activated complex. This might also be true in other enzymes but certainly does not hold for this particular one because the mechanism shown in C8.9C.IIIa–e does actually imply LBHB formation but, here, this bond should *increase* and not *lower* the activation barrier because it is *lost* and not *formed* along the reaction pathway. It is clear that things are more complicated than previously imagined and that, although it is certainly true that *highly energetic H-bonds* play a *functional role* in some enzymatic mechanisms, it may be uncertain how and why they do. In HIV-1 protease, for instance, the main role played by the $[O\cdots H\cdots O]^-$ LBHB could well be that of easing the deprotonation of the aspartic acid dimer to restore the reaction center at the end the reaction.

8.6 Σ-bond cooperativity and anticooperativity in PAHBs

8.6.1 *Cooperative and anticooperative water chains*

The concepts of *σ-bond cooperative* (Jeffrey and Saenger, 1991) and *polarization-assisted H-bond (PAHB)* (Gilli and Gilli, 2000) have already been introduced

CHART 8.10. PAHB & (+)CAHB: PROTON TRANSMISSION IN WATER

Σ-BOND COOPERATIVE OR POLARIZATION-ASSISTED HB (PAHB)

```
 R              R              R              R
  \              \              \              \
----O-H----O-H----O-H----O-H----O-H----O-H----O-H-----
          /              /              /
         R              R              R
```

(C8.10.Ia)

Homodromic σ-bond cooperative chain of water molecules

```
      H              H              H               H
     /              /                \               \
-----H—O----H—O----H—O----H    H----O-H----O-H----O-H-----
          \               \O/            /
           H                            H
```

(C8.10.Ib)

Breaking of the homodromic chain in two *antidromic* chains by action of a single point of σ-bond anticooperativity (in bold)

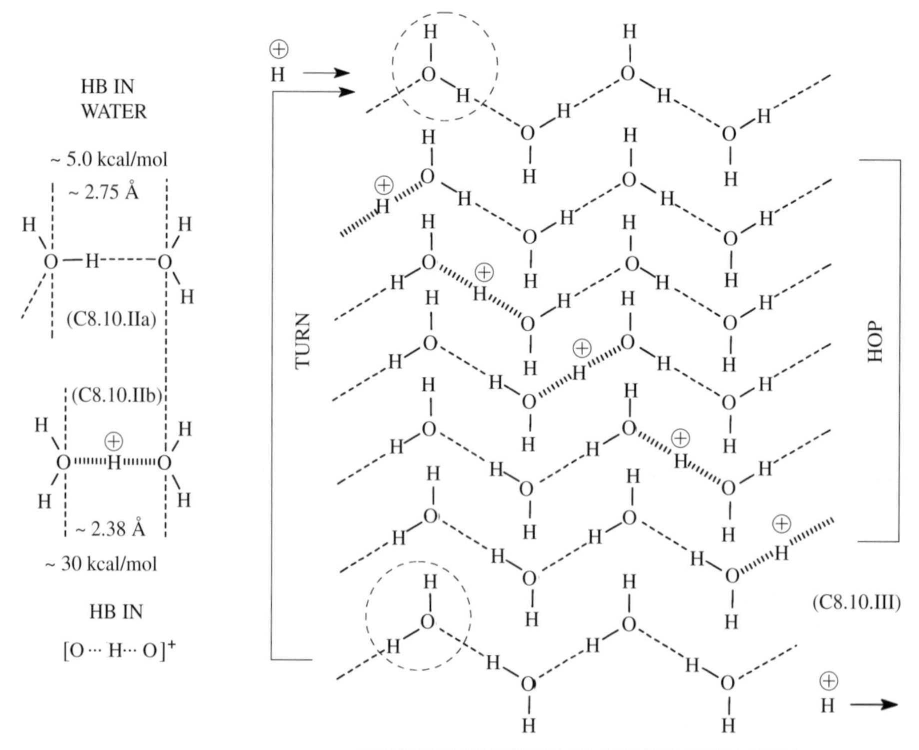

THE '*HOP AND TURN*' OR '*RELAY*' MECHANISM
C.J.T. de Grotthuss, *Ann. Chim.* 58: 54–74 (1806)

in Sections 3.2.1 and 3.3.1.1F and both make reference to the ability of water (or alcohols or phenols) to form infinite chains (or rings or segments of chains) of ···O–H···O–H··· bonds significantly stronger than usual because of the mutual σ-polarization induced on the O–H bond by the multiple H-bonds running along the chain. It has been estimated that PAHB chains can provide H-bond-energy enhancements of 20–45% with respect to the dimer, associated with an O···O shortening from 2.75 to 2.62 Å. The chains appear in two basic forms, the fully *homodromic* σ-bond cooperative chain of iso-oriented molecules (C8.10.Ia) and the sum of two *antidromic* half-chains (C8.10.Ib) that are σ-bond cooperative in opposite directions and linked by a single point of σ-bond anti-cooperativity marked in bold. In the following, it will be shown that both forms may play a definite functional role, though of opposite type.

It is intuitive, and was in fact remarked from the first H-bond studies, that homodromic chains of water molecules resemble electronic wires where conductivity is granted by proton (instead of electron) transfer and that this hypothesis might well explain the anomaly that protons have a limiting ionic conductivity in water ($\lambda = 34.96$ mS m^2 mol^{-1} at 298 K) six times greater than that of any other cation, an anomaly which seems to be due to a *naked proton* because it violates the general rule that ions of smaller radius are less conductive because of their greater hydration sphere.

^1H and ^{17}O NMR studies have shown that the time of proton transfer from a molecule to the next is about 1.5 ps, a value compatible with the time, measured by inelastic neutron scattering, taken by a water molecule to reorientate itself through about 1 rad. This observation contributes to focus the attention on the famous *Grotthuss mechanism* (C8.10.II and III) that derives its name from a very early hypothesis that chains of dipoles can explain the transport of charge in water (de Grotthuss, 1806). This mechanism (also called the *hop-and-turn* or *relay* mechanism) consists of a sequence of elementary proton hops that move the proton from left to right (the hop) followed by a reorientation of all molecules in the chain (the turn) that restores the original configuration of the chain. The need for a reorientation comes from the fact that the proton position is a point of σ-bond anticooperativity (in the sense of Scheme C8.10.Ib) whose movement to the right leaves back a completely inverted chain. Since the turning step is slower than the hopping one, the former comes to be the limiting factor of the global proton-transfer rate.

The Grotthuss mechanism is believed to be the essence of proton conduction in ice (Hückel, 1928; Bernal and Fowler, 1933). In liquid water the situation is more complex because the H-bonded network solvating the hydronium ion, H_3O^+, can exert a relevant influence on the rate of proton mobility. The subject has recently given rise to renewed interest (Agmon, 1995; Cukierman, 2006) particularly in view of the growing role played by *ab initio* molecular dynamics methods (Tuckerman *et al.*, 1995). It cannot be said, however, that the problem has found a unified solution, the debate being always focused on the nature of the solvated species that actually participate in the transfer of the proton, particularly the

'Zundel' $H_5O_2^+$ (Zundel, 1976) and 'Eigen' $H_9O_4^+$ (Eigen *et al.*, 1964) cations that could combine in Eigen–Zundel–Eigen or Zundel–Zundel transformations according to rules that become too specialist to be considered here.

8.6.2 *An example of cooperativity: The gramicidin A channel*

The idea that water wires can provide electric conductivity by fast proton transmission is not limited to ice and water and, in a recent review by Ball (2008), a considerable number of different phenomena are listed that can be interpreted in terms of this conduction mechanism. They include water wires in the

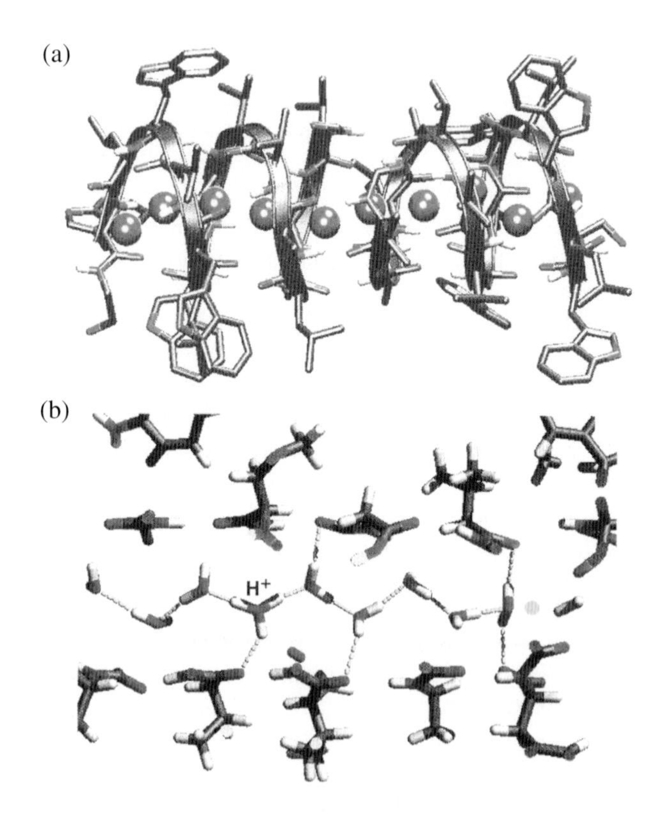

FIG. 8.2. (a) β-helix structure of the gramicidin A dimer. The narrow cylindrical pore formed by two molecules of the polypeptide can accommodate a single-file chain of water molecules depicted as larger spheres; (b) Snapshot from the molecular dynamics simulation showing a representative conformation of the H-bond network of the protonated water wire in gramicidin. The excess proton is in the form of a hydronium H_3O^+ ion whose motion causes the progressive inversion of the H-bonded chain in agreement with the Grotthuss mechanism. (Reproduced by permission from Pomès and Roux, 2002)

photosynthetic reaction center of *Rhodobacter sphaeroides*, proton conduction in the catalytic mechanism of carbonic anhydrase and cytochrome *c* oxidase, proton motion through the pore-forming membrane peptide gramicidin A, water wires in the transmembrane proton pump bacteriorhodopsin and, finally, proton mobilities along one-dimensional water chains inside carbon nanotubes that can become 40 times greater than in bulk water. Most of these contributions are quite recent, the field being still in rapid growth due to the concomitant advances of macromolecular crystallography and molecular-dynamics simulation methods.

The example discussed here concerns gramicidins, a class of peptide antibiotics that are active against gram-positive bacteria by forming ion-conducting channels through the bacterial membrane. Gramicidin S (Llamas-Saiz *et al.*, 2007)

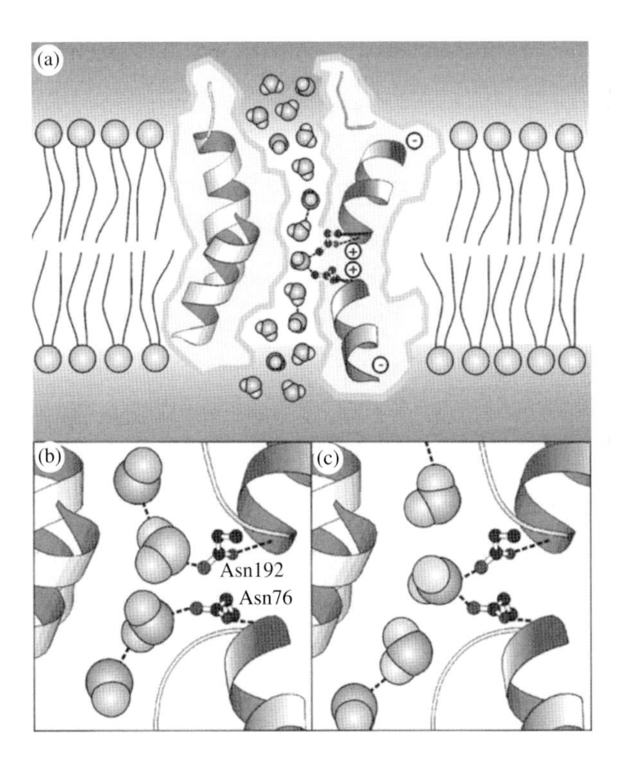

FIG. 8.3. Schematic representation of the structure of aquaporin-1 embedded in the cell membrane and its mechanism for blocking proton permeation. (a) Diagram showing how partial charges from the helix dipoles restrict the orientation of the chain of water molecules passing through the pore. (b, c) Reorientation of the water molecule at the center of the chain by simultaneous H-binding to the Asn76 and Asn192 residues. (Reproduced by permission from Murata *et al.*, 2000)

is a cyclodecapeptide, built up by two identical pentapeptides joined head to tail in a cyclic way, while gramicidin A (the compound considered here) is a pentadecapeptide consisting of alternating L- and D-aminoacids arranged in a β-helix. There has been considerable debate on the real structure of gramicidin A (Burkhart *et al.*, 1998; Duax *et al.*, 2003) and, to simplify the matter, Fig. 8.2(a) reports a drawing of such a structure by making reference to the rather idealized model used in its molecular-dynamics simulation (Pomès and Roux, 2002; Roux, 2002). The active form of gramicidin A is seen to be a head-to-head homodimer of two pentadecapeptides having a β-helix structure forming inside a cylindrical pore that can accommodate either a single-file chain of water molecules or a chain of alternating waters and cations (depicted in the figure as larger spheres). Fig. 8.2(b) is a step of the dynamic simulation showing a representative conformation of the H-bonded water chain while crossed by a single proton. The excess proton is in the form of a hydronium H_3O^+ ion that introduces a point of σ-bond anticooperativity in the chain and whose motion

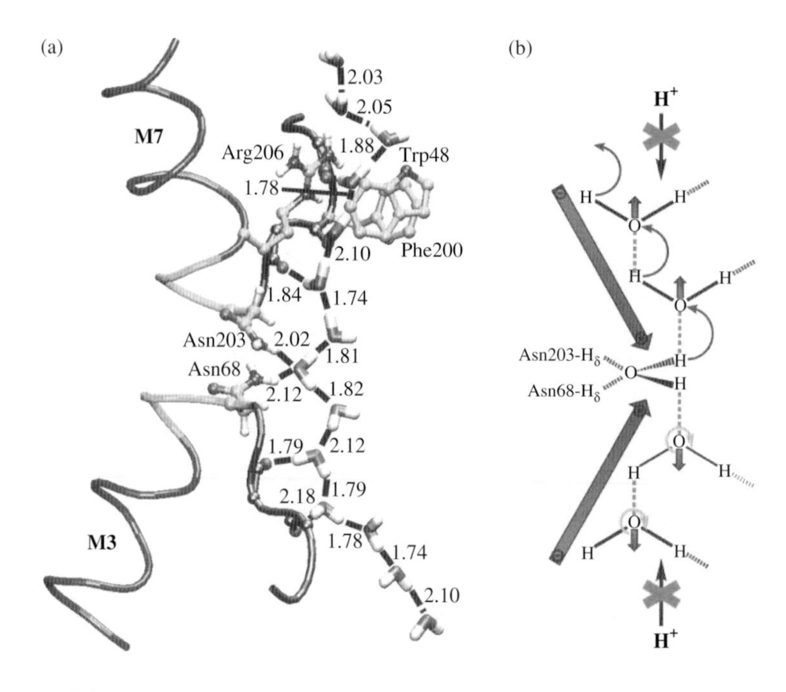

FIG. 8.4. (a) Snapshot from the molecular-dynamics simulation of *Escherichia coli* aquagliceroporin revealing the antidromic orientation of the two half-chains of H-bonded waters which precludes proton transmission. Water rotation is caused by the double H-bond donated by Asn68 and Asn203 to the central water molecule. (b) Schematic representation of the proton-transmission preclusion acting in both directions of the chain. (Reproduced by permission from Tajkhorshid *et al.*, 2002)

produces the progressive inversion of the H-bonded chain, in agreement with the Grotthuss mechanism.

8.6.3 *An example of anticooperativity: Water-without-proton transmission in aquaporin channels*

The existence of membrane proteins deputed to form water-specific membrane channels has been postulated for several decades but their identity remained unknown until the serendipitous discovery in red blood cells of the first molecular water channel, called *aquaporin*-1 *(AQP*1*)* (Preston *et al.*, 1992). In the following years several other aquaporin channels were identified in both animals and plants. According to a recent review (King *et al.*, 2004) eleven mammalian aquaporins are known, forming a complete family that can be divided on the basis of their permeability characteristics in true aquaporins, that are only permeable to water (7 members), and aquaglyceroporins that are permeable to water but also to other small molecules, particularly glycerol (4 members). Members of the aquaporin family have been so far found in archea, eubacteria and eucaryotes, including fungi, animals, and plants where they serve an astonishing variety of physiological functions and are easily identified by their sequence similarities across all kingdoms of life (Törnroth-Horsefield *et al.*, 2006).

The specific characteristic that all aquaporins must be endowed with is easily understood from the property of gramicidin A of killing gram-positive bacteria by injecting protons through their cell membrane. Conversely aquaporins, which act as water channels controlling the normal physiological flow of water molecules through cell membranes, must be endowed with a transmission mechanism that, though providing a complete water permeability, hinders the transfer of ions, and particularly protons, because any even small change of the electrochemical potential across the cell membrane would kill the cell itself. How to do that? The answer found by nature is straightforward: by forcing the introduction of a single point of σ-bond anticooperativity along the water chain and so breaking the chain into two antidromic halves, an arrangement illustrated in C8.10.Ib that allows water flux in either directions and proton flux in neither because charged particles can travel only in opposite directions along the two halves of the chain.

The recent X-ray crystal structures of *Escherichia coli* aquaglyceroporin (Fu *et al.*, 2000) and human aquaporin-1 (Sui *et al.*, 2001) show that both aquaporins are symmetric tetramers of four channels, each tetramer being circled by a hydrophobic surface having the thickness of the lipid bilayer in which the tetramer is vertically embedded in the cell wall. Each of the four parallel channels consists of six transmembrane and two half-membrane-spanning α-helices that form a right-handed helical bundle around the channel. A section of the channel of aquaporin-1 embedded in the cell membrane is shown in Fig. 8.3(a). The section is through a transmembrane α-helix (on the left) and the interrupted helix (on the right). The two halves of the latter meet in the center of the membrane at their NH_2-terminal ends in correspondence of residues Asn76 and 192 (Asn68 and 203 in aquaglyceroporin). These two groups orient a water molecule in anticooperative fashion creating the defect that inverts the direction of the cooperative H-bonds in the lower part of the chain (Fig. 8.3(b,c)). A more detailed representation of the H-bonded water chain that inverts its dromicity in correspondence of the Asp couple is shown in Fig. 8.4(a), which represents a snapshot from the molecular dynamics simulation of *Escherichia coli* aquaglyceroporin (Tajkhorshid *et al.*, 2002). Fig. 8.4(b) illustrates, in an even more schematic way, the resulting blockage of proton transmission while the water flux remains unaltered.

REFERENCES

Abdur-Rashid, K., D.G. Gusev, S.E. Landau, A.J. Lough, and R.H. Morris (1998). Organizing chain structures by use of proton-hydride bonding. The single-crystal X-ray diffraction structures of $[K(Q)][Os(H)_5(P^iPr_3)_2]$ and $[K(Q)][Ir(H)_4(P^iPr_3)_2]$, Q = 18-crown-6 and 1,10-diaza-18-crown-6. J Am Chem Soc 120: 11826–11827.

Abdur-Rashid, K., D.G. Gusev, A.J. Lough, and R.H. Morris (2000). Intermolecular proton-hydride bonding in ion pairs: Synthesis and structural properties of $[K(Q)][MH_5(P^iPr_3)_2]$ (M = Os, Ru; Q = 18-crown-6, 1-aza-18-crown-6, 1,10-diaza-18-crown-6). Organometallics 19: 834–843.

Agmon, A. (1995). The Grotthuss mechanism. Chem Phys Lett 244: 456–462.

Alder, R.W. (1990). Intrabridgehead chemistry. Tetrahedron 46: 683–713.

Alkorta, I., I. Rozas, and J. Elguero (1998). Non-conventional hydrogen bonds. Chem Soc Rev 27: 163–170.

Allen, F.H. (2002). The Cambridge Structural Database: a quarter of a million crystal structures and rising. Acta Crystallogr B58: 380–388.

Allen, F.H., S. Bellard, M.D. Brice, B.A. Cartwright, A. Doubleday, H. Higgs, T. Hummelink, B.G. Hummelink-Peters, O. Kennard, W.D.S. Motherwell, J.R. Rodgers, and D.G. Watson (1979). The Cambridge Crystallographic Data Centre: computer-based search, retrieval, analysis and display of information. Acta Crystallogr B35: 2331–2339.

Allen, F.H., C.M. Bird, R.S. Rowland, and P.R. Raithby (1997). Resonance-induced hydrogen bonding at sulfur acceptors in $R_1R_2C{=}S$ and $R_1CS_2^-$ systems. Acta Crystallogr B53: 680–695.

Allen, F.H., O. Kennard, and D.G. Watson (1994). Crystallographic databases: Search and retrieval of information from the Cambridge Structural Data Base. In: *Structure correlation*, Vol. 1. Edited by H.-B. Bürgi and J.D. Dunitz. VCH: Weinheim, D; Chapter 3, pp. 71–110.

Allen, F.H., O. Kennard, D.G. Watson, L. Brammer, A.G. Orpen, and R. Taylor (1987). Tables of bond lengths determined by X-ray and neutron diffraction. Part 1. Bond lengths in organic compounds. J Chem Soc Perkin Trans 2: S1–S19.

Anderson, P.W. (1972). More is different. Broken symmetry and the nature of the hierarchical structure of science. Science 177: 393–396.

Antipin, M.Yu., A.E. Kalinin, Yu.T. Struchkov, E.I. Matrosov, and M.I. Kabachnik (1980). Hydrogen bond in the $[Ph_3P{:}O\cdots H\cdots O{:}PPh_3]^+ClO_4^-$ crystal of a molecular complex of triphenylphosphine oxide with perchloric acid of 2:1 composition at room and low (-120°) temperatures. Kristallografiya (Russ.) 25: 514–525.

Applequist, J., and T.G. Mahr (1966). The conformation of poly-L-tyrosine in quinoline from dielectric dispersion studies. J Am Chem Soc 88: 5419–5429.

Åqvist, J., and A. Warshel (1993). Simulation of enzyme reactions using valence bond force fields and other hybrid quantum/classical approaches. Chem Rev 93: 2523–2544.

Arduengo, A.J. III, S.F. Gamper, M. Tamm, J.C. Calabrese, F. Davidson, and H.A. Craig (1995). A *bis*(carbene)-proton complex. Structure of a C–H–C hydrogen bond. J Am Chem Soc 117: 572–573.

Arnett, E.M., and E.J. Mitchell (1971). Hydrogen bonding. VI. A dramatic difference between proton transfer and hydrogen bonding. J Am Chem Soc 93: 4052–4053.

Asakawa, N., S. Kuroki, H. Kurosu, I. Ando, A. Shoji, and T. Ozaki (1992). Hydrogen-bonding effect on ^{13}C NMR chemical shifts of L-alanine residue carbonyl carbons of peptides in the solid state. J Am Chem Soc 114: 3261–3265.

Atwood, J.L., S.G. Bott, C.M. Means, A.W. Coleman, H. Zhang, and M.T. May (1990). Synthesis of salts of the hydrogen dichloride anion in aromatic solvents. 2. Syntheses and crystal structures of [K·18-crown-6][Cl–H–Cl], [Mg·18-crown-6] $[Cl–H–Cl]_2$, $[H_3O$·18-crown-6][Cl–H–Cl], and the related $[H_3O$·18-crown-6] [Br–H–Br]. Inorg Chem 29: 467–470.

Atwood, J.L., J.E.D. Davies, and D.D. MacNicol (1984). *Inclusion compounds*. Vols. 1–3. Academic Press: London, UK.

Atwood, J.L., J.E.D. Davies, and D.D. MacNicol (1991). *Inclusion compounds*. Vols. 4–5. Oxford University Press: Oxford, UK.

Atwood, J.L., G.A. Koutsantonis, F.-C. Lee, and C.L. Raston (1994). A thermally stable alane-secondary amine adduct: $[H_3Al$(2,2,6,6-tetramethylpiperidine)]. J Chem Soc Chem Commun 91–92.

Auf der Heyde, T.P.E., and H.-B. Bürgi (1989a) Molecular geometry of d^8 five-coordination. 1. Data search, description of conformation, and preliminary statistics. Inorg Chem 28: 3960–3969.

Auf der Heyde, T.P.E., and H.-B. Bürgi (1989b) Molecular geometry of d^8 five-coordination. 2. Cluster analysis, archetypal geometries, and cluster statistics. Inorg Chem 28: 3970–3981.

Auf der Heyde, T.P.E., and H.-B. Bürgi (1989c) Molecular geometry of d^8 five-coordination. 3. Factor analysis, static deformations, and reaction coordinates. Inorg Chem 28: 3982–3989.

Aullón, G., D. Bellamy, L. Brammer, E.A. Bruton, and A.G. Orpen (1998). Metal-bound chlorine often accepts hydrogen bonds. Chem Commun 653–654.

Ault, B.S., E. Steinback, and G.C. Pimentel (1975). Matrix isolation studies of hydrogen bonding. The vibrational correlation diagram. J Phys Chem 79: 615–620.

Ayala P.Y., and H.B. Schlegel (1997). A combined method for determining reaction paths, minima, and transition state geometries. J Chem Phys 107: 375–384.

Azarnia, N., H.M. Berman, and R.D. Rosenstein (1972). The crystal structure of D-*iso*-ascorbic acid. Acta Crystallogr B28: 2157–2161.

Bader, R.F.W. (1990). *Atoms in molecules: A quantum theory*. Oxford University Press: New York.

Bader, R.F.W., and C. Gatti (1998). A Green's function for the density. Chem Phys Lett 287: 233–238.

Bakhmutov, V.I. (2008). *Dihydrogen bond: Principles, experiments, and applications.* John Wiley and Sons: Hoboken, NJ.

Ball, P. (2008). Water as an active constituent in cell biology. Chem Rev 108: 74–108.

Banerjee, B., and D.K. Ganguly (1995). Thermodynamic studies with acetylthiocholine on nicotinic receptors of mammalian skeletal muscle *in vitro*. Biochem Pharmacol 49: 1713–1716.

Banerjee, B., and D.K. Ganguly (1996). Thermodynamics of the interaction of *d*-tubocurarine with nicotinic receptors of mammalian skeletal muscle *in vitro*. Eur J Pharmacol 310: 13–17.

Barnes, A.J. (1983). Molecular complexes of the hydrogen halides studied by matrix isolation infrared spectroscopy. J Mol Struct 100: 259–280.

Barnes, A.J. (2004). Blue-shifting hydrogen bonds – are they improper of proper? J Mol Stuct 704: 3–9.

Bartmess, J.E. (2005). Negative ion energetics data. In: *NIST Chemistry WebBook*, NIST Standard Reference Database Number 69. Edited by P.J. Linstrom and W.G. Mallard. National Institute of Standard and Technology: Gaithersburg, MD; http://webbook.nist.gov/chemistry.

Batsanov, A.S., M.G. Davidson, J.A.K. Howard, S. Lamb, and C. Lustig (1996). Phosphonium ylides as hydrogen bond acceptors: Intermolecular C–H···C interactions in the crystal structure of triphenylphosphonium benzylide. Chem Commun 1791–1792.

Baughcum, S.L., R.W. Duerst, W.F. Rowe, Z. Smith, and E.B. Wilson (1981). Microwave spectroscopic study of malonaldehyde (3-hydroxy-2-propenal). 2. Structure, dipole moment, and tunneling. J Am Chem Soc 103: 6296–6303.

Beauchamp, J.L. (1971). Ion cyclotron resonance spectroscopy. Ann Rev Phys Chem 22: 527–561.

Becke, A.D., and K.E. Edgecombe (1990). A simple measure of electron localization in atomic and molecular systems. J Chem Phys 92: 5397–5403.

Belkova, N.V., E.S. Shubina, and L.M. Epstein (2005). Diverse world of unconventional hydrogen bonds. Acc Chem Res 38: 624–631.

Bent, H.A. (1968). Structural chemistry of donor-acceptor interactions. Chem. Rev 68:, 587–648.

Bergerhoff, G., R. Hundt, R. Sievers, and I.D. Brown (1983). The Inorganic Crystal Structure Database. J Chem Inf Comput Sci 23: 66–69.

Berglund, B., and R.W. Vaughan (1980). Correlations between proton chemical shift tensors, deuterium quadrupole couplings, and bond distances for hydrogen bonds in solids. J Chem Phys 73: 2037–2043.

Berkovitch-Yellin, Z., S. Ariel, and L. Leiserowitz (1983). The comparative roles of the proton-acceptor properties of amide and carboxyl groups in influencing crystal packing patterns: Doubly vs. singly hydrogen-bonded systems in *N*-acyl amino acids and in other amide-acid crystals. J Am Chem Soc 105: 765–767.

Berkovitch-Yellin, Z., and L. Leiserowitz (1982). Atom-atom potential analysis of the packing characteristics of carboxylic acids. A study based on experimental electron density distributions. J Am Chem Soc 104: 4052–4064.

Bernal, J.D., and R.H. Fowler (1933). A theory of water and ionic solution, with particular reference to hydrogen and hydroxyl ions. J Chem Phys 1: 515–548.

Bernal, J.D., and H.D. Megaw (1935). Function of hydrogen in intermolecular forces. Proc Roy Soc London 151A: 384–420.

Bernstein, J. (2002). *Polymorphism in molecular crystals.* Oxford University Press: Oxford, UK.

Bernstein, J., M.D. Cohen, and L. Leiserowitz (1974). The structural chemistry of quinones. In: *The chemistry of the quinonoid compounds*, Part 1. Edited by S. Patai. John Wiley & Sons: Chichester, UK; pp. 37–110.

Bernstein, J., R.E. Davis, L. Shimoni, and N.-L. Chang (1995). Patterns in hydrogen bonding: Functionality and graph set analysis in crystals. Angew Chem Int Ed 34: 1555–1573.

Bernstein, F.C., T.F. Koetzle, G.J.B. Williams, E.F. Jr. Meyer, M.D. Brice, J.R. Rodgers, O. Kennard, T. Shimanouchi, M. Tasumi (1977). The Protein Data Bank: A computer-based archival file for macromolecular structures. J Mol Biol 112: 535–542.

Berry, R.S. (1960). Correlation of rates of intramolecular tunneling processes, with application to some group V compounds. J Chem Phys 32: 933–938.

Bertolasi, V. (2007). Private communication.

Bertolasi, V., V. Ferretti, P. Gilli, G. Gilli, Y.M. Issa, and O.E. Sherif (1993). Intramolecular N—H···O hydrogen bonding assisted by resonance. Part 2. Intercorrelation between structural and spectroscopic parameters for five 1,3-diketone arylhydrazones derived from dibenzoylmethane. J Chem Soc Perkin Trans 2: 2223–2228.

Bertolasi, V., P. Gilli, V. Ferretti, and G. Gilli (1991). Evidence for resonance-assisted hydrogen bonding. 2. Intercorrelation between crystal structure and spectroscopic parameters in eight intramolecularly hydrogen bonded 1,3-diaryl-1,3-propanedione enols. J Am Chem Soc 113: 4917–4925.

Bertolasi, V., P. Gilli, V. Ferretti, and G. Gilli (1994a). Intramolecular N–H···O hydrogen bonding assisted by resonance. III. Structural studies of 1-ketone-2-arylhydrazone derivatives. Acta Crystallogr B50: 617–625.

Bertolasi, V., P. Gilli, V. Ferretti, and G. Gilli (1995). Intermolecular N–H···O hydrogen bonds assisted by resonance. Heteroconjugated systems as hydrogen-bond-strengthening functional groups. Acta Crystallogr B51: 1004–1015.

Bertolasi, V., P. Gilli, V. Ferretti, and G. Gilli (1996). Resonance-assisted O–H···O hydrogen bonding: Its role in the crystalline self-recognition of β-diketone enols and its structural and IR characterization. Chem Eur J 2: 925–934.

Bertolasi, V., P. Gilli, V. Ferretti, and G. Gilli (1998). Intermolecular N–H···O hydrogen bonding assisted by resonance. II. Self-assembly of hydrogen-bonded secondary enaminones in supramolecular catemers. Acta Crystallogr B54: 50–65.

Bertolasi, V., P. Gilli, V. Ferretti, and G. Gilli (2001). General rules for the packing of hydrogen-bonded crystals as derived from the analysis of squaric acid anions: Aminoaromatic nitrogen base co-crystals. Acta Crystallogr B57: 591–598.

Bertolasi, V., P. Gilli, V. Ferretti, G. Gilli, and K. Vaughan (1999). Interplay between steric and electronic factors in determining the strength of intramolecular resonance-assisted N–H···O hydrogen bond in a series of β-ketoarylhydrazones. New J Chem 23: 1261–1267.

Bertolasi, V., P. Gilli, and G. Gilli (2009). Crystal chemistry and prototropic tautomerism in 2-(1-iminoalkyl)-phenols (or naphthols) and 2-diazenyl-phenols (or naphthols). Curr Org Chem 13: 250–268.

Bertolasi, V., L. Nanni, P. Gilli, V. Ferretti, G. Gilli, Y.M. Issa, and O.E. Sherif (1994b). Intramolecular N–H···O=C hydrogen bonding assisted by resonance. Intercorrelation between structural and spectroscopic data for six β-diketoarylhydrazones derived from benzoylacetone or acetylacetone. New J Chem, 18: 251–261.

Bertolasi, V., L. Pretto, G. Gilli, and P. Gilli (2006). π-Bond cooperativity and anticooperativity effects in resonance-assisted hydrogen bonds (RAHBs). Acta Crystallogr B62: 850–863.

Bian, L. (2003). Proton donor is more important than proton acceptor in hydrogen bond formation: A universal equation for calculation of hydrogen bond strength. J Phys Chem A107: 11517–11524.

Bideau, J.-P., G. Bravic, and A. Filhol (1977). Liaison hydrogène et cétoénolisation en série barbiturique à l'état cristallin. II. Structure cristalline par diffraction des neutrons de l'acide diéthyl-1,3-thio-2-barbiturique. Acta Crystallogr B33: 3847–3849.

Bock, H., R. Dienelt, H. Schödel, and Z. Havlas (1993). The C–H···O hydrogen bond adduct of two trinitromethanes to dioxane. J Chem Soc Chem Commun 1792–1793.

Bock, H., T. Vaupel, and H. Schödel (1997). Interactions in crystals. 96. Preparation and structures of salts $[R_nN^+–H\cdots NR_n][B^-(C_6H_5)_4]$ with prototype hydrogen bridges $N–H^+\cdots N$. J Prakt Chem – Chem Zeitung 339: 26–37.

Bock, H., T. Vaupel, H. Schödel, U. Koch, and E. Egert (1994). Interactions in crystals. 44. Trialkylammonium-trialkylamine-tetraphenylborates $[R_3N^+–H\cdots NR_3][B^-(C_6H_5)_4]$ with prototype $N^+–H\cdots N$ hydrogen bridges. Tetrahedron Lett 35: 7355–7358.

Boer, F.P. (1972). The crystal structure of clopidol (3,5-dichloro-2,6-dimethyl-4-pyridinol). Acta Crystallogr B28: 3200–3206.

Boese, R., M.Yu. Antipin, D. Bläser, and K.A. Lyssenko (1998). Molecular crystal structure of acetylacetone at 210 and 110 K: Is the crystal disorder static or dynamic? J. Phys. Chem. B102: 8654–8660.

Bolvig, S., and P.E. Hansen (2000). Isotope effects on chemical shifts as an analytical tool in structural studies of intramolecular hydrogen bonded compounds. Curr Org Chem 4: 19–54.

Bondi, A. (1964). van der Waals volumes and radii. J Phys Chem 68: 441–451.

Boorman, P.M., X. Gao, and M. Parvez (1992). X-ray structural characterization of a thiolate salt displaying a very strong S–H···S hydrogen bond. J Chem Soc Chem Commun 1656–1658.

Bordwell, F.G. (1988). Equilibrium acidities in dimethyl sulfoxide solution. Acc Chem Res 21: 456–463.

Borea, P.A.; A. Dalpiaz, S. Gessi, and G. Gilli (1996a). Thermodynamics of 5-HT_3 receptor binding discriminates agonistic from antagonistic behaviour. Eur J Pharmacol 298: 329–334.

Borea, P.A., A. Dalpiaz, K. Varani, S. Gessi, and G. Gilli (1996b). Binding thermodynamics at A_1 and A_{2A} adenosine receptors. Life Sci 59: 1373–1388.

Borea, P.A., A. Dalpiaz, K. Varani, L. Guerra, and G. Gilli (1995). Binding thermodynamics of adenosine A_{2a} receptor ligands. Biochem Pharmacol 49: 461–469.

Borea, P.A., K. Varani, S. Gessi, P. Gilli, and A. Dalpiaz (2001). Thermodynamics of adenosine (A_1 and A_{2A})-receptor interactions. In: *Drug-receptor thermodynamics: Introduction and applications.* Edited by R.B. Raffa. John Wiley & Sons: Chichester, UK; Chapter 11, pp. 221–244.

Borea, P.A., K. Varani, S. Gessi, P. Gilli, and G. Gilli (1998). Binding thermodynamics at the human neuronal nicotine receptor. Biochem Pharmacol 55: 1189–1197.

Borea, P.A., K. Varani, L. Guerra, P. Gilli, and G. Gilli (1992). Binding thermodynamics of A_1 adenosine receptor ligands. Mol Neuropharmacol 2: 273–281.

Bountis, T. (Ed.) (1992). *Proton transfer in hydrogen-bonded systems.* NATO-ASI Series B, Vol. 291. Plenum Press: New York.

Braga, D., and F. Grepioni (1997). Hydrogen-bonding interactions with the CO ligand in the solid state. Acc Chem Res 30: 81–87.

Braga, D., F. Grepioni, F. Biradha, and G.R. Desiraju (1996a). Agostic interactions in organometallic compounds. J Chem Soc Dalton Trans, 3925–3930.

Braga, D., F. Grepioni, K. Biradha, V.R. Pedireddi, and G.R. Desiraju (1995). Hydrogen bonding in organometallic crystals. 2. C–H···O hydrogen bonds in bridged and terminal first-row metal carbonyls. J Am Chem Soc 117: 3156–3166.

Braga, D., F. Grepioni, E. Tedesco, K. Birahda, and G.R. Desiraju (1996b). Hydrogen bonding in organometallic crystals. 4. M–H···O hydrogen bonding interactions. Organometallics 15: 2692–2699.

Braga, D., F. Grepioni, E. Tedesco, K. Birahda, and G.R. Desiraju (1997). Hydrogen bonding in organometallic crystals. 6. X–H···M hydrogen bonds and M···(H–X) pseudo-agostic bonds. Organometallics 16: 1846–1856.

Bragg, W.L. (1913). The structure of crystals as indicated by their diffraction of X-rays. Proc Roy Soc London A89: 248–277.

Bragg, W.H., and W.L. Bragg (1913). The structure of the diamond. Proc Roy Soc London A89: 277–291.

Brammer, L. (2003). Metals and hydrogen bonds. J. Chem. Soc. Dalton Trans. 3145–3157.

Brammer, L., E.A. Bruton, and P. Sherwood (1999). Fluoride ligands exhibit marked departures from the hydrogen bond acceptor behaviour of their heavier halogen congeners. New J Chem 23: 965–968.

Brammer, L., J.M. Charnock, P.L. Goggin, R.J. Goodfellow, A.G. Orpen, and T.F. Koetzle (1991). The role of transition metal atoms as hydrogen bond acceptors: A neutron diffraction study of $[N^nPr_4]_2[PtCl_4]\cdot$*cis*-$[PtCl_2(NH_2Me)_2]$ at 20 K. J Chem Soc Dalton Trans 1789–1798.

Brammer, L., J.C. Mareque Rivas, and C.D. Spilling (2000). An intramolecular N–H···Co hydrogen bond and a structure correlation study of the pathway for protonation of the $Co(CO)_3L^-$ anion (L = CO, PR_3). J Organomet Chem 609: 36–43.

Brammer, L., M.C. McCann, R.M. Bullock, R.K. McMullan, and P. Sherwood (1992). $Et_3NH^+Co(CO)_4^-$: Hydrogen-bonded adduct or simple ion pair? Single-crystal neutron diffraction study at 15K. Organometallics 11: 2339–2341.

Brammer, L., J.K. Swearingen, E.A. Bruton, and P. Sherwood (2002). Hydrogen bonding and perhalometallate ions: A supramolecular synthetic strategy for new inorganic materials. Proc Natl Acad Sci USA 99: 4956–4961.

Brammer, L., D. Zhao, F.T. Ladipo, and J. Braddock-Wilking (1995). Hydrogen bonds involving transition metal centers – A brief review. Acta Crystallogr B51: 632–640.

Braun-Sand, S., M.H.M. Olsson, and A. Warshel (2005). Computer modeling of enzyme catalysis and its relationship to concepts in physical organic chemistry. Adv Phys Org Chem 40: 201–245.

Bricklebank, N., S.M. Godfrey, C.A. McAuliffe, and R.G. Pritchard (1993). Triphenylphosphonium bromide. Acta Crystallogr C49: 1017–1018.

Brown, M.P., and R.W. Heseltine (1968). Coordinated BH_3 as a proton acceptor group in hydrogen bonding. J Chem Soc Chem Commun 1551–1552.

Brycki, B., and M. Szafran (1982). Infrared and ^1H nuclear magnetic resonance studies of hydrogen bonds in some pyridine *N*-oxide trifluoroacetates and their deuterated analogues in dichloromethane, J Chem Soc Perkin Trans 2: 1333–1338.

Brzozowski, A.M., A.C.W. Pike, Z. Dauter, R.E. Hubbard, T. Bonn, O. Engström, L. Öhman, G.L. Greene, J.-A. Gustafsson, and M. Carlquist (1997). Molecular basis of agonism and antagonism in the oestrogen receptor. Nature 389: 753–758.

Buemi G. (2006). Intramolecular hydrogen bonds. Methodologies and strategies for their strength evaluation. In: *Hydrogen bonding – New insights*. Edited by S.J. Grabowski. Springer: Dordrecht, NL; Chapter 2, pp. 51–107.

Burchell, C.J., C. Glidewell, A.J. Lough, and G. Ferguson (2001). Salts of 3,5-dinitrobenzoic acid with organic diamines: Hydrogen-bonded supramolecular structures in one, two and three dimensions. Acta Crystallogr B57: 201–212.

Bürgi, H.-B. (1973). Chemical reaction coordinates from crystal structure data. I. Inorg Chem 12: 2321–2325.

Bürgi, H.-B. (1975). Stereochemistry of reaction paths as determined from crystal structure data – A relationship between structure and energy. Angew Chem Int Ed 14: 460–473.

Bürgi, H.-B. (1992). Contributions of structural chemistry to the understanding of chemical reactivity. In: *Perspectives in coordination chemistry.* Edited by A.F. Williams, C. Floriani, and A.E. Merbach. Verlag Helvetica Chimica Acta: Basel, CH, pp. 1–27.

Bürgi, H.-B., and J.D. Dunitz (1983). From crystal statics to chemical dynamics. Acc Chem Res 16: 153–161.

Bürgi, H.-B., and J.D. Dunitz (1987). Fractional bonds: Relations among their lengths, strengths, and stretching force constants. J Am Chem Soc 109: 2924–2926.

Bürgi, H.-B., and J.D. Dunitz (Eds.) (1994). *Structure correlation.* Vols.1–2. VCH: Weinheim, D.

Bürgi, H.-B., J.D. Dunitz, and E. Shefter (1973). Geometrical reaction coordinates. II. Nucleophilic addition to a carbonyl group. J Am Chem Soc 95: 5065–5067.

Bürgi, H.-B., J.D. Dunitz, and E. Shefter (1974). Chemical reaction paths. IV. Aspects of O···C=O interactions in crystals. Acta Crystallogr B30: 1517–1527.

Burkhart, B.M., N. Li, D.A. Langs, W.A. Pangborn, and W.L. Duax (1998). The conducting form of gramicidin A is a right-handed double-stranded double helix. Proc Natl Acad Sci USA 95: 12950–12955.

Burnett, M.N., and C.K. Johnson (1996). *ORTEP-III: Oak Ridge Thermal Ellipsoid Plot Program for Crystal Structure Illustrations.* Oak Ridge National Laboratory Report ORNL-6895, Oak Ridge, TN.

Caldwell, G., and P. Kebarle (1984). Binding energies and structural effects in halide anion -ROH and -RCOOH complexes from gas-phase equilibria measurements. J Am Chem Soc 106: 967–969.

Caldwell, G., M.D. Rozeboom, J.P. Kiplinger, and J.E. Bartmess (1984). Anion-alcohol hydrogen bond strengths in the gas phase. J Am Chem Soc 106: 4660–4667.

Calvert, L.D. (1981). The metal data file. Acta Crystallogr A37: C343.

Calvert, L.D., and J.R. Rodgers (1984). The metal data file. Computer Phys Commun 33: 93–98.

Calvin, M., and K.W. Wilson (1945). Stability of chelate compounds. J Am Chem Soc 67: 2003–2007.

Campbell, J.P., J.-W. Hwang, V.G. Jr. Young, R.B. Von Dreele, C.J. Cramer, and W.L. Gladfelter (1998). Crystal engineering using the unconventional hydrogen bond. Synthesis, structure, and theoretical investigation of cyclotrigallazane. J Am Chem Soc 120: 521–531.

Cassidy, C.S., J. Lin, J.B. Tobin, and P.A. Frey (1998). Low-barrier hydrogen bonding in aqueous and aprotic solutions of dicarboxylic acids: Spectroscopic characterization. Bioorg Chem 26: 213–219.

Cavalli, A., P. Carloni, and M. Recanatini (2006). Target-related applications of first principles quantum chemical methods in drug design. Chem Rev 106: 3497–3519.

Chapman, N.B., and J. Shorter (Eds.) (1978). *Correlation analysis in chemistry. Recent advances.* Plenum Press: New York.

Cheeseman, J.R., M.T. Carroll, and R.F.W. Bader (1988). The mechanics of hydrogen bond formation in conjugated systems. Chem Phys Lett 143: 450–458.

Chemla, D.S., and J. Zyss (1987). *Nonlinear optical properties of organic molecules and crystals.* Vols. 1–2. Academic Press: Orlando, FL.

Cheng, Y.-C., and W.H. Prusoff (1973). Relationship between the inhibition constant (K_I) and the concentration of inhibitor which causes 50 per cent inhibition (IC_{50}) of an enzymatic reaction. Biochem Pharmacol 22: 3099–3108.

Cleland, W.W. (1992). Low-barrier hydrogen bonds and low fractionation factor bases in enzymatic reactions. Biochemistry 31: 317–319.

Cleland, W.W., and M.M. Kreevoy (1994). Low-barrier hydrogen bonds and enzymic catalysis. Science 264: 1887–1890.

Contreras, M.L., B.B. Wolfe, and P.B. Molinoff (1986). Thermodynamic properties of agonist interactions with the *beta*-adrenergic receptor-coupled adenylate cyclase system. I. High- and low-affinity states of agonist binding to membrane-bound *beta*-adrenergic receptors. J Pharmacol Exp Ther 237: 154–164.

Cookson, R.F. (1974). The determination of acidity constants. Chem Rev 74: 5–28.

Coppens, P. (1997). *X-ray charge densities and chemical bonding.* Oxford University Press: New York.

Coppens, P., and M.B. Hall (Eds.) (1981). *Electron distributions and the chemical bond.* Plenum Press: New York.

Cotton, F.A., C.K. Fair, G.E. Lewis, G.N. Mott, F.K. Ross, A.J. Schultz, and J.M. Williams (1984). Precise structural characterizations of the hexaaquovanadium(III) and diaquohydrogen ions. X-ray and neutron diffraction studies of $[V(H_2O)_6][H_5O_2](CF_3SO_3)_4$. J Am Chem Soc 106: 5319–5323.

Cotton, F.A., and G. Wilkinson (1988). *Advanced inorganic chemistry,* 5th edition. John Wiley & Sons: New York.

Coulson, C.A. (1959). The hydrogen bond. In: *Hydrogen bonding.* Edited by D. Hadži and H.W. Thompson. Pergamon Press: London, UK; pp. 339–360.

Coulson, C.A. (1961). *Valence,* 2nd edition. Oxford University Press: Oxford, UK.

Coulson, C.A., and U. Danielsson (1954a). Ionic and covalent contributions to the hydrogen bond. I. Ark Fys 8: 239–244.

Coulson, C.A., and U. Danielsson (1954b). Ionic and covalent contributions to the hydrogen bond. II. Ark Fys 8: 245–255.

Cowan, J.A., J.A.K. Howard, G.J. McIntyre, S.M.-F. Lo, and I.D. Williams (2003). Variable-temperature neutron diffraction studies of the short, strong N···O hydrogen bonds in the 1:2 co-crystal of benzene-1,2,4,5-tetracarboxylic acid and 4,4′-bipyridyl . Acta Crystallogr B59: 794–801.

Cowan, J.A., J.A.K. Howard, G.J. McIntyre, S.M.-F. Lo, and I.D. Williams (2005). Variable-temperature neutron diffraction studies of the short, strong hydrogen bonds in the crystal structure of pyridine-3,5-dicarboxylic acid. Acta Crystallogr B61: 724–730.

Crabtree, R.H. (1993). Transition metal complexation of σ bonds. Angew Chem Int Ed 32: 789–805.

Crabtree, R.H. (1998a). Recent advances in hydrogen bonding studies involving metal hydrides. J Organometal Chem 577: 111–115.

Crabtree, R.H. (1998b). A new type of hydrogen bond. Science 282: 2000–2001.

Crabtree, R.H., O. Eisenstein, G. Sini, and E. Peris (1998). New types of hydrogen bonds. J Organometal Chem 567: 7–11.

Crabtree, R.H., P.E.M. Siegbahn, O. Eisenstein, A.L. Rheingold, and T.F. Koetzle (1996). A new intermolecular interaction: Unconventional hydrogen bonds with element-hydride bonds as proton acceptors. Acc Chem Res 29: 348–354.

Cram, D.J., and J.M. Cram (1994). *Container molecules and their guests.* The Royal Society of Chemistry: Cambridge, UK.

Cramer, C.J., and W.L. Gladfelter (1997). Ab initio characterization of $[H_3N{\cdot}BH_3]_2$, $[H_3N{\cdot}AlH_3]_2$, and $[H_3N{\cdot}GaH_3]_2$. Inorg Chem 36: 5358–5362.

Cubero, E., M. Orozco, and F.J. Luque (1999). Electron density topological analysis of the C–H···O anti-hydrogen bond in the fluoroform-oxirane complex. Chem Phys Lett 310: 445–450.

Cukierman, S. (2006). Et tu, Grotthuss! and other unfinished stories. Biochim Biophys Acta 1757: 876-885.

Custelcean, R., and J.E. Jackson (2001). Dihydrogen bonding: Structures, energetics, and dynamics. Chem Rev 101: 1963–1980.

Dalpiaz, A., P.A. Borea, S. Gessi, and G. Gilli (1996). Binding thermodynamics of 5-HT_{1A} receptors ligands. Eur J Pharmacol 312: 107–114.

Dalpiaz, A., S. Gessi, P.A. Borea, and G. Gilli (1995). Binding thermodynamics of serotonin to rat-brain 5-HT_{1A}, 5-HT_{2A} and 5-HT_3 receptors. Life Sci 57: PL141–146.

Danil de Namor, A.F., M.-C. Ritt, M.-J. Schwing-Weill, F. Arnaud-Neu, and D.F.V. Lewis (1991). Solution thermodynamics of amino acid–18-crown-6 and amino acid–cryptand 222 complexes in methanol and ethanol. Linear enthalpy–entropy compensation effect. J Chem Soc Faraday Trans 87: 3231–3239.

Dannenberg, J.J., and R. Rios (1994). Theoretical study of the enolic forms of acetylacetone. How strong is the H-bond? J Phys Chem 98: 6714–6718.

Davidson, W.R., J. Sunner, and P. Kebarle (1979). Hydrogen bonding of water to onium ions. Hydration of substituted pyridinium ions and related systems. J Am Chem Soc 101: 1675–1680.

DeFrees, D.J., K. Raghavachavi, H.B. Schlegel, J.A. Pople, and P.v.R. Schleyer (1987). Binary association complexes of LiH, BeH_2, and BH_3. Relative isomer stabilities and barrier heights for their interconversion: Energy barriers in the dimerization reactions. J Phys Chem, 91: 1857–1864.

de Grotthuss, C.J.T. (1806). Mémoire sur la décomposition de l'eau et des corps qu'elle tient en dissolution à l'aide de l'électricité galvanique. Ann Chim 58: 54–74.

de Meester, P., M.V. Jovanovic, S.S.C. Chu, and E.R. Biehl (1986). Investigation of the structure and ^{13}C NMR spectrum of 5-phenylbarbituric acid. J Heterocycl Chem 23: 337–341.

Desiraju, G.R. (1991). The C–H···O hydrogen bond in crystals: What is it? Acc Chem Res 24: 290–296.

Desiraju, G.R. (2002). Hydrogen bridges in crystal engineering: Interactions without borders. Acc Chem Res 35: 565–573.

Desiraju, G.R., and Th. Steiner (1999). *The weak hydrogen bond in structural chemistry and biology.* Oxford University Press: Oxford, UK.

Destro, R. (1991). Proton transfer in the solid state: Thermodynamic parameters from an X-ray study in the temperature range 20–293 K. Chem Phys Lett 181: 232–236.

Destro, R., and R.E. Marsh (1984). Temperature dependence of tautomeric equilibria in the solid state: The case of citrinin. J Am Chem Soc 106: 7269–7271.

Dill, J.D., P.v.R. Schleyer, J.S. Binkley, and J.A. Pople (1977). Molecular orbital theory of the electronic structure of molecules. 34. Structures and energies of small compounds containing lithium or beryllium. Ionic, multicenter, and coordination bonding. J Am Chem Soc 99: 6159–6173.

Donohue, J. (1952). The hydrogen bond in organic crystals. J Phys Chem 56: 502–510.

Donohue, J. (1968). Selected topics in hydrogen bonding. In: *Structural chemistry and molecular biology.* Edited by A. Rich and N. Davidson. Freeman: San Francisco, CA; pp. 443–465.

Drück, U., and W. Littke (1980). The structures of two rubazoic acid derivatives. Acta Crystallogr B36: 3002–3007.

Duarte, E.P., C.R. Oliveira, and A.P. Carvalho (1988). Thermodynamic analysis of antagonist and agonist interactions with dopamine receptors. Eur J Pharmacol 147: 227–239.

Duax, W.L., V. Pletnev, and B.M. Burkhart (2003). Mechanism of ion transport and gating in gramicidin noanotubes. J Mol Struct 647: 97–111.

Dulmage, W.J., and W.N. Lipscomb (1951). The crystal structures of hydrogen cyanide, HCN. Acta Crystallogr 4: 330–334.

Dunitz, J.D. (1979). *X-ray analysis and the structure of organic molecules.* Cornell University Press: Ithaca, NY.

Dunitz, J.D., and R. Taylor (1997). Organic fluorine hardly ever accepts hydrogen bonds. Chem Eur J 3: 89–98.

Durocher, G., and C. Sandorfy (1967). A study of the effects of solvent on overtone frequencies. J Mol Spectrosc 22: 347–359.

Dziembowska, T., P.E. Hansen, and Z. Rozwadowski (2004). Studies based on deuterium isotope effect on ^{13}C chemical shifts. Prog Nucl Magn Reson Spectrosc 45: 1–29.

Edsall, J.T., and H. Gutfreund (1983). *Biothermodynamics: The study of biochemical processes at equilibrium.* John Wiley & Sons: Chichester, UK.

Eichhorn, K.D. (1991). Neutron structure analysis of 1-hydroxypyridinium trichloroacetate, $C_7H_6Cl_3NO_3$, at 120 K. Z Kristallogr 195: 205–220.

Eigen, M. (1964). Proton transfer, acid-base catalysis, and enzymatic hydrolysis. Part I: Elementary processes. Angew Chem Int Ed 3: 1–19.

Eigen, M., W. Kruse, G. Maass, and L. de Maeyer (1964). Rate constants of protolytic reactions in aqueous solution. Prog React Kinet 2: 285–318.

Eliard, P.H., and G.G. Rousseau (1984). Thermodynamics of steroid binding to the human glucocorticoid receptor. Biochem J 218: 395–404.

Elsaesser, Th., and H.J. Bakker (Eds.) (2002). *Ultrafast hydrogen bonding dynamics and proton transfer processes in the condensed phase.* Kluwer Academic Publishers: Dordrecht, NL.

Emsley, J. (1980). Very strong hydrogen bonding. Chem Soc Rev 9: 91–124.

Emsley, J. (1984). The composition, structure and hydrogen bonding of the β-diketones. Struct Bond 57: 147–191.

Epstein, L.M., A.N. Krylov, and E.S. Shubina (1994). Novel types of hydrogen bonds involving transition metal atoms and proton transfer ($XH\cdots M$, $[MH]^+\cdots B$, $[MH]^+\cdots A^-$). J Mol Struct 322: 345–352.

Epstein, L.M., and E.S. Shubina (2002). New types of hydrogen bonding in organometallic chemistry. Coord Chem Rev 231: 165–181.

Epstein, L.M., E.S. Shubina, E.V. Bakhmutova, L.N. Saitkulova, V.I. Bakhmutov, A.L. Chistyakov, and I.V. Stankevich (1998). Unusual hydrogen bonds with a hydride atom in boron hydrides acting as proton acceptor. Spectroscopic and theoretical studies. Inorg Chem 37: 3013–3017.

Epstein, L.M., E.S. Shubina, A.N. Krylov, A.Z. Kreindlin, and M.I. Rybinskaya (1993). Interaction between $[(\eta^5\text{-}C_5Me_5)_2OsH]^+PF_6^-$ and nitrogen or oxygen bases. Hydrogen bonds of the type $[OsH]^+\cdots B$. J Organometal Chem 447: 277–280.

Espinosa, E., I. Alkorta, J. Elguero, and E. Molins (2002). From weak to strong interactions: A comprehensive analysis of the topological and energetic properties of the electron density distribution involving $X\text{–}H\cdots F\text{–}Y$ systems. J Chem Phys 117: 5529–5542.

Espinosa, E., E. Molins, and C. Lecomte (1998). Hydrogen bond strengths revealed by topological analyses of experimentally observed electron densities. Chem Phys Lett 285: 170–173.

Espinosa, E., M. Souhassou, H. Lachekar, and C. Lecomte (1999). Topological analysis of the electron density in hydrogen bonds. Acta Cryst B55: 563–572.

Etter, M.C. (1982). A new role for hydrogen-bond acceptors in influencing packing patterns of carboxylic acids and amides. J Am Chem Soc 104: 1095–1096.

Etter, M.C. (1990). Encoding and decoding hydrogen-bond patterns of organic compounds. Acc Chem Res 23: 120–126.

Etter, M.C. (1991). Hydrogen bonds as design elements in organic chemistry. J Phys Chem 95: 4601–4610.

Etter, M.C., R.C. Hoye, and G.M. Vojta (1988). Solid-state NMR and X-ray crystallography: Complementary tools for structure determination. Cryst Rev 1: 281–338.

Etter, M.C., J.C. MacDonald, and J. Bernstein (1990). Graph-set analysis of hydrogen-bond patterns in organic crystals. Acta Crystallogr B46: 256–262.

Etter, M.C., Z. Urbańczyk-Lipkowska, D.A. Jahn, and J.S. Frye (1986). Solid-state structural characterization of 1,3-cyclohexanedione and of a 6:1 cyclohexanedione:benzene cyclamer, a novel host-guest species. J Am Chem Soc 108: 5871–5876.

Eyring, H., and M. Polanyi (1931). Simple gas reactions. Z Physik Chem B12: 279–311.

Fairhurst, S.A., R.A. Henderson, D.L. Hughes, S.K. Ibrahim and C.J. Pickett (1995). An intramolecular W–H···O=C hydrogen bond? Electosynthesis and X-ray crystallographic structure of $[WH_3(\eta^1\text{-}OCOMe)(Ph_2PCH_2CH_2PPh_2)_2]$. J Chem Soc Chem Commun 1569–1570.

Farnham, W.B., D.A. Dixon, W.J. Middleton, J.C. Calabrese, R.L. Harlow, J.F. Whitney, G.A. Jones, and L.J. Guggenberger (1987). The crystal and molecular structure of tris(substituted amino) sulfonium ions. J Am Chem Soc 109: 476–483.

Ferguson, G., W.C. Marsh, R.J. Restivo, and D. Lloyd (1975). Conformational studies of 2,3-diacyl-5-nitrocyclopentadienes. Delocalized systems with very short intramolecular O···H···O hydrogen bonds. Crystal and molecular structures of 2,3-diacetyl- and 2,3-dibenzoyl-5-nitrocyclopentadiene. J Chem Soc Perkin Trans 2: 998–1004.

Fermi, E. (1956). *Thermodynamics.* Dover Publications: New York.

Ferretti, V., V. Bertolasi, G. Gilli, and P.A. Borea (1985). Structures of two 2-arylpyrazolo[4,3-*c*]quinolin-3-ones: CGS8216, $C_{16}H_{11}N_3O$, and CGS9896, $C_{16}H_{10}ClN_3O$, Acta Crystallogr C41: 107–110.

Ferretti, V., V. Bertolasi, P. Gilli, and G. Gilli (1993). Out-of-plane deformation pathways of the $R(X{=})C\text{–}NR_2$ fragment present in amides, thioamides, amidines, enamines and anilines. A concerted study making use of structural data, molecular mechanics, and *ab initio* calculations. J Phys Chem 97: 13568–13574.

Ferretti, V., K.C. Dubler-Steudle, and H.-B. Bürgi (1992). Structure correlations, reaction pathways, and energy surfaces for chemical reactions. In: *Accurate molecular structures: Their determination and importance.* Edited by A. Domenicano and I. Hargittai. Oxford University Press: Oxford, UK; Chapter 17, pp. 412–436.

Ferretti, V., P. Gilli, V. Bertolasi, and G. Gilli (1996). Structure correlation methods in chemical crystallography. Cryst Rev 5: 3–104.

Ferstanding, L.L. (1962). Carbon as a hydrogen bonding base and carbon-hydrogen-carbon bonding. J Am Chem Soc 84: 3553–3557.

Filarowski, A., T. Glowiak, and A. Koll (1999). Strengthening of the intramolecular O···H···N hydrogen bonds in Schiff bases as a result of steric repulsion. J Mol Struct 484: 75–89.

Filarowski, A., A. Koll, T. Glowiak, E. Majewski, and T. Dziembowska (1998). Proton transfer reaction in N-methyl-2-hydroxy-Schiff bases. Ber Bunsen-Ges Phys Chem 102: 393–402.

Filippini, G., and A. Gavezzotti (1993). Empirical intermolecular potentials for organic crystals: the '6-exp' approximation revisited. Acta Crystallogr B49: 868–880.

Fillaux, F., M.H. Limage, and F. Romain (2002). Quantum proton transfer and interconversion in the benzoic acid crystal: vibrational spectra, mechanism and theory. Chem Phys 276: 181–210.

Finar, I.L. (1973). *Organic chemistry*. Vols. 1–2. Longman: Harlow, UK.

Fitzgerald, P.M.D., B.M. McKeever, J.F. VanMiddlesworth, J.P. Springer, J.C. Heimbach, C.-T. Leu, W.K. Herber, R.A.F. Dixon, and P.L. Darke (1990). Crystallographic analysis of a complex between human immunodeficiency virus type 1 protease and acetyl-pepstatin at 2.0 Å resolution. J Biol Chem 265: 14209–14219.

Foster, R., and C.A. Fyfe (1969). Nuclear magnetic resonance of organic charge-transfer complexes. Prog Nucl Magn Reson Spectrosc 4: 1–89.

Frank, H.S., and M.W. Evans (1945). Free volume and entropy in condensed systems. III. Entropy in binary liquid mixtures; partial molar entropy in dilute solutions; structure and thermodynamics in aqueous electrolytes. J Chem Phys 13: 507–532.

Franks, F. (Ed.) (1972–82). *Water: A comprehensive treatise*. Vols. 1–7. Plenum Press: New York, London.

Frey, P.A. (2001). Strong hydrogen bonding in molecules and enzymatic complexes. Magn Reson Chem 39: S190–S198.

Frey, P.A. (2002). Characterization of a low barrier hydrogen bond in the active site of chymotrypsin. J Mol Struct 615: 153–161.

Frey, P.A., S.A. Whitt, and J.B. Tobin (1994). A low-barrier hydrogen bond in the catalytic triad of serine protease. Science 264: 1927–1930.

Frisch, M.J., A.C. Scheiner, H.F. III Schaefer, and J.S. Binkley (1985). The malonaldehyde equilibrium geometry: A major structural shift due to the effects of electron correlation. J Chem Phys 82: 4194–4198.

Frisch, M.J., G.W. Trucks, H.B. Schlegel, G.E. Scuseria, M.A. Robb, J.R. Cheeseman, V.G. Zakrzewski, J.A. Jr. Montgomery, R.E. Stratmann, J.C. Burant, S. Dapprich, J.M. Millam, A.D Daniels, K.N. Kudin, M.C. Strain, O. Farkas, J. Tomasi, V. Barone, M. Cossi, R. Cammi, B. Mennucci, C. Pomelli, C. Adamo, S. Clifford, J. Ochterski, G.A. Petersson, P.Y. Ayala, Q. Cui, K. Morokuma, N. Rega, P. Salvador, J.J. Dannenberg, D.K. Malick, A.D. Rabuck, K. Raghavachari, J.B. Foresman, J. Cioslowski, J.V. Ortiz, A.G. Baboul, B.B. Stefanov, G. Liu, A. Liashenko, P. Piskorz, I. Komaromi, R. Gomperts, R.L. Martin, D.J. Fox, T. Keith, M.A. Al-Laham, C.Y. Peng, A. Nanayakkara, M. Challacombe, P.M.W. Gill, B. Johnson, W. Chen, M.W. Wong, J.L. Andres, C. Gonzalez, M. Head-Gordon, E.S. Replogle, and J.A. Pople (2002). GAUSSIAN 98 (Revision A.11.3); Gaussian, Inc.: Pittsburgh, PA.

Fu, D., A. Lisbon, L.J.W. Miercke, C. Weitzman, P. Nollert, J. Krucinski, and R.M. Stroud (2000). Structure of a glycerol-conducting channel and the basis for its selectivity. Science 290: 481–486.

Fuster, F., and B. Silvi (2000). Does the topological approach characterize the hydrogen bond? Theor Chem Acc 104: 13–21.

Fyfe, C.A. (1983). *Solid state NMR for chemists.* CFC Press: Guelph, Ontario, Canada.

Gatti, C., F. Cargnoni, and L. Bertini (2003). Chemical information from the source function. J Comput Chem 24: 422–436.

Gavezzotti, A., and G. Filippini (1994). Geometry of the intermolecular X–H···Y (X, Y = N, O) hydrogen bond, and the calibration of empirical hydrogen-bond potentials. J Phys Chem 98: 4831–4837.

Gerlt, J.A., and P.G. Gassman (1993). An explanation for rapid enzyme-catalyzed proton abstraction from carbon acids: Importance of late transition states in concerted mechanisms. J Am Chem Soc 115: 11552–11568.

Gerstein, B.C., R.G. Pembleton, R.C. Wilson, and L.M. Ryan (1977). High resolution NMR in randomly oriented solids with homonuclear dipolar broadening: Combined multiple pulse NMR and magic angle spinning. J Chem Phys 66: 361–362.

Giglio, E. (1969). Calculation of Van der Waals interactions and hydrogen bonding in crystals. Nature 222: 339–341.

Gill, S.J., S.F. Dec, G. Olofsson, and I. Wadsö (1985). Anomalous heat capacity of hydrophobic solvation. J Phys Chem 89: 3758–3761.

Gilli, G., F. Bellucci, V. Ferretti, and V. Bertolasi (1989). Evidence for resonance-assisted hydrogen bonding from crystal-structure correlations on the enol form of the β-diketone fragment. J Am Chem Soc 111: 1023–1028.

Gilli, G., and V. Bertolasi (1990). Structural chemistry of enols. In: *The chemistry of enols.* Edited by Z. Rappoport. John Wiley & Sons: Chichester, UK; Chapter 13, pp. 713–764.

Gilli, G., V. Bertolasi, F. Bellucci, and V. Ferretti (1986). Stereochemistry of the $R_1(X{=})C(sp^2){-}N(sp^3)R_2R_3$ fragment. Mapping of the *cis-trans* isomerization path by rotation around the C–N bond from crystallographic structural data. J Am Chem Soc 108: 2420–2424.

Gilli, G., V. Bertolasi, V. Ferretti, and P. Gilli (1993). Resonance-assisted hydrogen bonding. III. Formation of intermolecular hydrogen-bonded chains in crystals of β-diketone enols and its relevance to molecular association. Acta Crystallogr B49: 564–576.

Gilli, G., V. Bertolasi, P. Gilli, and V. Ferretti (2001). Associations of squaric acid and its anions as multiform building blocks of hydrogen-bonded molecular crystals. Acta Crystallogr B57:859–865.

Gilli, G., and P. Gilli (2000). Towards a unified hydrogen-bond theory. J Mol Struct 552: 1–15.

Gilli, P., V. Bertolasi, V. Ferretti, and G. Gilli (1994a). Covalent nature of the strong homonuclear hydrogen bond. Study of the O–H···O system by crystal structure correlation methods. J Am Chem Soc 116: 909–915.

Gilli, P., V. Bertolasi, V. Ferretti, and G. Gilli (2000). Evidence for intramolecular N–H···O resonance-assisted hydrogen bonding in β-enaminones and related heterodienes. A combined crystal-structural, IR and NMR spectroscopic, and quantum-mechanical investigation. J Am Chem Soc 122: 10405–10417.

Gilli, P., V. Bertolasi, L. Pretto, L. Antonov, and G. Gilli (2005a). Variable-temperature X-ray crystallographic and DFT computational study of the N–H···O/N···H–O tautomeric competition in 1-(arylazo)-2-naphthols. Outline of a transition-state hydrogen-bond theory. J Am Chem Soc 127: 4943–4953.

Gilli, P., V. Bertolasi, L. Pretto, V. Ferretti, and G. Gilli (2004). Covalent versus electrostatic nature of the strong hydrogen bond: Discrimination among single, double, and asymmetric single-well hydrogen bonds by variable-temperature X-ray crystallographic methods in β-diketone enol RAHB systems. J Am Chem Soc 126: 3845–3855.

Gilli, P., V. Bertolasi, L. Pretto, and G. Gilli (2006). Outline of a transition-state hydrogen-bond theory. J Mol Struct 790: 40–49.

Gilli, P., V. Bertolasi, L. Pretto, A. Lyčka, and G. Gilli (2002). The nature of solid-state N–H···O/O–H···N tautomeric competition in resonant systems. Intramolecular proton transfer in low-barrier hydrogen bonds formed by the ···O=C–C=N–NH··· $\rightleftarrows$ ···HO–C=C–N=N··· ketohydrazone-azoenol system. A variable-temperature X-ray crystallographic and DFT computational study. J Am Chem Soc 124: 13554–13567.

Gilli, P., V. Ferretti, V. Bertolasi, and G. Gilli (1996a). A novel approach to hydrogen bonding theory. In: *Advances in molecular structure research*, Vol. 2. Edited by M. Hargittai and I. Hargittai. JAI Press: Greenwich, CT; pp. 67–102.

Gilli, P., V. Ferretti, and G. Gilli (1996b). Hydrogen bonding models: Their relevance to molecular modeling. In: *Fundamental principles of molecular modeling.* Edited by W. Gans, A. Amann, and J.C.A. Boeyens. Plenum Press: New York; pp. 119–141.

Gilli, P., V. Ferretti, G. Gilli, and P.A. Borea (1994b). Enthalpy-entropy compensation in drug-receptor binding. J Phys Chem 98: 1515–1518.

Gilli, P., and G. Gilli (1992). *LSHB. A Computer program for performing Lippincott and Schroeder HB calculations.* Department of Chemistry, University of Ferrara, Ferrara, Italy.

Gilli, P., and G. Gilli (2002). Hydrogen bond at the dawn of the XXI century. New methods, new results, new ideas. In: *Strength from weakness: Structural consequences of weak interactions in molecules, supermolecules, and crystals.* NATO Science Series II, Vol. 68. Edited by A. Domenicano and I. Hargittai. Kluwer Academic Publishers: Dordrecht, NL; Chapter 14, pp. 261–280.

Gilli, P., G. Gilli, P.A. Borea , K. Varani, A. Scatturin, and A. Dalpiaz (2005b). Binding thermodynamics as a tool to investigate the mechanisms of drug-receptor interactions: Thermodynamics of cytoplasmic steroid/nuclear receptors in comparison with membrane receptors. J Med Chem 48: 2026–2035.

Gilli, P., L. Pretto, V. Bertolasi, and G. Gilli (2009). Predicting hydrogen-bond strengths from acid-base molecular properties. The pK_a slide rule: Toward the solution of a long-lasting problem. Acc Chem Res 42: 33–44.

Gilli, P., L. Pretto, and G. Gilli (2007). PA/pK_a equalization and the prediction of the hydrogen-bond strength: A synergism of classical thermodynamics and structural crystallography. J Mol Struct 844–845: 328–339.

Glasstone, S., K.J. Laidler, and H. Eyring (1941). *The theory of rate processes.* McGraw-Hill: New York.

Glendening, E.D., and A. Streitwieser (1994). Natural energy decomposition analysis: An energy partitioning procedure for molecular interactions with application to weak hydrogen bonding, strong ionic, and moderate donor-acceptor interactions. J Chem Phys 100: 2900–2909.

Grabowski, S.J. (2001). π-Electron delocalization for intramolecular resonance assisted hydrogen bonds. J Phys Org Chem 16: 797–802.

Grabowski, S.J. (2003). A new measure of hydrogen bonding strength – *ab initio* and atoms in molecules studies. Chem Phys Lett 338: 361–366.

Grabowski, S.J. (Ed.) (2006). *Hydrogen bonding – New insights.* Springer: Dordrecht, NL.

Grell, J., J. Bernstein, and G. Tinhofer (2002). Investigation of hydrogen bond patterns: A review of mathematical tools for the graph set approach. Cryst Rev 8: 1–56.

Grunwald, E. (1985). Structure-energy relations, reaction mechanism, and disparity of progress of concerted reaction events. J Am Chem Soc 107: 125–133.

Grunwald, E. (1997). *Thermodynamics of molecular species.* John Wiley & Sons: New York.

Grunwald, E. (2001). Enthalpy-entropy compensation and solvent reorganization. In: *Drug-receptor thermodynamics: Introduction and applications.* Edited by R.B. Raffa. John Wiley & Sons: Chichester, UK; Chapter 25, pp. 553–573.

Grunwald, E., and C. Steel (1995). Solvent reorganization and thermodynamic enthalpy-entropy compensation. J Am Chem Soc 117: 5687–5692.

Gu, Y., T. Kar, and S. Scheiner (1999). Fundamental properties of the CH···O interaction: Is it a true hydrogen bond? J Am Chem Soc 121: 9411–9422.

Guo, H., S. Sirois, E.I. Proynov, and D.R. Salahub (1997). Density functional theory and its applications to hydrogen-bonded systems. In: *Theoretical treatments of hydrogen bonding.* Edited by D. Hadži. John Wiley & Sons: Chichester, UK; Chapter 3, pp. 49–74.

Haddon, R.C. (1980). Symmetrical hydrogen bonding: Molecular orbital theory of the π-electron component. J Am Chem Soc 102: 1807–1811.

Hadži, D. (Ed.) (1997). *Theoretical treatments of hydrogen bonding.* John Wiley & Sons: Chichester, UK.

Hadži, D., and H.W. Thompson (Eds.) (1959). *Hydrogen bonding.* Papers presented at the Symposium on Hydrogen Bonding held at Ljubljana, Slovenia, 1957. Pergamon Press: London, UK.

Haisa, M., S. Kashino, T. Ueno, N. Shinozaki, and Y. Matsuzaki (1980). The structures of *N*-aromatic amides: *p*-acetanisidide, *N*-2-naphthylacetamide and *N*-2-fluorenylacetamide. Acta Crystallogr B36: 2306–2311.

Hall, H.E. (1974). *Solid state physics.* John Wiley & Sons: Chichester, UK.

Hallam, H.E. (1976). Matrix-isolation studies of hydrogen bonding. In: *The hydrogen bond: Recent developments in theory and experiments,* Vol. III. Edited by P. Schuster, G. Zundel, and C. Sandorfy. North-Holland: Amsterdam, NL; Chapter 22, pp. 1065–1105.

Hamilton, A.D. (1990). Molecular recognition – Design and synthesis of artificial receptors employing directed hydrogen bonding interactions. J Chem Educ 67: 821–828.

Hamilton, W.C., and Ibers, J.A. (1968). *Hydrogen bonding in solids: Methods of molecular structure determination*. Benjamin: New York.

Hammond, G.S. (1955). A correlation of reaction rates. J Am Chem Soc 77: 334–338.

Hansen, N.K., and P. Coppens (1978). Testing aspherical atom refinements on small-molecule data sets. Acta Crystallogr A34: 909–921.

Hansen, P.E. (1996). *Isotope effects on chemical shifts as a tool in structural studies*. Roskilde University Press: Roskilde, DK.

Hantzsch, A. (1910). Über die Isomerie-Gleichgewichte des Acetessigesters und die sogenannte Isorrhopesis seiner Salze. Chemische Berichte 43: 3049–3076.

Harris, T.K., and A.S. Mildvan (1999). High-precision measurement of hydrogen bond lengths in proteins by nuclear magnetic resonance methods. Proteins 35: 275–282.

Hatton, J.V., and R.E. Richards (1961). Solvent effects in the high-resolution hydrogen resonance spectra of some acetylenes. Trans Faraday Soc 57: 28–33.

Heldt, I., and U. Behrens (2005). Structures of alkali metal salts of aromatic, heterocyclic amides: Synthesis and structure of crown ether adducts of the alkali metal pyrrolides. Z Anorg Allg Chem 631: 749–758.

Hendricks, S.B., O.R. Wulf, G.E. Hilbert, and U. Liddel (1936). Hydrogen bond formation between hydroxyl groups and nitrogen atoms in some organic compounds. J Am Chem Soc 58: 1991–1996.

Herbstein, F.H., M. Kapon, G.M. Reisner, M.S. Lehman, R.B. Kress, R.B. Wilson, W.-I. Shiau, E.N. Duesler, I.C. Paul, and D.Y. Curtin (1985). Polymorphism of naphthazarin and its relation to solid-state proton-transfer – Neutron and X-ray diffraction studies on naphthazarin C. Proc Roy Soc London, A399: 295–319.

Hibbert, F., and J. Emsley (1990). Hydrogen bonding and chemical reactivity. Adv Phys Org Chem 26: 255–411.

Hilbert, G.E., O.R. Wulf, S.B. Hendricks, and U. Liddel (1936). The hydrogen bond between oxygen atoms in some organic compounds. J Am Chem Soc 58: 548–555.

Hinze, J., and H.H. Jaffé (1962). Electronegativity. I. Orbital electronegativity of neutral atoms. J Am Chem Soc 84: 540–546.

Hinze, J., and H.H. Jaffé (1963). Electronegativity. IV. Orbital electronegativities of the neutral atoms of the periods three A and four A and of positive ions of periods one and two. J Phys Chem 67: 1501–1506.

Hinze, J., M.A. Whitehead, and H.H. Jaffé (1963). Electronegativity. II. Bond and orbital electronegativities. J Am Chem Soc 85: 148–154.

Hirshfeld, F.L. (1971). Difference densities by least-squares refinement: Fumaramic acid. Acta Crystallogr B27: 769–781.

Hirshfeld, F.L. (1976). Can X-ray data distinguish bonding effects from vibrational smearing? Acta Crystallogr A32: 239–244.
Hobza, P., and Z. Havlas (2000). Blue-shifting hydrogen bonds. Chem Rev 100: 4253–4264.
Hobza, P., V. Špirko, H.L. Selzle, and E.W. Schlag (1998). Anti-hydrogen bond in the benzene dimer and other carbon proton donor complexes. J Phys Chem A102: 2501–2504.
Hohenberg, P., and W. Kohn (1964). Inhomogeneous electron gas. Phys Rev 136: B864–B871.
Hollander, F.J., D. Semmingsen, and T.F. Koetzle (1977). The molecular and crystal structure of squaric acid (3,4-dihydroxy-3-cyclobutene-1,2-dione) at 121°C: A neutron diffraction study. J Chem Phys 67: 4825–4831.
Holm, R.H., and F.A. Cotton (1958). Spectral investigations of metal complexes of β-diketones. I. Nuclear magnetic resonance and ultraviolet spectra of acetylacetonates. J Am Chem Soc 80: 5658–5663.
Holmes, R.R. (1972). Spectroscopy and structure of pentacoordinated molecules. Acc Chem Res 5: 296–303.
Holmes, R.R. (1975). Conformational energy calculations on five-membered rings in pentacoordinate phosphorus. J Am Chem Soc 97: 5379–5385.
Holmes, R.R. (1979). Structure of cyclic pentacoordinated molecules of main group elements. Acc Chem Res 12: 257–265.
Houk, K.N., A.G. Leach, S.P. Kim, and X. Zhang (2003). Binding affinities of host-guest, protein-ligand, and protein-transition-state complexes. Angew Chem Int Ed 42: 4872–4897.
Howard, J.A.K., V.J. Hoy, D. O'Hagan, and G.T. Smith (1996). How good is fluorine as a hydrogen bond acceptor? Tetrahedron 52: 12613–12622.
Hückel, E. (1928). Theory of the mobilities of the hydrogen and hydroxyl ions in aqueous solution. Z Elektrochem 34: 546–562.
Huggins, C.M., G.C. Pimentel, and J.N. Shoolery (1955). Proton magnetic resonance studies of chloroform in solution: Evidence for hydrogen bonding. J Chem Phys 23: 1244–1247.
Huggins, C.M., G.C. Pimentel, and J.N. Shoolery (1956). Proton magnetic resonance studies of the hydrogen bonding of phenol, substituted phenols and acetic acid. J Phys Chem 60: 1311–1314.
Huggins, M.L. (1931). The role of hydrogen bonds in conduction by hydrogen and hydroxyl ions. J Am Chem Soc 53: 3190–3191.
Huggins, M.L. (1936a). Hydrogen bridges in ice and liquid water. J Phys Chem 40: 723–731.
Huggins, M.L. (1936b). Hydrogen bridges in organic compounds. J Org Chem 1: 407–456.
Huggins, M.L. (1971). 50 Years of hydrogen bond theory. Angew Chem Int Ed 10: 147–152.
Huheey, J.E., E.A. Keiter, and R.L. Keiter (1993). *Inorganic chemistry: Principles of structure and reactivity*, 4th edition. HarperCollins College Publishers: New York.

Humbel, S. (2002). Short strong hydrogen bonds: A valence bond analysis. J Phys Chem A106: 5517–5520.

Hunter, E.P., and S.G. Lias (2005). Proton affinity data. In: *NIST Chemistry WebBook*, NIST Standard Reference Database Number 69. Edited by P.J. Linstrom and W.G. Mallard. National Institute of Standard and Technology: Gaithersburg, MD; http://webbook.nist.gov/chemistry.

Hunter, L. (1944). The H bond. Chem Ind (London) 155–157.

Hunter, L.(1945). Mesohydric tautomerism. J Chem Soc 806–809.

Hunter, L. (1946). The hydrogen bond. Annu Rep Prog Chem (Chem Soc London) 43: 141–155.

Hunter, L. (1954). The stereochemistry of the hydrogen bond. Prog Stereochem 1: 223–249.

Hurley, W.J., I.D. Kuntz, and G.E. Jr. Leroi (1966). Far-infrared studies of hydrogen bonding. J Am Chem Soc 88: 3199–3202.

Hussain, M.S., and E.O. Schlemper (1980). A combined neutron and X-ray diffraction study of hydrogenbis(*NN*-dimethylacetamide) tetrachloroaurate(III) revealing a short hydrogen bond. J Chem Soc Dalton Trans 750–755.

Hussain, M.S., and E.O. Schlemper (1982). Crystal structure of hydrogenbis(pyridine *N*-oxide) tetrachloroaurate(III), revealing a short hydrogen bond. J Chem Soc Dalton Trans 751–755.

Huta, O.M., I.O. Patsaj, A. Konitz, J. Meszko, and J. Blazejowski (2002). 9-Cyano-10-methylacridinium hydrogen dinitrate. Acta Crystallogr C58: o295–o297.

Huyskens, P.L., W.A.P. Luck, and Th. Zeegers-Huyskens (Eds.) (1991). *Intermolecular forces: An introduction to modern methods and results.* Springer Verlag: Heidelberg, D.

Huyskens, P., L. Sobczyk, and I. Majerz (2002). On a hard/soft hydrogen bond interaction. J Mol Struct 615: 61–72.

Huyskens, P.L., and Th. Zeegers-Huyskens (1964). Associations moléculaires et équilibres acide-base. J Chim Phys Phys-Chim Biol 61: 81–86.

Iijima, K., A. Ohnogi, and S. Shibata (1987). The molecular structure of acetylacetone as studied by gas-phase electron diffraction. J Mol Struct 156: 111–118.

Jakobsen, R.J., and J.W. Brasch (1965). Far-infrared studies of the hydrogen bond of phenols. Spectrochim Acta 21: 1753–1763.

Jaskólski. M., I. Olovsson, R. Tellgren, and D. Mickiewicz-Wichlacz (1982). A neutron diffraction study of 1,l'-ethylenebis(piperidine 1-oxide) monoperchlorate. Acta Crystallogr B38: 291–294.

Jaswal, J.S., S.J. Rettig, and B.R. James (1990). Ruthenium(III) complexes containing dimethylsulfoxide or dimethylsulfide ligands, and a new route to trans-dichlorotetrakis(dimethylsulfoxide)ruthenium(II). Can J Chem 68: 1808–1817.

Jeffrey, G.A. (1984). The structures of some small molecules. Ab initio molecular orbital calculations versus low temperature neutron diffraction crystal structure analyses. J Mol Struct – Theochem 108: 1–15.

Jeffrey, G.A. (1985). The structures of some small molecules. Ab initio molecular orbital calculations versus low temperature neutron diffraction crystal structures. Part II. J Mol Struct 130: 43–53.

Jeffrey, G.A. (1992). Accurate crystal structure analysis by neutron diffraction. In: *Accurate molecular structures: Their determination and importance.* Edited by A. Domenicano and I. Hargittai. Oxford University Press: Oxford, UK: Chapter 11, pp. 270–298.

Jeffrey, G.A. (1995). Hydrogen-bonding: An update. Cryst Rev 4: 213–259.

Jeffrey, G.A. (1997). *An introduction to hydrogen bonding.* Oxford University Press: Oxford, UK.

Jeffrey, G.A., and W. Saenger (1991). *Hydrogen bonding in biological structures.* Springer Verlag: Berlin, D.

Jeffrey, G.A., and Y. Yeon (1986). The correlation between hydrogen-bond lengths and proton chemical shifts in crystals. Acta Crystallogr B42: 410–413.

Jerzykiewicz, L.B., Z. Malarski, L. Sobczyk, T. Lis, and E. Grech (1998). The quasi-symmetric OHN and ODN bridges in the adducts of 3,5-dimethylpyridine with 3,5-dinitrobenzoic acid. J Mol Struct 440: 175–185.

Joesten, M.D., and Schaad, L.J. (1974). *Hydrogen bonding.* Marcel Dekker: New York.

Johnson, S.L., and K.A. Rumon (1965). Infrared spectra of solid 1:1 pyridine-benzoic acid complexes: The nature of the hydrogen bond as a function of the acid-base levels in the complex. J Phys Chem 69: 74–86.

Johnston, H.S., and C. Parr (1963). Activation energies from bond energies. I. Hydrogen transfer reactions. J Am Chem Soc 85: 2544–2551.

Jones, R.D.G. (1976). The crystal structure of the enol tautomer of 1,3-diphenyl-1,3-propanedione (dibenzoylmethane) by neutron diffraction. Acta Crystallogr B32: 1807–1811.

Kaplan, I.G. (1986). *Theory of molecular interactions.* Elsevier: Amsterdam, NL.

Katrusiak, A. (1990). High-pressure X-ray diffraction study on the structure and phase transition of 1,3-cyclohexanedione crystals. Acta Crystallogr B46: 246–256.

Katrusiak, A. (1991). Structure and phase transition of 1,3-cyclohexanedione crystals as a function of temperature. Acta Crystallogr B47: 398–404.

Katrusiak, A. (2003). Macroscopic and structural effects of hydrogen-bond transformations. Cryst Rev 9: 91–133.

Kavanau, J.L. (1964). *Water and solute-water interactions.* Holden-Day: San Francisco, CA.

Kazarian, S.G., P.A. Hamley, and M. Poliakoff (1993). Is intermolecular hydrogen-bonding to uncharged metal centers of organometallic compounds widespread in solution? A spectroscopic investigation in hydrocarbon, noble gas, and supercritical fluid solutions of the interaction between fluoro alcohols and (η^5-C_5R_5)ML_2 (R = H, Me; M = CO, Rh, Ir; L = CO, C_2H_4, N_2, PMe_3) and its relevance to protonation. J Am Chem Soc 115: 9069–9079.

Kebarle, P. (1977). Ion thermochemistry and solvation from gas phase ion equilibria. Ann Rev Phys Chem 28: 445–476.

Kilpatrick, G.J., N. El Tayar, H. van de Waterbeemd, P. Jenner, B. Testa, and C.D. Marsden (1986). The thermodynamics of agonist and antagonist binding to dopamine D_2 receptor. Mol Pharmacol 30: 226–234.

Kim, S.W., S.-S. Cha, H.-S. Cho, J.-S. Kim, N.-C. Ha, M.-J. Cho, S. Joo, K.K. Kim, K.Y. Choi, and B.-H. Oh (1997). High-resolution crystal structures of Δ^5-3-ketosteroid isomerase with and without a reaction intermediate analogue. Biochemistry 36: 14030–14036.

King, L.D., D. Kozono, and P. Agre (2004). From structure to disease: The evolving tale of aquaporin biology. Nature Rev 5: 687–698.

Kitaura, K., and K. Morokuma (1976). A new energy decomposition scheme for molecular interactions within the Hartree-Fock approximation. Int J Quant Chem 10: 325–340.

Klapotke, Th.M., H. Noth, H. Schwenk-Kircher, W.-H. Walther, and G. Holl (1999). Synthesis and X-ray structure of 1,1-dimethylhydrazinium azide. Polyhedron 18: 717–719.

Klebe, G. (1994). The use of composite crystal-field environments in molecular recognition and the *de novo* design of protein ligands. J Mol Biol 237: 212–235.

Klein, C.L., and E.D. Stevens (1988). Experimental measurements of electron density distributions and electrostatic potentials. In: *Molecular structure and energetics, Vol. 1: Structure and reactivity.* Edited by J.F. Liebman and A. Greenberg. VCH: New York; pp. 25–64.

Klyne, W., and V. Prelog (1960). Description of steric relationships across single bonds. Experientia 16: 521–523.

Kobko, N., and J.J. Dannenberg (2003). Cooperativity in amide hydrogen bonding chains. Relation between energy, position, and H-bond chain length in peptide and protein folding models. J Phys Chem A107: 10389–10395.

Koch, U., and P.L.A. Popelier (1995). Characterization of C–H–O hydrogen bonds on the basis of the charge density. J Phys Chem 99: 9747–9754.

Kochman, R.L., and J.D. Hirsch (1982). Thermodynamic changes associated with benzodiazepine and alkyl β-carboline-3-carboxylate binding to rat brain homogenates. Mol Pharmacol 22: 335–341.

Koetzle, Th.F., and M.S. Lehmann (1976). Neutron diffraction studies of hydrogen bonding in α-amino acids. In: *The hydrogen bond: Recent developments in theory and experiments,* Vol. II. Edited by P. Schuster, G. Zundel, and C. Sandorfy. North-Holland: Amsterdam, NL; Chapter 9, pp. 457–469.

Kohn, W., and L.J. Sham (1965). Self-consistent equations including exchange and correlation effects. Phys Rev 140: A1133–A1138.

Kolesnikov, S.P., I.V. Lyudkovskaya, M.Yu. Antipin, Yu.T. Struchkov, and O.M. Nefedov (1985). Etherates of Friedel-Crafts acids with a short hydrogen bond: the symmetrical cation ethyl ether-hydrogen-ethyl ether ($[Et_2O \cdots H \cdots OEt_2]^+$) in the crystal structure of the etherate bis(ethylether) hydrogen trichlorozincate ($(Et_2O)_2.HZnCl_3$). Izv Akad Nauk SSSR Ser Khim (Russ.) 79–86.

Kollman, P.A., and L.C. Allen (1972). The theory of the hydrogen bond. Chem Rev 72: 283–303.

Koritsanszky, T.S. (2006). Topology of X-ray charge density of hydrogen bonds. In: *Hydrogen bonding – New insights.* Edited by S.J. Grabowski. Springer: Dordrecht, NL; Chapter 12, pp. 441–470.

Koritsanszky, T.S., and P. Coppens (2001). Chemical applications of X-ray charge-density analysis. Chem Rev 101: 1583–1627.

Krebs, B. (1983). Thio- and seleno-compounds of main group elements – Novel inorganic oligomers and polymers. Angew Chem Int Ed 22: 113–134.

Krebs, B., and G. Henkel (1981). Investigations on compounds containing S–H···S hydrogen bonds. Crystal structure of diphenyldithiophosphinic acid at 140 and 293 K. Z Anorg Allg Chem 475: 143–155.

Kreevoy, M.M., and T.M. Liang (1980). Structures and isotopic fractionation factors of complexes, $A_1HA_2^{-1}$. J Am Chem Soc 102: 3315–3322.

Kristjansdottir, S.S., J.R. Norton, A. Moroz, R.L. Sweany, and S.L. Whittenburg (1991). Absence of hydrogen bonding between cobalt carbonyl $HCo(CO)_4$ and nitrogen and oxygen bases. IR and Raman study. Organometallics 10: 2357–2361.

Krogh Andersen, E., and I.G. Krogh Andersen (1975). α-Methyltetronic acid. A refinement of the structure based on new experimental data. Acta Crystallogr B31: 394–398.

Kroon, J., and J.A. Kanters (1974). Non-linearity of hydrogen bonds in molecular crystals. Nature 248: 667–669.

Kroon, J., J.A. Kanters, J.G.C.M. van Duijneveldt-van de Rijdt, F. B. van Duijneveldt, and J. A. Vliegenhart (1975). O–H···O hydrogen bonds in molecular crystals. A statistical and quantum-chemical analysis. J Mol Struct 24: 109–129.

Krug, R.R., W.G. Hunter, and R.A. Grieger (1976). Statistical interpretation of enthalpy-entropy compensation. Nature 261: 566–567.

Krygowski, T.M., R. Anulewicz, and J. Kruszewski (1983). Crystallographic studies and physicochemical properties of π-electron compounds. III. Stabilization energy and the Kekulé structure contributions derived from experimental bond lengths. Acta Crystallogr B39: 732–739.

Krygowski, T.M., K. Woźniak, R. Anulewicz, D. Pawlak, W. Kolodziejski, E. Grech, and A. Szady (1997). Through-resonance assisted ionic hydrogen bonding in 5-nitro-*N*-salicylideneethylamine. J Phys Chem A101: 9399–9404.

Kubas, G.J. (2001). *Metal dihydrogen and σ-bond complexes: Structure, theory, and reactivity.* Kluwer Academic/Plenum Publishers: New York.

Kuhn, P., M. Knapp, S.M. Soltis, G. Ganshaw, M. Thoene, and R. Bott (1998). The 0.78 Å structure of a serine protease: *Bacillus lentus* subtilisin. Biochemistry 37: 13446–13452.

Kuhs, W.F., and M.S. Lehmann (1986). The structure of ice I_h. Water Science Rev 2: 1–65.

Kuleshova, L.N., and P.M. Zorky (1980). Graphical enumeration of hydrogen-bonded structures. Acta Crystallogr B36: 2113–2115.

Landau, L., and E. Lifshitz (1967). *Physique statistique.* MIR: Moscow.

Larson, J.W., and T.B. McMahon (1982). Gas-phase bifluoride ion. An ion cyclotron resonance determination of the hydrogen bond energy in FHF^- from gas-phase fluoride transfer equilibrium measurements. J Am Chem Soc 104: 5848–5849.

Larson, J.W., and T.B. McMahon (1983). Strong hydrogen bonding in gas-phase anions. An ion cyclotron resonance determination of fluoride binding energetics to Brønsted acids from gas-phase fluoride exchange equilibria measurements. J Am Chem Soc 105: 2944–2950.

Larson, J.W. and T.B. McMahon (1984). Hydrogen bonding in gas-phase anions. An experimental investigation of the interaction between chloride ion and Brønsted acids from ion cyclotron resonance chloride exchange equilibria. J Am Chem Soc 106: 517–521.

Latimer, W.M., and W.H. Rodebush (1920). Polarity and ionization from the standpoint of the Lewis theory of valence. J Am Chem Soc 42: 1419–1433.

Lee, B. (1985) The physical origin of the low solubility of nonpolar solutes in water. Biopolymers 24: 813–823.

Lee, B. (1991). Solvent reorganization contribution to the transfer thermodynamics of small nonpolar molecules. Biopolymers 31: 993–1008.

Lee, B., and G. Graziano (1996). A two-state model of hydrophobic hydration that produces compensating enthalpy and entropy changes. J Am Chem Soc 118: 5163–5168.

Lee, J.C., A.L. Rheingold, B. Muller, P.S. Pregosin, and R.H. Crabtree (1994). Complexation of an amide to iridium *via* an iminol tautomer and evidence for an Ir–H···H–O hydrogen-bond. J Chem Soc Chem Commun 1021–1022.

Leffler, J.E. (1953). Parameters for the description of transition states. Science 117: 340–341.

Leffler, J.E., and E. Grunwald (1963). *Rates and equilibria of organic reactions.* John Wiley & Sons: New York.

Legon, A.C., P.D. Aldrich, and W.H. Flygare (1981). The rotational spectrum and molecular structure of the acetylene-HCl dimer. J Chem Phys 75: 625–630.

Legon, A.C., and D.J. Millen (1987a). Hydrogen bonding as a probe of electron densities: Limiting gas-phase nucleophilicities and electrophilicities of B and HX. J Am Chem Soc 109: 356–358.

Legon, A.C., and D.J. Millen (1987b). Directional character, strength, and nature of the hydrogen bond in gas-phase dimers. Acc Chem Res 20: 39–46.

Lennard-Jones, J., and J.A. Pople (1951). Molecular association in liquids. I. Molecular association due to lone-pair electrons. Proc Roy Soc London A205: 155–162.

Lewis, G.N. (1923). *Valence and the structure of atoms and molecules.* Chemical Catalog Co.: New York.

Li, Q., W.H. Yip, and T.C.W. Mak (1995). Hydrogen-bonded urea-anion host lattices. Part 2. Crystal structures of inclusion compounds of urea with tetraalkylammonium bicarbonates. J Inclus Phen Mol Recogn Chem 23: 233–244.

Liddel, U., and N.F. Ramsey (1951). Temperature dependent magnetic shielding in ethyl alcohol. J Chem Phys 19: 1608.

Liddel, U., and O.R. Wulf (1933). The character of the absorption of some amines in the near infra-red. J Am Chem Soc 55: 3574–3583.

Lide, D.R. (Ed.) (2006). *CRC Handbook of chemistry and physics,* 87th edition. CRC Press, Taylor and Francis: Boca Raton, FL.

Linstrom, P.J., and W.G. Mallard (Eds.) (2005). *NIST Chemistry WebBook,* NIST Standard Reference Database Number 69. National Institute of Standard and Technology: Gaithersburg, MD; http://webbook.nist.gov/chemistry.

Lippincott, E.R. (1955). Derivation of an internuclear potential function from a quantum mechanical model. J Chem Phys 23: 603.

Lippincott, E.R., and R. Schroeder (1955). One-dimensional model of the hydrogen bond. J Chem Phys 23: 1099–1106.

Lipscomb, W.N. (1954). Structures of the boron hydrides. J Chem Phys 22: 985–988.

Llamas-Saiz, A.L., C. Foces-Foces, and J. Elguero (1994). Proton sponges. J Mol Struct 328: 297–323.

Llamas-Saiz, A.L., G.M. Grotenbreg, M. Overhand, and M.J. van Raaij (2007). Double-stranded helical twisted β-sheet channels in crystals of gramicidin S grown in the presence of trifluoroacetic and hydrochloric acids. Acta Crystallogr D63: 401–407.

Lough, A.J., S. Park, R. Ramachandran, and R.H. Morris (1994). Switching on and off a new intramolecular hydrogen-hydrogen interaction and the heterolytic splitting of dihydrogen. Crystal and molecular structure of $[Ir\{(H(\eta^1\text{-}SC_5H_4NH)\}_2(PCy_3)_2]BF_4 \cdot 2.7CH_2Cl_2$. J Am Chem Soc 116: 8356–8357.

Low, J.N., P. Tollin, E. Brand, and C.C. Wilson (1986). Structure of 3-methylxanthine. Acta Crystallogr C42: 1447–1448.

Lowe, P.R., C.H. Schwalbe, and G.J.B. Williams (1987). Structure of 6-methylisocytosine. Acta Crystallogr C43: 330–333.

Lucas, M. (1976). Size effect in transfer of nonpolar solutes from gas or solvent to another solvent with a view on hydrophobic behavior. J Phys Chem 80: 359–362.

Luck, W.A.P. (1991). IR-overtone vibration spectroscopy. In: *Intermolecular forces: An introduction to modern methods and results.* Edited by P. L. Huyskens, W.A.P. Luck, and Th. Zeegers-Huyskens. Springer Verlag: Heidelberg, D; Chapter VII, pp. 157–193.

Luck, W.A.P., and W. Ditter (1968). Determination of hydrogen bonding in the harmonic range. J Mol Struct 1: 261–282.

Lumry, R., and S. Rajender (1970). Enthalpy-entropy compensation phenomena in water solutions of proteins and small molecules: A ubiquitous property of water. Biopolymers 9: 1125–1227.

Lyssenko, K.A., and M.Yu. Antipin (2006). The nature and energy characteristics of intramolecular hydrogen bonds in crystals. Russ Chem Bull Int Ed 55: 1–15.

Macdonald, A.L., J.C. Speakman, and D. Hadzi (1972). Crystal structures of the acid salts of some monobasic acids. Part XIV. Neutron-diffraction studies of potassium hydrogen bis(trifluoroacetate) and potassium deuterium bis(trifluoroacetate): Crystals with short and symmetrical hydrogen bonds. J Chem Soc Perkin Trans 2: 825–832.

MacDonald, J.C., G.W. Whitesides (1994). Solid-state structures of hydrogen-bonded tapes based on cyclic secondary diamines. Chem Rev 94: 2383–2420.

Madsen, G.K.H., C. Wilson, Th.M. Nymand, G.J. McIntyre, F.K. Larsen (1999). The structure of nitromalonamide: A combined neutron-diffraction and computational study of a very short hydrogen bond. J Phys Chem A103: 8684–8690.

Maes, G. (1991). Intermolecular interactions at low temperature. Matrix isolation spectroscopy applied to hydrogen-bonded complexes and charge transfer complexes. In: *Intermolecular forces: An introduction to modern methods and results.* Edited by P.L. Huyskens, W.A.P. Luck, and Th. Zeegers-Huyskens. Springer Verlag: Heidelberg, D; Chapter VIII, pp. 195–216.

Maguire, P.A., H.O. Villar, M.F. Davies, and G.H. Loew (1992). Thermodynamic analysis of binding to the cerebellar type I benzodiazepine receptor. Eur J Pharmacol, Mol Pharmacol Sect 226: 233–238.

Maksay, G. (1994). Thermodynamics of γ-aminobutyric acid type A receptor binding differentiate agonists from antagonists. Mol Pharmacol 46: 386–390.

Maksay, G. (1996). Distinct thermodynamic parameters of serotonin 5-HT_3 agonists and antagonists to displace [^{3}H]granisetron binding. J Neurochem 67: 407–412.

Malarski, Z., M. Rospenk, and L. Sobczyk (1982). Dielectric and spectroscopic studies of pentachlorophenol-amine complexes. J. Phys. Chem. 86: 401–406.

Malone, J.F., C.M. Murray, M.H. Charlton, R. Docherty, and A.J. Lavery (1997). X–H$\cdots\pi$(phenyl) interactions. Theoretical and crystallographic observations. J Chem Soc Faraday Trans 93: 3429–3436.

Marcus, R.A. (1960). Exchange reactions and electron transfer reactions including isotopic exchange. Theory of oxidation-reduction reactions involving electron transfer. Part 4. A statistical-mechanical basis for treating contributions from solvent, ligands, and inert salt. Discuss Faraday Soc 29: 21–31.

Marcus, R.A. (1968). Theoretical relations among rate constants, barriers, and Brønsted slopes of chemical reactions. J Phys Chem 72: 891–899.

Maréchal, Y. (2007). *The hydrogen bond and the water molecule: The physics and chemistry of water, aqueous and bio media.* Elsevier: Amsterdam, NL.

Mareque Rivas, J.C., and L. Brammer (1998). Self-assembly of l-D chains of different topologies using the hydrogen-bonded inorganic supramolecular synthons N–H$\cdots$$Cl_2$M or N–H$\cdots$$Cl_3$M. Inorg Chem 37: 4756–4757.

Martell, A.E., R.M. Smith, and R.J. Motekaitis (2004). *NIST critically selected stability constants of metal complexes.* NIST Standard Reference Database Number 46, Version 8.0, National Institute of Standard and Technology: Gaithersburg, MD.

Maskill, H. (1985). *The physical basis of organic chemistry.* Oxford University Press: New York.

Mastropaolo, D., A. Camerman, and N. Camerman (1991). Structure of 4-deoxyphenobarbital. Acta Crystallogr C47: 1050–1054.

Mautner, F.A., C. Gspan, K. Gatterer, M.A.S. Goher, M.A.M. Abu-Youssef, E. Bucher, and W. Sitte (2004). Synthesis and characterization of three 5-(4-pyridyl)tetrazolato complexes obtained by reaction of 4-cyanopyridine with metal azides from aqueous solutions. Polyhedron 23: 1217–1224.

McMullan, R.K., and B.M. Craven (1989). Crystal structure of 1-methyluracil from neutron diffraction at 15, 60 and 123 K. Acta Crystallogr B45: 270–276.
McWeeny, R. (1979). *Coulson's valence.* Oxford University Press: Oxford, UK.
Meot-Ner (Mautner) M. (1984). The ionic hydrogen bond and ion solvation. 1. $NH^+\cdots O$, $NH^+\cdots N$, and $OH^+\cdots O$ bonds. Correlations with proton affinity. Deviations due to structural effects. J Am Chem Soc 106: 1257–1264.
Meot-Ner (Mautner), M. (1987). Ionic hydrogen bonds. Part I. Thermochemistry, structural implications, and role in ion solvation. In: *Molecular structure and energetics, Vol. 4. Biophysical aspects.* Edited by J.F. Liebman and A. Greenberg. VCH Publishing: New York; pp. 71–103.
Meot-Ner (Mautner), M. (1988). Models for strong interactions in proteins and enzymes. 2. Interactions of ions with the peptide link and with imidazole. J Am Chem Soc 110: 3075–3080.
Meot-Ner (Mautner), M. (2005). The ionic hydrogen bond. Chem Rev 105: 213–284.
Meot-Ner (Mautner), M., and S.G. Lias (2005). Thermochemistry of cluster ion data. In: *NIST Chemistry WebBook*, NIST Standard Reference Database Number 69. Edited by P.J. Linstrom and W.G. Mallard. National Institute of Standard and Technology: Gaithersburg, MD; http://webbook.nist.gov/chemistry.
Meot-Ner (Mautner), M., and L.W. Sieck (1985). The ionic hydrogen bond and ion solvation. 4. $SH^+\cdots O$ and $NH^+\cdots S$ bonds. Correlations with proton affinity. Mutual effects of weak and strong ligands in mixed clusters. J Phys Chem 89: 5222–5225.
Meot-Ner (Mautner), M., and L.W. Sieck (1986). The ionic hydrogen bond and ion solvation. 5. $OH\cdots O^-$ bonds. Gas-phase solvation and clustering of alkoxide and carboxylate anions. J Am Chem Soc 108: 7525–7529.
Merighi, S., K. Varani, S. Gessi, K.-N. Klotz, E. Leung, P.G. Baraldi, and P.A. Borea (2002). Binding thermodynamics at the human A_3 adenosine receptor. Biochem Pharmacol 63: 157–161.
Mildvan, A.S., M.A. Massiah, T.K. Harris, G.T. Marks, D.H.T. Harrison, C. Viragh, P.M. Reddy, and I.M. Kovach (2002). Short, strong hydrogen bonds on enzymes: NMR and mechanistic studies. J Mol Struct 615: 163–175.
Minshall, P.C., and G.M. Sheldrick (1978). Pyridinium 2,2,5,5-tetrathio-*cyclo*-di (phosphadithianate), $[(\text{pyridine})_2H]_2^{2+}\,[P_2S_8]^{2-}$. Acta Crystallogr B34: 1378–1380.
Moers, O., I. Lange, K. Wijaya, A. Blaschette, and P.G. Jones (2001). Polysulfonylamines, CXLII. A supramolecular monomer-dimer pair: Strong and weak hydrogen bonding in the crystal structures of pyridinium dimesylamide and 4,4′-bipyridinediium bis(dimesylamide). Z Naturforsch B56: 1041–1051.
Moore, T.S., and T.F. Winmill (1912). The state of amines in aqueous solution. J Chem Soc Trans 101: 1635–1676.
Mootz, D., and K. Bartmann (1991). Zur Kristallchemie von Supersäuren: Bildung und Struktur der tiefschmelzenden Addukte $SbF_5{\cdot}CF_3SO_3H$ und $SbF_5{\cdot}7HF$. Z Naturforsch B46: 1659–1663.

Mootz, D., and A. Deeg (1992). Poly(hydrogen chlorides). Formation and crystal structure of the low-melting adducts $Me_2S.4HCl$ and $Me_2S.5HCl$. Z Anorg Allg Chem 615: 109–113.

Morokuma, K. (1971). Molecular orbital studies of hydrogen bonds. III. C=O···H–O hydrogen bond in $H_2CO \cdots H_2O$ and $H_2CO \cdots 2H_2O$. J Chem Phys 55: 1236–1244.

Morokuma, K. (1977). Why do molecules interact? The origin of electron donor-acceptor complexes, hydrogen bonding, and proton affinity. Acc Chem Res 10: 294–300.

Morse, P.M. (1929). Diatomic molecules according to the wave mechanics. II. Vibrational levels. Phys Rev 34: 57–64.

Moulton, W.G., and R.A. Kromhout (1956). Nuclear magnetic resonance: Structure of the amino group. J Chem Phys 25: 34–37.

Muller, N. (1990). Search for a realistic view of hydrophobic effects. Acc Chem Res 23: 23–28.

Munn, R.W., and C.J. Eckhardt (2001). A model for resonance-assisted hydrogen bonding in crystals and its graph set analysis. J Phys Chem A105: 6938–6942.

Munshi, P., and T.N. Guru Row (2005). Evaluation of weak intermolecular interactions in molecular crystals *via* experimental and theoretical charge densities. Cryst Rev 11: 199–241.

Murata, K., K. Mitsuoka, T. Hirai, T. Walz, P. Agre, J.B. Heymann, A. Engel, and Y. Fujiyoshi (2000). Structural determinants of water permeation through aquaporin-1. Nature 407: 599–605.

Murphy, K.M.M., and S.H. Snyder (1982). Heterogeneity of adenosine A_1 receptor binding in brain tissue. Mol Pharmacol 22: 250–257.

Murray-Rust, P., H.-B. Bürgi, and J.D. Dunitz (1975) Chemical reaction paths. V. The S_N1 reaction of tetrahedral molecules. J Am Chem Soc 97: 921–922.

Murray-Rust, P., and J. P. Glusker (1984). Directional hydrogen bonding to sp^2- and sp^3-hybridized oxygen atoms and its relevance to ligand-macromolecule interactions. J Am Chem Soc 106: 1018–1025.

Murray-Rust, P., W.C. Stallings, C.T. Monti, R.K. Preston, and J.P Glusker (1983). Intermolecular interactions of the C–F bond. The crystallographic environment of fluorinated carboxylic acids and related structures. J Am Chem Soc 105: 3206–3214.

Murthy, A.S.N., and C.N.R. Rao (1968). Spectroscopic studies of the hydrogen bond. Appl Spectrosc Rev 2: 69–191.

Nakagawa, N., and S. Fujiwara (1961). Solvent shift of methyl proton nuclear magnetic resonance. Bull Chem Soc Japan 34: 143–146.

Nakamoto, K., M. Margoshes, and R.E. Rundle (1955). Stretching frequencies as a function of distances in hydrogen bonds. J Am Chem Soc 77: 6480–6486.

Nishio, M., M. Hirota, and Y. Umezawa (1998). *The CH/π interaction. Evidence, nature, and consequences.* Wiley-VCH: New York.

Noble, P.N., and R.N. Kortzeborn (1970). LCA-MO-SCF studies of HF_2^- and the related unstable systems HF_2^0 and HeF_2. J Chem Phys 52: 5375–5387.

Norman, I., and G. Porter (1954). Trapped atoms and radicals in a glass "cage". Nature 174: 508–509.

Northrop, D.B. (2001). Follow the protons: A low-barrier hydrogen bond unifies the mechanisms of the aspartic proteases. Acc Chem Res 34: 790–797.

Novak, A. (1974). Hydrogen bonding in solids. Correlation of spectroscopic and crystallographic data. Struct Bond 18: 177–216.

Nyburg, S.C. (1961). *X-ray analysis of organic structures.* Academic Press: New York.

Nygren, C.L., C.C. Wilson, and J.F.C. Turner (2005). Electron and nuclear positions in the short hydrogen bond in urotropine-*N*-oxide·formic acid. J Phys Chem A109: 1911–1919.

Oddo, G., and E. Puxeddu (1906). Sui 5-azoeugenoli e la costituzione dei cosidetti *o*-ossiazocomposti. Gazz Chim It 36(II): 1–48.

Ohms, U., H. Guth, E. Hellner, H. Dannöhl, and A. Schweig (1984). Comparison of observed and calculated electron-densities. 22. 2-Pyridone, C_5H_5NO, crystal structure refinements at 295 K and 120 K, experimental and theoretical deformation density studies. Z Kristallogr 169: 185–200.

Olivieri, A.C., R.B. Wilson, I.C. Paul, D.Y. Curtin (1989). ^{13}C NMR and X-ray structure determination of 1-(arylazo)-2-naphthols. Intramolecular proton transfer between nitrogen and oxygen atoms in the solid state. J Am Chem Soc 111: 5525–5532.

Olovvson, I., and P.-G. Jönsson (1976). X-ray and neutron diffraction studies of hydrogen bonded systems. In: *The hydrogen bond: Recent developments in theory and experiments,* Vol. II. Edited by P. Schuster, G. Zundel, and C. Sandorfy. North-Holland: Amsterdam, NL; Chapter 8, pp. 393–456.

Page, M.I., and W.P. Jencks (1971). Entropic contributions to rate accelerations in enzymic and intramolecular reactions and the chelate effect. Proc Natl Acad Sci USA 68: 1678–1683.

Park, S., R. Ramachandran, A.J. Lough, and R.H. Morris (1994). A new type of intramolecular H···H···H interaction involving N–H···H(Ir)···H–N atoms. Crystal and molecular structure of $[IrH(\eta^1\text{-}SC_5H_4NH)_2(\eta^2\text{-}SC_5H_4N)(PCy_3)]BF_4 \cdot 0.72CH_2Cl_2$. J Chem Soc Chem Commun 2201–2202.

Parr, R.G., and W. Yang (1989). *Density-functional theory of atoms and molecules.* Oxford University Press: New York.

Pauling, L. (1931). The nature of the chemical bond. Application of results obtained from the quantum mechanics and from a theory of paramagnetic susceptibility to the structure of molecules. J Am Chem Soc 53: 1367–1400.

Pauling, L. (1939, 1940, 1960). *The nature of the chemical bond and the structure of molecules and crystals: An introduction to modern structural chemistry.* Cornell University Press: Ithaca, NY.

Pauling, L. (1947). The nature of the interatomic forces in metals. II. Atomic radii and interatomic distances in metals. J Am Chem Soc 69: 542–553.

Pearce, P.J., and R.J.J. Simkins (1968). Acid strengths of some substituted picric acids. Can. J. Chem. 46: 241–248.

Pearson, R.G. (1969). A symmetry rule for predicting molecular structures. J Am Chem Soc 91: 4947–4955.

Pedireddi, V.R., and G.R. Desiraju (1992). A crystallographic scale of carbon acidity. J Chem Soc Chem Commun 988–990.

Perchard, J.P. (1981). Hydrogen bonding in matrices. In: *Matrix isolation spectroscopy.* NATO-ASI Ser C, Vol. 76. Edited by A.J. Barnes, W.J. Orville-Thomas, A. Müller, and R. Gaufrès. Reidel: Dordrecht, NL; Chapter 24, pp. 551–563.

Peris, E., and R.H. Crabtree (1995). Infrared detection of M–H···$OPPh_3$ hydrogen bonds. J Chem Soc Chem Commun 2179–2180.

Peris, E., J.C. Lee, J.R. Rambo, O. Eisenstein, and R.H. Crabtree (1995). Factors affecting the strength of X–H···H–M hydrogen bonds. J Am Chem Soc 117: 3485–3491.

Perrin, C.L., and J.B. Nielson (1997). "Strong" hydrogen bonds in chemistry and biology. Annu Rev Phys Chem 48: 511–544.

Perrin, C.L., and B.K. Ohta (2001). Symmetry of the N–H–N hydrogen bonds in 1,8-bis(dimethylamino)naphthalene $\cdot H^+$ and 2,7-dimethoxy-1,8-bis(dimethylamino)naphthalene $\cdot H^+$. J Am Chem Soc 123: 6520–6526.

Pfeiffer, P., Ph. Fischer, J. Kunter, P. Monti, and Z. Pros (1913). Zur Theorie der Farblacke. Justus Liebig Ann Chem 398: 137–196.

Piana, S., and P. Carloni (2000). Conformational flexibility of the catalytic Asp dyad in HIV-1 protease: An *ab initio* study on the free enzyme. Proteins: Struct Funct Genet 39: 26–36.

Pimentel, G.C., and A.L. McClellan (1960). *The hydrogen bond.* Freeman: San Francisco, CA.

Pimentel, G.C., and A.L. McClellan (1971). Hydrogen bonding. Ann Rev Phys Chem 22: 347–385.

Pimentel, G.C., and C.H. Sederholm (1956). Correlation of infrared stretching frequencies and hydrogen bond distances in crystals. J Chem Phys 24: 639–641.

Platt, J.R. (1949). Classification of spectra of cata-condensed hydrocarbons. J Chem Phys 17: 484–495.

Politzer, P., and D.G. Truhlar (1981). *Chemical applications of atomic and molecular electrostatic potentials. Reactivity, structure, scattering, and energetics of organic, inorganic, and biological systems.* Plenum Press: New York.

Pomès, R., and B. Roux (2002). Molecular mechanism of H^+ conduction in the single-file water chain of the gramicidin channel. Biophys J 82: 2304–2316.

Popelier, P., A.T.H. Lenstra, C. Van Alsenoy, and H.J. Geise (1989). An ab initio study of crystal field effects: Solid-state and gas-phase geometry of acetamide. J Am Chem Soc 111: 5658–5660.

Pople, J.A. (1951). Molecular association in liquids. II. A theory of the structure of water. Proc Roy Soc London A205: 163–178.

Pople, J.A. (1956). Proton magnetic resonance of hydrocarbons. J Chem Phys 24: 1111.

Popper, K.R. (1959). *The logic of scientific discovery.* Hutchinson: London, UK.

Prasad, P.N., and D.J. Williams (1991). *Introduction to nonlinear optical effects in molecules and polymers.* John Wiley & Sons: New York.

Preston, G.M., T.P. Carroll, W.B Guggino, and P. Agre (1992). Appearance of water channels in *Xenopus* oocytes expressing red cell CHIP28 protein. Science 256: 385–387.

Primas, H. (1983). *Chemistry, quantum mechanics and reductionism. Perspectives in theoretical chemistry,* 2nd edition. Springer Verlag: Heidelberg, D.

Raffa, R.B. (Ed.) (2001). *Drug-receptor thermodynamics: Introduction and applications.* John Wiley & Sons: Chichester, UK.

Rappoport, Z. (Ed.) (1990). *The chemistry of enols.* John Wiley & Sons: Chichester, UK.

Ratajczak, H., and W.J. Orville-Thomas (Eds.) (1980, 1981, 1982). *Molecular interactions.* Vols. 1–3. John Wiley & Sons: Chichester, UK.

Ratajczak, H., and L. Sobczyk (1969). Dipole moments of hydrogen-bonded complexes and proton-transfer effect. J. Chem. Phys. 50: 556–557.

Read, W.G., E.J. Campbell, and G. Henderson (1983). The rotational spectrum and molecular structure of the benzene-hydrogen chloride complex. J Chem Phys 78: 3501–3508.

Rebek, J. Jr. (1990). Molecular recognition with model systems. Angew Chem Int Ed 29: 245–255.

Reed, A.E., L.A. Curtiss, and F. Weinhold (1988). Intermolecular interactions from a natural bond orbital, donor-acceptor viewpoint. Chem Rev 88: 899–926.

Reeves, L.W., and W.G. Schneider (1957). Nuclear magnetic resonance measurements of complexes of chloroform with aromatic molecules and olefins. Can J Chem 35: 251–261.

Reich, H.J. (Ed.) (2008). Bordwell pK_a table (Acidity in DMSO). Available from http://www.chem.wisc.edu/areas/reich/pkatable/index.htm.

Reid, C. (1959). Semiempirical treatment of the hydrogen bond. J Chem Phys 30: 182–190.

Richardson, Th.B., S. de Gala, R.H. Crabtree, and P.E.M. Siegbahn (1995). Unconventional hydrogen bonds: Intermolecular B–H···H–N interactions. J Am Chem Soc 117: 12875–12876.

Robillard, G., and R.G. Shulman (1974). High resolution nuclear magnetic resonance studies of the active site of chymotrypsin. I. The hydrogen bonded protons of the "charge relay" system. J Mol Biol 86: 519–540.

Rodrigues, B.L., R. Tellgren, and N.G. Fernandes (2001). Experimental electron density of urea-phosphoric acid (1/1) at 100 K. Acta Crystallogr B57: 353–358.

Rohlfing, C.M., L.C. Allen, and R. Ditchfield (1983). Proton chemical shift tensors in hydrogen-bonded dimers of RCOOH and ROH. J Chem Phys 79: 4958–4966.

Roof, R.B. Jr. (1956). The crystal structure of ferric acetylacetonate. Acta Crystallogr 9: 781–786.

Rosenfield, R.E. Jr., K.N. Trueblood, and J.D. Dunitz (1978). A test for rigid-body vibrations, based on a generalization of Hirshfeld's 'rigid bond' postulate. Acta Crystallogr A34: 828–829.

Roux, B. (2002). Computational studies of the gramicidin channel. Acc Chem Res 35: 366–375.

Rowlinson, J.S. (1951). The lattice energy of ice and the second virial coefficient of water vapor. Trans Faraday Soc 47: 120–129.

Ruiz-Gomez, A., M. Garcìa-Calvo, J. Vaszquez, J.C.G. Marvizón, F. Valdivieso, and F. Jr. Mayor (1989). Thermodynamics of agonist and antagonist interaction with the strychnine-sensitive glycine receptor. J. Neurochem. 52: 1775–1780.

Ryan, L.M., R.E. Taylor, A.J. Paff, and B.C. Gerstein (1980). An experimental study of resolution of proton chemical shifts in solids: Combined multiple pulse NMR and magic-angle spinning. J Chem Phys 72: 508–515.

Saenger, W. (1984). *Principles of nucleic acid structure.* Springer Verlag: New York.

Saenger, W., and M. Mikolajczyk (1973). Crystal and molecular structure of 4-methyl-1,3,2-dioxaphosphorinane 2-oxide. A P–H···O hydrogen bond. Chem Ber 106: 3519–3523.

Sanderson, R.T. (1967). *Inorganic Chemistry.* Van Nostrand-Reinhold: New York.

Scheiner S. (1997). *Hydrogen bonding: A theoretical perspective.* Oxford University Press: New York.

Schiøtt, B., B.B. Iversen, G.K.H. Madsen, and T.C. Bruice (1998). Characterization of the short strong hydrogen bond in benzoylacetone by ab initio calculations and accurate diffraction experiments. Implications for the electronic nature of low-barrier hydrogen bonds in enzymatic reactions. J Am Chem Soc 120: 12117–12124.

Schleyer, P.v.R., and A. Allerhand (1962). Strong hydrogen bonds to carbon in isocyanides. J Am Chem Soc 84: 1322–1323.

Schmalle, H.W., G. Hänggi, and E. Dubler (1988). Structure of hypoxanthine. Acta Crystallogr C44: 732–736.

Schmutzler, R., D. Schomburg, R. Bartsch, and O. Stelzer (1984). Reactions of (2-trimethylsiloxyphenyl)diphenylphosphine with phosphorous halides – Crystal and molecular structure of (2-hydroxyphenyl)diphenylphosphonium bromide. Z Naturforsch B39: 1177–1184.

Schneider, H.-J. (1991). Mechanisms of molecular recognition: Investigations of organic host-guest complexes. Angew Chem Int Ed 30: 1417–1436.

Schroeder, R., and E.R. Lippincott (1957). Potential function model of hydrogen bonds. II. J Phys Chem 61: 921–928.

Schuster, P., G. Zundel, and C. Sandorfy (Eds.) (1976). *The hydrogen bond: Recent developments in theory and experiments.* Vols. I–III. North-Holland: Amsterdam, NL.

Schweber, S.S. (1993). Physics, community and the crisis in physical theory. Phys Today 46: 34–40.
Semmingsen, D. (1974). The crystal structure of triose reductone. Acta Chem Scand B28: 141- 146.
Semmingsen, D. (1977). The crystal structure of phenylmalondialdehyde at –162 °C. Acta Chem Scand B31: 114–118.
Semmingsen, D., and J. Feder (1974). A structural phase transition in squaric acid. Solid State Commun 15: 1369–1372.
Semmingsen, D., and P. Groth (1987). Deltic acid, a novel compound. J Am Chem Soc 109: 7238–7239.
Semmingsen, D., F.J. Hollander, and T.F. Koetzle (1977). A neutron diffraction study of squaric acid (3,4-dihydroxy-3-cyclobutene-1,2-dione). J Chem Phys 66: 4405–4412.
Sennikov, P.G. (1994). Weak H-bonding by second-row (PH_3, H_2S) and third-row (AsH_3, H_2Se) hydrides. J Phys Chem 98: 4973–4981.
Shaik, S.S., H.B. Schlegel, and S. Wolfe (1992). *Theoretical aspects of physical organic chemistry. The S_N2 mechanism.* John Wiley & Sons: New York.
Shaik, S., and A. Shurki (1999). Valence bond diagrams and chemical reactivity. Angew Chem Int Ed 38: 586–625.
Shimoni, L., and J.P. Glusker (1994). The geomerty of intermolecular interactions in some crystalline fluorine-containing organic compounds. Struct Chem 5: 383–397.
Shubina, E.S., N.V. Belkova, and L.M. Epstein (1997). Novel types of hydrogen bonding with transition metal π-complexes and hydrides. J Organomet Chem 536–537: 17–29.
Shubina, E.S., N.V. Belkova, A.N. Krylov, E.V. Vorontsov, L.M. Epstein, D.G. Gusev, M. Niederman, and H. Berke (1996). Spectroscopic evidence for *intermolecular* M–H···H–OR hydrogen bonding: Interaction of $WH(CO)_2(NO)L_2$ hydrides with acidic alcohols. J Am Chem Soc 118: 1105–1112.
Shubina, E.S., A.N. Krylov, A.Z. Kreindlin, M.I. Rybinskaya, and L.M. Epstein (1993). Intermolecular hydrogen bonds with *d*-electrons of transition metal atoms. H-complexes with metallocenes of the iron subgroup. J Mol Struct 301: 1–5.
Silvestre, J.-P., I. Bkouche-Waksman, G. Heger, and N.Q. Dao (1990). A neutron diffraction study of the rubidium trihydrogen 1-hydroxy-1, 1-ethanedi(phosphonate) dihydrate. New J Chem 14: 29–35.
Silvestre, J.-P., N.Q. Dao, G. Heger, and A. Cousson (2002). Refinement by neutron diffraction of the crystal structure of hydroxyethylidene bisphosphonic acid monohydrate: $C(CH_3)(OH)(PO_3H_2)_2 \cdot H_2O$. Phosphorus Sulfur Silicon Relat Elem 177: 277–288.
Singh, T.R., and J.L. Wood (1968). Two-dimensional double-minimum model of hydrogen bonding: The symmetric case. J Chem Phys 48: 4567–4581.
Smith, D.A. (Ed.) (1994). *Modeling the hydrogen bond.* ACS Symp Ser 569, American Chemical Society: Washington DC.

Smith, K.M., F.W. Bobe, O.M. Minnetian, H. Hope, and M.D. Yanuck (1985). Novel substituent orientation in Reimer-Tiemann reactions of pyrrole-2-carboxylates. J Org Chem 50: 790–792.

Smith, M.B., and J. March (2001). *March's advanced organic chemistry*, 5th edition. John Wiley & Sons: New York.

Sobczyk, L. (1998). X-ray diffraction, IR, UV and NMR studies on proton transfer equilibrating phenol-N-base systems. Ber Bunsen-Ges Phys Chem 102: 377–383.

Sokolov, N.D., and V.A. Savel'ev (1977). Dynamics of the hydrogen bond: Two-dimensional model and isotope effects. Chem Phys 22: 383–399.

Southall, N.T., K.A. Dill, and A.D.J. Haymet (2002). A view of the hydrophobic effect. J Phys Chem B106: 521–533.

Spackman, M.A., and A.S. Brown (1994). Charge densities from X-ray diffraction data. Annu Rep Prog Chem C91: 175–212.

Speakman, J.C. (1972). Acid salts of carboxylic acids, crystals with some "very short" hydrogen bonds. Struct Bond 12: 141–199.

Speller, C.V., and M. Meot-Ner (Mautner) (1985). The ionic hydrogen bond and ion solvation. 3. Bonds involving cyanides. Correlations with proton affinities. J Phys Chem 89: 5217–5222.

Srinivasan, R., J.S. Feenstra, S.T. Park, S. Xu, and A.H. Zewail (2004). Direct determination of hydrogen-bonded structures in resonant and tautomeric reactions using ultrafast electron diffraction. J Am Chem Soc 126: 2266–2267.

Staab, H.A., and Th. Saupe (1988). "Proton sponges" and the geometry of hydrogen bonds: Aromatic nitrogen bases with exceptional basicities. Angew Chem Int Ed 27: 865–879.

Steele, D., E.R. Lippincott, and J.T. Vanderslice (1962). Comparative study of empirical internuclear potential functions. Rev Mod Phys 34: 239–251.

Steiner, Th. (1998a). Lengthening of the covalent X–H bond in heteronuclear hydrogen bonds quantified from organic and organometallic neutron crystal structures. J Phys Chem A102: 7041–7052.

Steiner, Th. (1998b). Structural evidence for resonance-assisted O–H···S hydrogen bonding. Chem Commun 411–412.

Steiner, Th. (2002). The hydrogen bond in the solid state. Angew Chem Int Ed 41: 48–76.

Steiner, Th., I. Majerz, and C.C. Wilson (2001). First O–H–N hydrogen bond with a centered proton obtained by thermally induced proton migration. Angew Chem Int Ed 40: 2651–2654.

Steiner, Th., and W. Saenger (1993). Role of C–H···O hydrogen bonds in the coordination of water molecules. Analysis of neutron diffraction data. J Am Chem Soc 115: 4540–4547.

Stewart, R.F. (1976). Electron population analysis with rigid pseudoatoms. Acta Crystallogr A32: 565–574.

Stone, A.J. (2000). *The theory of intermolecular forces.* Clarendon Press: Oxford, UK.

Streitwieser, A. Jr. (1961). *Molecular orbital theory for organic chemists.* John Wiley & Sons: New York.

Sturtevant, J.M. (1977). Heat capacity and entropy changes in processes involving proteins. Proc Natl Acad Sci USA 74: 2236–2240.

Sui, H., B.-G. Han, J.K. Lee, P. Walian, and B.K. Jap (2001). Structural basis of water-specific transport through the AQP1 water channel. Nature 414: 872–878.

Sutor, D.J. (1962). The C–H···O hydrogen bond in crystals. Nature 195: 68–69.

Sutor, D.J. (1963). Evidence for the existence of C–H···O hydrogen bonds in crystals. J Chem Soc 1105–1110.

Swann, R.T., A.W. Hanson, and V. Boekelheide (1984). Conversion of η^4,η^6-bis(arene)ruthenium(0) complexes to cyclohexadienyl analogues of ruthenocene. J Am Chem Soc 106: 818–819.

Tajkhorshid, E., P. Nollert, M.Ø. Jensen, L.J.W. Miercke, J. O'Connell, R.M. Stroud, and K. Schulten (2002). Control of the selectivity of the aquaporin water channel family by global orientational tuning. Science 296: 525–530.

Tanenbaum, D.M., Y. Wang, S.P. Williams, and P.B. Sigler (1998). Crystallographic comparison of the estrogen and progesterone receptor's ligand binding domains. Proc Natl Acad Sci USA 95: 5998–6003.

Taylor, R., and O. Kennard (1982). Crystallographic evidence for the existence of C–H···O, C–H···N, and C–H···Cl hydrogen bonds. J Am Chem Soc 104: 5063–5070.

Taylor, R., and O. Kennard (1984). Hydrogen-bond geometry in organic crystals. Acc Chem Res 17: 320–326.

Tellgren, R. (1975). Theoretical and experimental aspects of the neutron diffraction technique and its application to the study of ferroelectric and related phenomena. Doctoral Thesis. Acta Univ Upsaliensis 344: 1–59.

Tomlinson, E. (1983). Enthalpy-entropy compensation analysis of pharmaceutical, biochemical and biological systems. Int J Pharm 13: 115–144.

Törnroth-Horsefield, S., Y. Wang, K. Hedfalk, U. Johanson, M. Karlsson, E. Tajkhorshid, R. Neutze, and P. Kjellbom (2006). Structural mechanism of plant aquaporin gating. Nature 439: 688–694.

Trifonov, L., J.H. Bieri, R. Prewo, A.S. Dreiding, D.M. Rast, and L. Hoesch (1982). The constitution of vertinolide, a new derivative of tetronic acid, produced by *Verticillium intertextum*. Tetrahedron 38: 397–403.

Trotter, J. (1960). A three-dimensional analysis of the crystal structure of *p*-benzoquinone. Acta Crystallogr 13: 86–95.

Truter, M.R. (1967). Comparison of photographic and counter observations for the X-ray crystal structure analysis of thiourea. Acta Crystallogr 22: 556–559.

Tuckerman, M., K. Laasonen, M. Sprik, and M. Parrinello (1995). *Ab initio* molecular dynamics simulation of the solvation and transport of H_3O^+ and OH^- ions in water. J Phys Chem 99: 5749–5752.

Vajda, E., and I. Hargittai (1992) Molecular structure of tetrafluorohydroquinone and geometrical indication of intramolecular hydrogen bonding from gas-phase electron diffraction. J Phys Chem 96: 5843–5846.

Vanquickenborne, L.G. (1991). Quantum chemistry of the hydrogen bond. In: *Intermolecular forces: An introduction to modern methods and results.* Edited by P.L. Huyskens, W.A.P. Luck, and Th. Zeegers-Huyskens. Springer Verlag: Heidelberg, D; Chapter II, pp. 31–53.

van Thiel, M., E.D. Becker, and G.C. Pimentel (1957a). Infrared studies of hydrogen bonding of methanol by the matrix isolation technique. J Chem Phys 27: 95–99.

van Thiel, M., E.D. Becker, and G.C. Pimentel (1957b). Infrared studies of hydrogen bonding of water by the matrix isolation technique. J Chem Phys 27: 486–490.

Vinogradov, S.N., and R.H. Linnell (1971). *Hydrogen bonding.* Van Nostrand-Reinhold: New York.

Vogt, J., and N. Vogt (2004). Structure searching in the MOGADOC database. J Mol Struct 695–696: 237–241.

Weiland, G.A., K.P. Minneman, and P.B. Molinoff (1979). Fundamental difference between the molecular interactions of agonists and antagonists with the β-adrenergic receptor. Nature 281: 114–117.

Werner, A. (1902). Über Haupt- und Nebenvalenzen und die Constitution der Ammoniumverbindungen. Justus Liebig Ann Chem 322: 261–296.

Whitaker, A. (1980). The crystal structure of a second polymorph (β) of C.I. Pigment Red 1, 1-[(4-nitrophenyl)azo]-2-naphthol. Z Kristallogr 152: 227–238.

Whittle, E., D.A. Dows, and G.C. Pimentel (1954). Matrix isolation method for the experimental study of unstable species. J Chem Phys 22: 1943.

Wijaya, K., D. Henschel, O. Moers, A. Blaschette, and P.G. Jones (1997). Polysulfonylamines, XCVI. Homoconjugates formed from di(organosulfonyl) amines and their conjugate anions: Four crystal structures featuring asymmetric $[N\cdots H\cdots N]^-$ hydrogen bonds. Z Naturforsch B52: 1219–1228.

Wijaya, K., O. Moers, A. Blaschette, and P.G. Jones (1999). Polysulfonylamines, CXII. A novel solid-state aspect of 2-pyridones: Isolation and X-ray structures of bis(2-pyridone)hydrogen(I) and bis(6-methyl-2-pyridone)hydrogen(I) dimesylamides. Z Naturforsch B54: 643–648.

Williams, D.E., W.L. Dumke, and R.E. Rundle (1962). The crystal structure of bis(*m*-bromobenzoyl)methane. Acta Crystallogr 15: 627–635.

Wilson, C.C. (2000). *Single crystal neutron diffraction from molecular materials.* World Scientific: Singapore.

Wilson, C.C. (2007). The evolution of hydrogen atom parameters under changing external conditions by time-of-flight single crystal neutron diffraction. Cryst Rev 13: 143–198.

Wilson, C.C., and A.E. Goeta (2004). Towards designing proton-transfer systems – Direct imaging of proton disorder in a hydrogen-bonded carboxylic acid dimer by variable-temperature X-ray diffraction. Angew Chem Int Ed 43: 2095–2099.

Wilson, C.C., N. Shankland, A.J. Florence (1996). A single-crystal neutron diffraction study of the temperature dependence of hydrogen-atom disorder in benzoic acid dimers. J Chem Soc Faraday Trans 92: 5051–5057.

Wright, W.B., and E.A. Meyers (1980). Tris(acetylacetonato)cobalt(III)· 2-Selenourea, $C_{17}H_{29}CoN_4O_6Se_2$. Cryst Struct Commun 9: 1173–1180.

Wulf, O.R., U. Liddel, and S.B. Hendricks (1936). The effect of *ortho* substitution on the absorption of the OH group of phenol in the infrared. J Am Chem Soc 58: 2287–2293.

Yamdagni, R., and P. Kebarle (1971). Hydrogen-bonding energies to negative ions from gas-phase measurements of ionic equilibria. J Am Chem Soc 93: 7139–7143.

Yamdagni R., and P. Kebarle (1976). Gas-phase basicities and proton affinities of compounds between water and ammonia and substituted benzenes from a continuous ladder of proton transfer equilibria measurements. J Am Chem Soc 98: 1320–1324.

Yao, W., O. Eisenstein, and R.H. Crabtree (1997). Interactions between C–H and N–H bonds and d^8 square planar metal complexes: Hydrogen bonded or agostic? Inorg Chim Acta 254: 105–111.

Yoshida, Z., and E. Osawa (1966). Hydrogen bonding of phenol to π electrons of aromatics, polyolefins, heteroaromatics, fulvenes and azulenes. J Am Chem Soc 88: 4019–4026.

Zeegers-Huyskens Th. (1986). Energies of $OH\cdots X^-$ hydrogen bonds in the gas phase and proton affinities. Chem Phys Lett 129: 172–175.

Zeegers-Huyskens Th. (1988). Importance of the proton affinities in hydrogen bond studies. A new exponential expression. J Mol Struct 177: 125–141.

Zeegers-Huyskens, Th. (1991). Experimental vibrational characteristics of the hydrogen bond. In: *Intermolecular forces: An introduction to modern methods and results.* Edited by P.L. Huyskens, W.A.P. Luck, and Th. Zeegers-Huyskens. Springer Verlag: Heidelberg, D; Chapter VI, pp. 123–155.

Zeegers-Huyskens, Th. (1999). Are the enol-enolate hydrogen bonds at matched PA really symmetrical? J Org Chem 64: 4946–4948.

Zhang, Z.-Y., M.-C. Shao, X.-J. Xu, Y.-Q. Tang, and Y.-Q. Tu (1981). The crystal structure of urea fluorosilicate. Huaxue Tongbao (Chin.) 142–143.

Zhao, Q., C. Abeygunawardana, A.G. Gittis, and A.S. Mildvan (1997). Hydrogen bonding at the active site of Δ^5-3-ketosteroid isomerase. Biochemistry 36: 14616–14626.

Zundel, G. (1976). Easily polarizable hydrogen bonds – Their interactions with the environment – IR continuum and anomalous large proton conductivity. In: *The hydrogen bond: Recent developments in theory and experiments,* Vol. II. Edited by P. Schuster, G. Zundel, and C. Sandorfy. North-Holland: Amsterdam, NL; Chapter 15, pp. 683–766.

INDEX